Irreversibility in the Many-Body Problem

Irreversibility in the Many-Body Problem

Sitges International School of Physics, May 1972

Director: L. M. Garrido

Director, Institute of Theoretical Physics
National Council of Research, Spain

Edited by

J. Biel

University of Madrid

and

J. Rae

Université Libre de Bruxelles

ℚℙ SPRINGER SCIENCE+BUSINESS MEDIA, LLC

Library of Congress Catalog Card Number 72-87519

ISBN 978-1-4899-2671-5   ISBN 978-1-4899-2669-2 (eBook)
DOI 10.1007/978-1-4899-2669-2

© 1972 Springer Science+Business Media New York
Originally published by Plenum Press, New York in 1972
Softcover reprint of the hardcover 1st edition 1972

PREFACE

The Sitges International School of Physics
is the second one to be held in Spain on the Many
Body Problem. The first one took place on Mallorca
during the summer 1969.

The aim of the school was mainly to direct
the interest of professors and students of Spanish
Universities towards this concrete field of research. For this purpose 55 especially prepared
lectures were given by an eminent collection of
lecturers. Besides, a school of this class contributes to the scientific formation of many students
from other countries. Also, in a meeting of this
kind, personal contacts are born that favour future
collaboration between scientists.

In view of the success of the first two
schools, we intend to foster future international
meetings on this subject until interest in it is
consolidated in Spain.

All the lectures given are published here
except those of Professor P.C. Martin whose lectures have previously been published.

I would like to thank all those people who
helped to make this school a success, and in
particular:

Prof. J.L. Villar-Palasi, Minister of Education of Spain for sponsoring the school.

Dr. R. Diez-Hochleitner, Undersecretary of
the Ministry of Education for receiving the project
of this school with such enthusiasm.

Prof. E.Costa-Novella, Director General of Universities in Spain and Dr. F.Arias-Salgado who showed such interest and patience while assuring the necessary finance would be found for the school.

To several people who in different ways have contributed to transform the project of the school into a reality my most sincere thanks:

Dr. J.M. de Müller y Abadal, President of Diputación de Barcelona; Dr. E. Bassols, Director General of Popular Culture; Dr. E. Thomas de Carranza, Director General of Cultural Relations; Prof. A. Romañá, of the National Council of Research; Dr. A. Martinez Sardá, Mayor of Sitges.

Also I wish to extend my thanks to Prof. I. Prigogine (Université Libre de Bruxelles) who provided much useful information.

Dr. L. Navarro (University of Barcelona) for continuous help in the organization of the school.

Dr. J. Biel (University of Madrid) and Dr. J. Rae (Université Libre de Bruxelles) co-editors of the Proceedings.

Mrs. R. Chester (University of Edinburgh) and Mrs. H. Meier-Loy who acted as secretaries.

And finally to my wife for her patience and understanding during the preparation of the school and the typing of the manuscript.

L.M. GARRIDO
Director, Institute of Theoretical Physics
National Council of Research, Spain

LECTURERS

Prof. R. Balescu, Brussels
Prof. C.P. Enz, Geneva
Prof. I.E. Farquhar, St. Andrews
Prof. P.C. Martin, Harvard
Prof. S. Miracle, Zaragoza
Prof. P. Résibois, Brussels
Prof. L. Rosenfeld, NORDITA
Prof. E. Santos, Valladolid
Prof. N.G. van Kampen, Utrecht
Prof. M.G. Velarde, Madrid
Prof. H. Wergeland, Trondheim

PARTICIPANTS

Dr. P. Allen, Brussels
Dr. F. Barocchi, Florence
Prof. A. Bernalte, Barcelona
Dr. K. Binder, Munich
Prof. B. Bosco, Florence
Mr. J.J. Brey, Sevilla
Dr. A. Coniglio, Naples
Dr. J. Dancz, Manchester
Mr. T. Dreyfus, Geneva
Mr. R. Fleckinger, Toulouse
Dr. A. Gomes, Toulouse
Dr. R. Giachetti, Florence
Mr. J. Graells, Barcelona
Mr. G.P. Hagedorn, Kiel
Dr. H. Hahn, Tübingen
Mr. I. Hernando, Valladolid
Mr. P. v. Hoyningen, London

Mr. M.J. Kahn, London

Dr. P. Kleban, Grenoble

Dr. J. Lam, Munich

Dr. R. Lima, Marseille

Mr. A. Malaspines, Geneva

Mr. J. Marro, Barcelona

Mr. F. Mauricio, Barcelona

Mr. J. Navarro, Valencia

Dr. L. Navarro, Barcelona

Dr. K. Nishioka, Oslo

Dr. B. Novak, Tübingen

Dr. W. Peier, Zürich

Mr. A. Perez-Machado, Valencia

Prof. V. Perez-Villar, Valladolid

Mr. A. Robledo, St. Andrews

Dr. M.T. Sacchi, Cagliari

Prof. J. de la Rubia, Sevilla

Prof. E. Santos, Valladolid

Mr. A. Schenzle, Stuttgart

Mr. P. Seglar, Barcelona

Prof. J. Sesma, Valencia

Dr. Y. Soulet, Toulouse

Mr. W.H. Steeb, Kiel

Dr. R. Swendsen, Köln

Prof. F. Tejerina, Valencia

Dr. V. Tognetti, Florence

Prof. E. Tuttle, Oxford

Mrs. I. Veretennicoff, Brussels

Mr. A. Villaeys, Strasbourg

Mr. B. Wilkinson, Oxford

Dr. J.G. Williams, Birmingham

DIRECTOR

Prof. L.M. Garrido, Barcelona

EDITORS

Dr. J. Biel, Madrid

Dr. J. Rae, Brussels

SECRETARIAT

Mrs. R. Chester, Edinburgh

Mrs. H. Meier-Loy, Zürich

CONTENTS

GENERAL INTRODUCTION TO IRREVERSIBILITY

L. Rosenfeld

NORDITA, Copenhagen

1. Historical Background
2. The "arrow of time"
3. Irreversibility of large quantal systems

1.   Historical background

Although a naive feeling of a privileged direction of time has always been present in human thought (Fugit irreparabile tempus, Humpty Dumpty etc.), the scientific concept of irreversibility is relatively modern (19th century).

Modern physics can be said to have started with Stevin and Galileo in the 16th century, but at that time no attention was paid to irreversible phenomena because the primitive state of technology could permit only crude quantitative considerations. The main topic of scientific investigation was the analysis of planetary motions, which were characterized by periodicities exhibiting no irreversibility, and the definition and measurement of time also involved periodic motions, such as that of a pendulum.

Moreover, even though conservation laws were implicit in the laws of mechanics (Kepler's second law, for instance, expresses the conservation of angular momentum), they were not explicitly recognized by Newton and his immediate followers. It was only one century later that Laplace pointed out the relation between Kepler's second law and the conservation of angular momentum. The conservation law of mechanical energy was not formulated in full generality by the 18th century physicists. The growing importance of the use of machines prompted Lazare Carnot to discuss the efficiency of such mechanical systems. His main

conclusion was that maximum efficiency could be
achieved by avoiding abrupt changes of velocity,
so as to minimize the loss of "living force".
This concept of "living force" had been introduced by Leibnitz as a measure of "motion" in
opposition to Descartes' view, according to which
this measure would be the absolute value of the
momentum. The underlying idea was that all
interactions between the particles of matter
were contact interactions, i.e. transmission of
motion by collisions, and that the total amount
of "motion" should be conserved, i.e. that the
collisions were elastic. From this point of view,
the "living force", correctly defined by Leibnitz
as the product of the mass and the square of the
velocity of the particles, is indeed conserved
in all dynamical processes. However, it could
be "lost" in the cases considered by Carnot, involving production of heat.

    Sadi Carnot, no doubt influenced by his
father's work, tried to find the condition of
maximum efficiency for the new source of power
developed in the meantime, the steam engine. His
masterly analysis showed that this condition was
a reversible cycle of processes, consisting in
an alternation of the only two reversible modes
of transformation - isothermal and adiabatic. The
Carnot principle of maximum efficiency could be
applied quite generally to any transformation
involving heat transfer. It made possible the

analysis of experimental data concerning the
thermal properties of gases and vapours and also
made possible the definition of an absolute scale
of temperature. Carnot's book was addressed to
engineers and remained unnoticed by the physicists,
who continued to regard friction and heat losses
as a nuisance spoiling the regularity of the
purely dynamical behaviour. It was only after
W. Thomson had become aware of Carnot's work
(through a paper by Clapeyron based on it) that
the fruitfulness of Carnot's ideas was appreciated.

However, Thomson and Clausius were soon faced
with an apparent contradiction between Carnot's
principle and the discovery by Joule of the law
of equivalence between heat and mechanical work.
Indeed, Joule's principle implied the transformability ·of heat into motion, whereas Carnot's
argument was based on the assumption that heat was
was a substance, the amount of which remained
constant in all transformations. Clausius recon-
ciled the two principles by regarding them as
logically independent, and abandoning the idea
of the substantiality of heat. Thomson came independently to the same conclusion about a year
later.

Thus the two fundamental laws of thermodynamics were established in full generality.
Clausius introduced the concept of entropy as a
"measure of the transformability (Verwandlungs-
mass) of a system: reversible transformations of

an isolated system leave the entropy constant,
irreversible transformations lead to an increase
of its entropy.

   The question then arose of deriving these
principles of thermodynamics from the laws of
mechanics. Indeed, since Descartes' time it had
been quite generally accepted that material
bodies were mechanical systems, consisting of
elementary particles in motion. From this point
of view, heat had to be regarded as a particular
mode of motion of the particles and the thermal
properties of bodies must appear as consequences
of the purely dynamical behaviour of these particles. It was natural to assume that the motion
of the particles was governed by the same laws as
those that governed the bulk motions of macroscopic bodies. The first law of thermodynamics
then resulted from an immediate application of
the conservation law of mechanical energy. As
to the second law, Clausius and Boltzmann, in
an early attempt to derive it on a mechanical
basis, both made the same mistake; they replaced at one stage of the argument the individual
velocity of an atom by its average velocity in the
steady state, thus unwittingly introducing a
statistical element into the derivation. Maxwell was the first to realize the essentially
statistical character of the second law; he was
apologetic about the introduction of statistical
considerations, which he excused by our ignorance
of the exact initial conditions of the atomic

motions. Boltzmann adopted the statistical method
without at first realizing its full implications.
Thus, his proof of the second law ("H-theorem")
was based on the well-known "Stosszahlansatz",
which is not valid when the system is far away
from the steady state. He only reached a clear
understanding of the situation when he had to
face two serious objections. The first was put
forward by Loschmidt. He argued that if a system
is started in some definite state, and if at some
later time during its evolution the velocities
of all particles were inverted, the system must
retrace its steps back to the initial state. This
implies that in this process the entropy should
decrease, and therefore the H-theorem appears to
be in contradiction with the laws of mechanics.
Another paradox was pointed out by Zermelo, who
referred to Poincare's theorem stating that the
motion of any isolated mechanical system of
finite extension is quasi-periodic: such cyclic
behaviour is at variance with the irreversibility
expressed by the second law. The gist of these
two objections was to emphasize the logical im-
possibility of deriving any form of irreversible
behaviour from the strictly reversible laws of
dynamics. In refuting the objections, Boltzmann
did not deny this logical point, but argued that
it was irrelevant, inasmuch as the derivation of
the H-theorem, owing to its statistical character,
was not strictly dynamical. Thus the essential
link between the occurrence of irreversibility and

a statistical mode of description was clearly
recognized. Boltzmann's views received further
elaboration by Ehrenfest, who showed that the
non-dynamical element was introduced by a passage
from a strictly continuous to a "coarse-grained"
distribution of the states of the system, and by
Gibbs, who compared the coarse-grained evolution
of the system to a "mixing" process.

In modern terms, we have here a typical
relationship of complementarity between two modes
of description of the phenomena, corresponding to
conditions of observation which are mutually ex-
clusive. Contradictions in the use of such complementary modes of description are avoided by
specifying appropriate limitations of their
validity. In the present case, these limitations
refer essentially to the size of the system, more
precisely to the number of degrees of freedom.
The atomic mode of description requires an
immense number of dynamical variables; the
thermodynamic, macroscopic mode of description
uses a very limited number of variables. The
thermodynamic description applies to "large"
systems, i.e. to systems of spatial dimensions
much larger than the atomic dimensions: such
systems behave in general irreversibly, while
small systems do not (as the case of Brownian
motion shows). The critical size corresponds
to typical dimensions of our sense organs, compared to those of single atoms. With respect
to time, the evolution of the system is

irreversible on a macroscopic time scale, long
with respect to the atomic time scale defined
by the average time of atomic interactions, and
again corresponding to the relaxation times of
the processes of reception of signals by our
sense organs. Such reference to the process of
observation does not imply (as has sometimes been
stated, e.g. by G.N. Lewis, who called the
entropy a "subjective" concept) any departure
from objectivity - if this word is used in its
only sensible meaning as denoting the consensus
of all observers placed in the same conditions.

The preceding considerations apply just as
well to quantal systems, since the laws of
quantum mechanics are also reversible. In this
case, we meet a statistical element even at the
level of individual atomic processes: this
statistical element is logically independent of
the one introduced in the description of large
systems of atoms. Nevertheless, there is an
indirect connection between the two owing to the
fact that the variables describing the behaviour
of single atoms are defined by measuring operations performed with the help of macroscopic
apparatus.

2. The "arrow of time" and the irreversibility
   of individual processes

It was Boltzmann who first raised the possibility of associating the sense of the flow of time
from past to future, the "arrow of time" (the expression is due to Eddington), with the second

law, by defining the future as that direction of
time in which the entropy of the universe increases. Although this procedure is logically acceptable
if the direction of time is referred to a standard
physical phenomenon (i.e. heat transfer), it is
unsatisfactory because it does not express the
real origin of the distinction between past and
future. From a historical and a psychological point
of view, human beings refer past and future to the
evolution of their own life, and thus attach to the
direction of time a much more profound significance than an increase of entropy of material systems
can express. The direction of time is associated
with the endeavour to achieve desirable aims, common to men and all animals, to orientate action towards such aims. Scientific observation, combined
with rational thinking, is just a sophisticated
example of this activity, inasmuch as knowledge
is acquired in view of its use in orientating our
behaviour. One ought therefore rather try to associate a direction of time with the process of observation. This can be done by realizing that an
observation requires the reception of a signal,
emitted from the outside world, and propagated
with finite velocity. Hence, one can <u>define</u> a direction of time by stating that the instant of reception is <u>later</u> than the instant of emission.
Having done this, the second law is given a well
defined physical content, where entropy becomes in
every case a measure of the transformability of the
system. Moreover, this point of view makes it

possible to avoid the paradoxes and confusion found
in the literature. In fact, the invariance of the
mechanical laws for time reversal is not contradic-
tory with the distinction between past and future
just introduced. When performing an observation the
physicist normally prepares an experimental arrange-
ment in order to observe what is going to happen.
The consequences of such an observation are naturally expressed by solutions of the time-reversal invariant, but describe the evolution oriented towards the future (e.g. the retarded electromagnetic
potentials).

However, the observer can also use the time-
inverse solutions (e.g. advanced potentials) to
infer by a "retrodictive" process the events which
may have led to the situation revealed by a present
observation.

By establishing a direct connection between the
sense of time and the process of observation, we
not only satisfy the epistemological requirement
of going back as far as possible to the immediate
psychological roots of our scientific concepts,
but we also emphasize the fact that irreversibility
has a much wider scope than its occurrence in the
second law of thermodynamics: an irreversible element may be attached to an individual atomic process, obeying a reversible law of evolution, by
the conditions under which it is observed. As a
typical example, we shall discuss the "decaying
states" of atomic systems, and more specifically

those of atomic nuclei, which appear in the study
of nuclear reactions. For simplicity, we shall consider reactions involving neutrons, in order to
deal only with finite range interactions. This
allows us to introduce the concept of <u>channel</u>, in
the following way: outside the finite region of
interaction, the fragments (projectile and target,
or end products of the reaction) can be regarded
as moving freely, and the asymptotic form of their
wave-function can accordingly be factorized and
written down explicitly - such an asymptotic wave-
function defines an entrance or exit channel. If
we treat the process as stationary, i.e. occurring
in both directions at a given energy out of the
continuous energy spectrum, the channel wave-function plays the part of a boundary condition selec-
ting a particular wave-function among the manifold
of those belonging to the given energy.

Let us first consider the simple case of "potential scattering" (i.e. scattering of the particle by a fixed potential of finite range): only
one channel is possible, namely elastic scattering.
We may decompose the wave function into "partial
waves", i.e. components corresponding to definite
spherical harmonics with respect to the centre of
the potential: the total cross-section will be the
sum of the partial cross-sections. The partial
wave components of the motion of the particle have
the asymptotic form $I_c(k)\exp(-ikr)+O_c(k)\exp(ikr)$,
with incoming and outgoing amplitudes $I_c(k), O_c(k)$
depending on the wave-number k and a set of

"channel quantum numbers" c. In the absence of the
potential, these amplitudes would be the same (if
the phases are chosen appropriately); the effect
of the scattering is then represented by the
amplitude difference

$$1-S_{cc}(k) = 1 - \frac{O_c(k)}{I_c(k)} ,$$

and the cross-section is proportional to the absolute square of this quantity. The ratio $S_{cc}(k)$ expresses the asymptotic relation established by the
potential between an incoming and an outgoing wave
in the only available channel c.

   In the more general case in which several channels c, c' are available we may still write channel
wave-functions, of the same form as above, for each
of them, with an extra factor, consisting of the
wave-functions representing the internal states
of the interacting fragments. The partial ampli
tudes will then be of the form

$$\delta_{cc'} - S_{cc'}(k) , \qquad S_{cc'}(k) = \frac{O_{c'}(k)}{I_c(k)} ,$$

where k is one of the wave-numbers, on which all
the others depend by the condition of energy
conservation $k_c^2 - \varepsilon_c = k_{c'}^2 - \varepsilon_{c'} = E$ ($\varepsilon_c, \varepsilon_{c'}$ being the internal energies in the various channels). The S-ma-
trix thus introduced must satisfy two conditions:
it must be unitary - this expresses the conservation of the total flux of particles, and it must be
symmetrical, $S_{cc'}(k) = S_{c'c}(k)$, as a consequence
of the invariance of the stationary processes ocur-
ring at a given energy for time-reversal: all
reactions occur equally in both directions.

As a function of the wave-number k or the
energy E, the cross-sections $\sigma_{cc'}(E)$ present experimentally for definite energies E_n a number of
peaks which, when well separated, have definite
"widths" Γ_n: these are called "resonances". These
can be interpreted physically in the following way:
Let us introduce a fictitious infinite barrier
around the interaction region. Within the barrier
the energy spectrum is discrete, whereas outside
it is continuous. Removing the barrier couples the
two regions and allows the particle, which was in
a perfectly sharp stationary state, to escape to
the continuum. This process has associated with it
a characteristic decay time: this is the physical
origin of the "resonance" phenomena. Therefore, an
asymmetry in time is introduced even at this elementary level.

The decaying of the wave-function in time can
be expressed by adding to the energy of a stationary state E_n, an imaginary term $-\frac{i}{2}\Gamma_n$; this
gives a time factor

$$e^{-\frac{i}{\hbar}(E_n - \frac{i}{2}\Gamma_n)t}$$

to the wave-function. How can one relate this concept of a "decaying state" with the S-matrix? If
we extend the definition of the functions $S_{cc'}(E)$
to the complex energy plane, a pole of $S_{cc'}(E)$ in
the lower half plane will give rise to a term in
its expansion of the form

$$\frac{R_n^{1/2}}{E-(E_n - \frac{i\Gamma_n}{2})}$$

which gives a contribution to the cross-section

of the form $\dfrac{R_n}{(E-E_n)^2 - \frac{1}{2}\,\Gamma_n^2}$, correctly represen-

ting an isolated resonance peak. Of course, there
must also exist poles in the upper half-plane,
corresponding to unphysical situations which would
be the time-irreverses of the decaying states. We
thus have no contradiction with reversibility. The
observation process selects from all the poles
those corresponding to the decaying states.

This interpretation is only valid if the S-matrix actually has the assumed analytic properties.
The theory can be developed in full generality for
the case of finite range potential scattering.
There, one can prove that S(E) is indeed meromor-
phic in the complex E plane. In the general case,
no such proof can be given. In the first place,
the wave-numbers are non-analytic functions of the
selected independent variable k or E, and the S-matrix elements are therefore non-analytic functions
with branch points on the real axis, corresponding
to the "opening" of the channels. These branch
points necessitate the introduction of cuts in
the complex E plane. However, it is then possible
to extend the theory of elastic scattering by assuming a distribution of poles on the different
Riemann sheets.

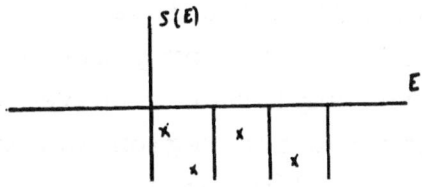

The idea of decaying states was first introduced by Gamow in the early days of quantum mechanics, to describe the spontaneous emission of
α -particles from nuclei. It was received with
such scepticism at the time, since the meaning of
solutions of the Schrödinger equation with complex
eigenvalues was unclear. Born elucidated their
physical meaning by the method outlined above. Now,
we realize that the complex poles are in fact eigenvalues of a non-hermitian generator, which has
been explicitly written down by Claude Bloch. Its
non-hermitian part, related to the irreversibility
of the decaying state concept exhibits a direct dependence on the boundary conditions.

We meet here a situation for which the regular
framework of quantum mechanics is obviously too
narrow (as the early critics of Gamow's work correctly felt): this breakdown of the usual rules of
quantum mechanics is shown by the fact that the
wave-function corresponding to a decaying state is
not normalizable. In fact, besides the time-factor
expressing exponential decay in time, this function has a radial factor exp $(ik_n r)$, with the
complex wave-number $k_n = K_n - i\gamma_n$ $(\gamma_n > 0)$, such that
$k_n^2 = E_n - \frac{1}{2}i\, \Gamma_n$. This means that the amplitude of the
wave-function increases exponentially with increa-
sing distance from the interaction region. The physical meaning of this property was already explained
by Gamow: the amplitude received at a given distance
r of the interaction region was emitted at an ante-
rior time $t-(r/v)$ (where v is the velocity of the

outgoing motion), its variation with r must thus
reflect the inverse of the variation with t .The
non-renormalizability of the decay state wave-
function is accordingly a direct consequence of
the irreversibility in time of the process it
describes. We expect, therefore, that the incorpo-
ration of irreversible processes in a quantal description will necessitate a widening of the formalism of quantum mechanics.

3. Irreversibility of large quantal systems

If one attempts a general theory of large
systems with the aim of analysing the origin of
their irreversible behaviour, two different points
of view offer themselves, both initiated by Boltzmann: the ergodic and the kinetic one. In the ergodic approach, one considers the evolution of finite
systems, confined in the classical case in a finite
manifold of their phase space. The cyclic character
of the phase space trajectories makes it necessary
to consider the whole motion in a global way, and
the irreversibility appears in a "coarse-grained"
description, in which the ideally precise locali-
zation in phase-space is replaced by a statistical
distribution over finite cells. Two conditions -
"metrical transitivity" and "mixing" - must be
satisfied in order to ensure an irreversible trend
of the coarse-grained distribution towards an equilibrium distribution. However, these conditions
have little practical value, since they are not
directly related to the Hamiltonian and therefore

do not provide us with any criterium to decide
whether a given system satisfies them. Moreover,
the quantal generalization of the coarse-graining
concept gives rise to great formal complications.

The kinetic approach circumvents these difficulties if it applied to an infinite system in the
"thermodynamic limit" (in which the extension and
the number of degrees of freedom both become infinite in such a way that the density remains finite):
for in this limit, the Poincaré cycle becomes of
infinite duration, and the dynamical behaviour
must therefore be expected to exhibit (in general)
an irreversible trend in the asymptotic limit of
sufficiently large, but finite, times. Thus, one
obtains at the macroscopic level of description,
a smoothed-out representation of the time evolution of the system, showing an approach to equilibrium on the macroscopic time scale: this is the
content of Boltzmann's H-theorem. In the usual
treatment, the definition of the asymptotic limit
implies a departure from the strict dynamical description, taking the form of some statistical
assumption - Stosszahlansatz, time averaging, etc.
It is possible, however, as we shall see, to dis-
pense with such explicit assumptions and obtain
asymptotic solutions satisfying the dynamical
equation of evolution exactly, but exhibiting an
asymmetry in time which only results from the
property of incoherent atomic correlations of limited range of having "no memory",i.e. of dying
out after times of atomic order of magnitude. This

method, due to the Brussels group led by I. Prigogine, has the further advantage of yielding a general criterium of "dissipativity", directly expressed in terms of the Hamiltonian, and allowing
one to decide whether a given system will show
any irreversible approach to equilibrium on the
macroscopic time-scale. The theory has recently
been given by Prigogine, George and myself [1] a
quite general form, based on the formalism of
quantum mechanics, but representing a non-trivial
extension of this formalism, enabling it to embody
the treatment of irreversible processes. Let us now
proceed to a brief account of this theory, valid
for at any rate an extensive class of very large
quantal systems considered in the thermodynamic
limit.

  In a superspace defined as the direct product
of the Hilbert space and its dual, the density
operator $\rho(t)$ appears as a supervector, varying
in time according to a Liouville equation of the
form

$$i \, \dot{\rho}(t) = L \, \rho(t).$$

The Liouville superoperator L may be expressed in
terms of the Hamiltonian H of the system. For this
purpose, we may use a convenient notation for the
special class of "factorizable" superoperators
$O \equiv M \times N$ depending on a pair of supervectors M, N
according to the definition $O\rho = M \rho N$; we may then
write $L = H \times 1 - 1 \times H$. Our aim being to find out the
long time effect of correlations, we must, to begin
with, compare our system, defined by the Hamiltonian

H, with a "model" system H_o, from which the interaction energy V, responsible for the correlations,
is removed, in such a way that $H=H_o+V$. The eigenstates of H_o form a complete orthogonal basis of
representation in Hilbert space, from which we
construct a similar basis in superspace: the latter may be divided into two classes of supervectors,
those built up of pairs of identical (or physically
equivalent eigenstates, and those built up of pairs
of different eigenstates; they belong, respectively,
to two orthogonal subspaces of superspaces of superspace, characterized by projection superoperators P_o, P_c. Then, the projections $\rho_o=P_o\rho$ and
$\rho_c=P_c\rho$ of the density supervector correspond,
respectively, to the average distribution densities
and the correlation amplitudes. Putting $L_{oo}=P_oLP_o$,
$L_{oc}=P_oLP_c$, etc. we obtain for ρ_o and ρ_c the
coupled Liouville equations

$$i\,\dot{\rho}_o=L_{oo}\,\rho_c - L_{oc}\,\rho_c \qquad , \qquad i\dot{\rho}_c=L_{cc}\rho_c-L_{co}\rho_o. \qquad (1)$$

The next step is to extract from these equations
the asymptotic forms $\tilde{\rho}_o(t)$, $\tilde{\rho}_c(t)$ of ρ_o and ρ_c
for large positive values of the time variable:
these are expected to express our possibilities
of prediction of the future evolution of the system
on the macroscopic time scale.

We must here restrict the generality of the
Liouville superoperator in order to characterize
the class of systems which we expect to exhibit
the "normal" asymptotic behaviour, i.e. an approach
to a state of equilibrium. To this end, we observe
that the time evolution of the correlation density

$_c$ is essentially governed by the superoperator
$T_c = \boxed{\exp}\,(-iL_{cc}t)$, depending on the part of the
Liouville superoperator which acts entirely in the
correlation subspace. We assume accordingly that
the asymptotic effect of this superoperator $T_c(t)$
upon any regular supervector A which is not an
invariant is to reduce this supervector to zero:
$\boxed{\lim}_{t \to \infty} T_c(t)A=o$; we express by this assumption
the fading of the system's "memory" of its corre-
lations. This condition, which may also be expressed
as an analyticity condition on the Laplace transform of $T_c(t)$, has first been verified in this
form by a perturbation expansion, for infinite
systems in whose description there enters a "small"
physical parameter such as the coupling constant
or the density [3]. More recently, it has been shown
that for soluble models, such as the Friedrichs
model, the analyticity assumption is satisfied
regorously (i.e. independently of any perturbative
approach) for a large class of interactions [4]. By
means of this assumption, we readily derive from
the second Liouville equation (1) the following
relation between the asymptotic densities:

$$\tilde{\rho}_c(t)=\int_0^\infty \boxed{d}\,\tau\,\boxed{e}^{\,-iL_{cc}\tau}\,(-iL_{co})\,\tilde{\rho}_o(t-\tau)\ ; \qquad (2)$$

it has the form of an integral equation, showing
the asymptotic correlations built up by sequences
of processes starting from the average situations
through which the system passes in the course of
time.

Let us introduce at this state an asymptotic

time evolution operator by writing $\tilde{\rho}_o(t)$ in the form

$$\rho_o(t) = e^{-i\Theta t}\,\tilde{\rho}_o(o). \tag{3}$$

The advantage of the representation (3) is to reduce the integral equation (2) to a simple linear relation between $\hat{\rho}_o(t)$ and $\tilde{\rho}_c(t)$ taking at the same time:

$$\tilde{\rho}_c(t)=C\,\tilde{\rho}_o(t) \;,\;\; C=C(\Theta)=\int_o^\infty \boxed{d}\tau\,\boxed{e}^{-iL_{cc}\tau}(-iL_{co})\,\boxed{e}^{\,i\Theta\tau} \tag{4}$$

The first Liouville equation (1) then yields a functional equation for Θ:

$$\Theta = L_{oo} + L_{oc}\,C\,(\Theta), \tag{5}$$

which can be solved by iteration.

The total asymptotic density $\tilde{\rho} = \tilde{\rho}_o - \tilde{\rho}_c$ thus obtained has the remarkable property of being an <u>exact</u> solution of the Liouville equation. According to Eq. (4), it may be written in the form $\tilde{\rho} = P_a\tilde{\rho}$, with $P_a = P_o - C$. It is again remarkable that this superoperator P_a has the characteristic properties of a projection operator in superspace, indempotency and "adjoint symmetry" (i.e. it is such that the projection $P_a A$ of a self-adjoint supervector A is self-adjoint). The projector P_a defines a subspace in which the asymptotic density is confined. This subspace differs from the average subspace P_o by the adjunction of a part of the correlation subspace P_c, namely that part which is specified by the superoperator C; the latter may be interpreted as representing the building up of correlations from <u>asymptotic</u> average situations (we call it the superoperator of correlation

creation) - it sorts out those correlation processes which have a long time effect and accordingly
manifest themselves at the macroscopic level. Thus,
the asymptotic density   is not, as one might have
expected, governed by a "kinetic equation" different
from the dynamical Liouville equation: it is an
exact solution of the latter, and its asymptotic
character is conferred upon it by its confinement
to a subspace defined by the projector P_a.

The expression for the superoperator of correlation creation C, which enters in the definition
of P_a, is clearly unsymmetrical in time, and gives
the projector P_a the expected bias towards a pre-
ferred direction of the time evolution. In fact,
time inversion transforms P_a into a different
projector $P_a = P_0 + D$, where the superoperator

$$D = \int_0^\infty d\tau\, e^{i\eta\tau} (-iL_{oc})\, e^{-iL_{cc}\tau}$$

is the time-inverse of C; it contains the superoperator η which is the time-inverse of Θ and
obeys the equation $\eta = L_{oo} - DL_{co}$ derived from Eq.
(5) by time-inversion. In contrast to C, the superoperator D describes sequences of "destructions" of
correlations leading to asymptotic average situations.

An important element is still missing in the
picture: we must establish a link between the asymptotic density supervector $\tilde{\rho}$ (t) and the arbitrarily
chosen dynamical density supervector ρ (o) from
which the time-evolution is assumed to start. This
is readily supplied, however, on the basis of a
further remarkable property (easily derived) of the

superoperator Θ and its time-inverse   :

$$P_a \Theta = L P_a, \qquad \eta \overline{P}_a = \overline{P}_a L. \qquad (6)$$

With the notation $N_0 = 1 + DC$, it follows from Eqs. (6) and (3) that

$$N_0 \tilde{\rho}_0(t) = N_0 \boxed{e}^{-i\Theta t} \tilde{\rho}_0(o) = \boxed{e}^{-i\eta t} N_0 \tilde{\rho}_0(o)$$

and, on the other hand,

$$\overline{P}_a \rho(t) = \overline{P}_a \boxed{e}^{-iLt}(o) = \boxed{e}^{-i\eta t} \overline{P}_a \rho(o).$$

This shows that $N_0 \tilde{\rho}_0(t)$ and $\overline{P}_a \rho(t)$ have the same time-evolution, governed by the superoperator $\boxed{\exp}(-i\eta t)$: we may therefore equate them at any instant and in this way fix the correspondence between the dynamical and the asymptotic density. This gives the quite fundamental relation, valid at any time,

$$\tilde{\rho}(t) = \tilde{\pi}\rho(t) \quad \boxed{\text{with}} \quad \tilde{\pi} = P_a N_0^{-1} \overline{P}_a \qquad (7)$$

from which follows

$$\tilde{\rho}(t) = \tilde{\sum}(t)\rho(o) \boxed{\text{with}} \tilde{\sum}(t) = \tilde{\pi}\boxed{e}^{-iLt}$$

the answer to our last question, completing the theory.

The striking feature about the superoperator $\tilde{\pi}$ occurring in Eq. (7) is that it is also a projector in the extended sense defined above (which does not include the property of self-adjointness); moreover, it is time-reversal invariant: it defines a time-symmetrical subspace of the superspace in which the asymptotic part of the time-evolution, starting from any given situation, remains confined, exhibiting the features observed at the macroscopic level; whereas the irregular fluctuations occurring

on the atomic time scale are contained in the complementary subspace orthogonal to the asymptotic
one. That such a clean separation between the two
aspects of the atomic system could be effected is
an entirely unexpected property specific to the
density supervector representation: it could never
have been found by a study of the evolution of the
system in Hilbert space, for it can only be formulated in terms of the superspace formalism.

The superoperator $\widetilde{\pi}$ is not factorizable: one
cannot ascribe any state vector to an asymptotic
situation as we have defined it, but only a density
supervector $\widetilde{\rho}$. In fact, as appears from Eq. (4),
the correlation part $\widetilde{\rho}_c$ of the density $\widetilde{\rho}$ is directly derived from the part $\widetilde{\rho}_o$ expressing the average
probability distributions of the system; owing to
this remarkable structure of the projector $\widetilde{\pi}$, the
evolution in $\widetilde{\pi}$ = space may be entirely described
in terms of probabilities only. In particular, the
"reduction of the wave-packet" of an atomic system
after its interaction with a measuring apparatus
is a direct consequence of this property of the
$\widetilde{\pi}$ -space description: the essential point being
that the apparatus must necessarily belong to the
macroscopic level of quantum mechanics, and its
behaviour accordingly be described in terms of its
$\widetilde{\pi}$ -space variables (loosely speaking, the behaviour
of the apparatus has "thermodynamical" character,
inasmuch as variables pertaining to thermodynamic
equilibrium or near-equilibrium states all belong
to $\widetilde{\pi}$ -space). Any phase relations in the initial

state of the atomic system are therefore wiped
out (i.e. are rejected into the orthogonal subspace): this is the only meaning of the "reduction"
of the initial state of the atomic system resulting
from the measurement. As to the human observer,
his interaction with the apparatus is also entirely
described in the $\widetilde{\pi}$ -subspace, and therefore without
any influence whatsoever on whatever goes on in
the orthogonal subspace.

One further point should be mentioned. The
theory gives us a simple criterium to decide
whether the system shows the normal macroscopic
behaviour described by thermodynamics. The superoperator of asymptotic time-evolution Θ is closely
related to a "collision superoperator" defined as
the Laplace transform of the superoperator
$L_{oc} T_c(\tau) L_{co}$:

$$\gamma(z) = \int_0^\infty d\tau\, L_{oc}\, T_c(\tau)\, L_{co}\, e^{-z\tau} ;$$

indeed, Eq. (5) may be written as a functional
relation in terms of $\gamma(z)$:

$$i\, \Theta = i\, L_{oo} + \gamma(-i\, \Theta). \tag{8}$$

Now, a homogeneous system (i.e. a system for which
$L_{oo} = 0$) will obviously not exhibit any tendency
towards equilibrium if the collision operator
vanishes identically. Eq. (8) then shows that it
will exhibit an irreversible tendency towards equilibrium provided that Θ itself does not vanish identically. This "condition of dissipativity" is a
practical one: it can be tested in concrete cases[4]
by actual computation of Θ.

REFERENCES

1. C. George, I. Prigogine and L. Rosenfeld:
    Det kgl. Danske Videnskabernes Selskab,
    mat.-fys. Meddelelser 38 (1972) no. 12

2. I. Prigogine, C. George and F. Henin:
    Physica 45,418 (1969); see also R. Balescu
    and L. Brenig: Physica 54, 504 (1971)

3. I. Prigogine: Non-Equilibrium Statistical
    Mechanics (Interscience Publ., 1962).

4. A. Grecos and I. Prigogine: Physica (1972);
    Proc. Nat. Acad. Sci. (1972).

ERGODICITY AND RELATED TOPICS

I.E. Farquhar

Department of Theoretical Physics
University of St. Andrews
St. Andrews, Scotland

1. The Irrelevance of Ergodicity to Irreversibility.

2. Abstract Dynamical Systems.

3. Application of Abstract Dynamical Theory to Physical Systems.

4. Irreversibility and Infinite Systems.

5. Integrable Systems.

1.   The Irrelevance of Ergodicity to
Irreversibility

1.1   In these lectures I have been asked to
review ergodic theory, whilst the subject matter
of this school is irreversibility. Our natural
programme would seem to be first to say what we
mean by ergodicity and then to see what relevance
this concept has for irreversibility.

So let us do this.

First, what do we mean by ergodicity? It is
the property of an ergodic system, a system for
which the ergodic theorem is valid. And in
physics the ergodic theorem is the assertion that
time averages and ensemble averages are equal for
all phasefunctions of physical interest in a
classical system and for expectation values of
all physical observables in a quantal system.

We shall concentrate attention on classical
systems for the simple reason that ergodic theory
in quantum mechanics has not given rise to any
results of much significance.

1.2   In classical mechanics we are familiar with
the concept of a phase space Γ , such that the
state of the system at time t is determined by
a point $x_t \in \Gamma$. The evolution of the system is
described by the motion of this phase point in
Γ under the action of a set of transformations
T_t that for any t map Γ onto itself; thus

$$x_t = T_t x \qquad (1.1)$$

for all $x \in \Gamma$. These transformations are induced by the equations of motion of the system.
Since the transformations have the properties

$$T_{t_1 + t_2} = T_{t_1} T_{t_2} \; ; \quad T_o = 1 \; ;$$
$$T_t \, T_t^{-1} = 1 = T_t^{-1} \, T_t \quad , \qquad (1.2)$$

they form a one-parameter group of transformations
$T_t (-\infty < t < \infty)$ of Γ onto itself, that is, a
one-parameter group of automorphisms. Such a
group is termed a **flow**.

A basic feature of this flow is that it is
<u>measure-preserving</u>. For a classical system of
n degrees of freedom the usual choice of phase
space Γ is R^{2n}, the 2n-dimensional real
Euclidean space. The natural choice of measure
in this space is the Lebesgue measure μ , which
is just the measure that corresponds to the
'volume' in phase space. That this measure is
invariant under the flow engendered by the
Hamiltonian equations of motion is precisely the
content of Liouville's theorem.

If we suppose that the measure of Γ is
finite, we may for convenience take $\mu(\Gamma) = 1$.
(Should $\mu(\Gamma)$ be unbounded, we restrict attention to some subset of Γ of finite measure and
use this subset in place of Γ .)

Consider now any integrable phasefunction
f on Γ . The phase average of f is defined

as

$$\langle f \rangle_\Gamma \equiv \int_\Gamma f(x)d\mu ; \quad \mu(\Gamma) = 1. \tag{1.3}$$

The time average of f is defined as

$$\langle f \rangle_J \equiv \lim_{T \to \infty} \frac{1}{T} \int_{t_0}^{t_0+T} f(T_t\, x)dt \tag{1.4}$$

for all $x \in \Gamma$ for which this time average exists.

The ergodic theorem may be regarded as asserting that for all integrable phasefunctions f - or, less strongly, for all integrable phase-functions that are of physical significance - this time average (i) exists almost everywhere in Γ (that is, for all $x \in \Gamma$ except at most on a set of measure zero), and (ii) whenever it exists, is equal to the corresponding phase average.

1.3   Very often in classical mechanics we have to deal with a conservative Hamiltonian system. The system is both closed and isolated - not only is there no exchange of particles between system and surroundings, but there is also no exchange of energy. Such a system of n degrees of freedom is describable by a set of conjugate variables $(q,p) \equiv (q_1, q_2, \ldots, q_n; p_1, p_2, \ldots, p_n)$ that obey Hamiltonian equations of motion

$$\dot{q}_k = \frac{\partial H}{\partial p_k} \; ; \quad \dot{p}_k = -\frac{\partial H}{\partial q_k} \quad (k = 1, \ldots, n)$$
$$(1.5)$$

where the Hamiltonian is such that

$$H = H(q,p) \neq H(q,p;\, t) \quad , \qquad (1.6)$$

in accordance with the system's being isolated.
The motion of a given phase point in Γ is now
constrained by the relation

$$H(q,\, p) = E \quad . \qquad (1.7)$$

We may envisage a family of (hyper-) surfaces of
constant energy in Γ , phase trajectories for
a given system being confined to a particular
surface S_E characterised by the value of the
energy parameter E.

In terms of the invariant measure on the
2n-dimensional phase space Γ any energy surface, supposedly (2n-1)-dimensional, has measure
zero. Thus we cannot obtain useful results
concerning motion on an energy surface if we
utilise the 2n-dimensional measure μ on Γ .
On the other hand, the Lebesgue measure on the
energy surface, corresponding to the 'area' of
the hypersurface, is also unsuitable because it
is not invariant under the flow on Γ , —
according to Liouville's theorem it is the 2n-
dimensional measure μ that is invariant under
the motion of Γ . Accordingly, we introduce
on an energy surface S_E the measure given by

$$d\mu_E \equiv d\sigma/|\text{grad } H| \, , \qquad (1.8)$$

where $d\sigma$ denotes the (2n-1)-dimensional Lebesgue measure on S_E ; it is easy to see that this measure has the required property of invariance.

On the supposition that the energy surface S_E has finite measure we now define the phase average of any integrable phasefunction f on S_E as

$$\langle f \rangle_{S_E} \equiv \int_{S_E} f(x) \, d\mu_E \, ; \quad \mu_E(S_E) = 1 \, . \qquad (1.9)$$

Ergodicity then has the meaning that for all phasefunctions of interest time and phase averages are equal almost everywhere on S_E.

1.4   Now that we have stated what we mean by an ergodic system we may turn to the question of what such ergodicity has to do with irreversibility.

Let us look first at a very simple system, namely, the <u>linear harmonic oscillator.</u> For this system we have a Hamiltonian

$$H = \frac{1}{2} \frac{p^2}{m} + \frac{1}{2} m\omega^2 q^2 \qquad (1.10)$$

and energy surfaces

$$H(q,p) = E \qquad (1.11)$$

in the two-dimensional phase space; for a
given energy $E = E_0$ the energy surface is thus
an ellipse.

The equations of motion

$$\dot{q} = \frac{\partial H}{\partial p} \quad ; \quad \dot{p} = -\frac{\partial H}{\partial q} \qquad (1.12)$$

have the solution

$$q = q_0 \sin \omega t \quad ; \quad p = m\, q_0 \omega \cos \omega t, \qquad (1.13)$$

which are parametric equations of an ellipse.
Thus the trajectory is periodic and traces out
the entire (one-dimensional) energy surface. For
such periodic motion the time average of any
integrable phasefunction is the average taken over
one period of the motion and clearly this is equal
to the phase average of the phasefunction taken
also over the ellipse. This system is thus
<u>ergodic</u>, the ergodicity being a consequence of
the trajectory's covering the entire energy surface.

So in the linear harmonic oscillator we have
an ergodic system. The question now arises of
what is irreversible about the motion of such an
oscillator - and the answer would appear to be
'nothing whatsoever'.

So the property of ergodicity in itself is
insufficient to give rise to irreversibility.

1.5   It may be objected that the system we have
considered is of a very special nature in that
the phase trajectory is identical with the energy
surface, both here being one-dimensional.

Let us look then at another simple system
free from this property. The two-dimensional
anisotropic oscillator would serve, this having
a four-dimensional phase space and three-
dimensional energy surfaces; however, we should
have to take into account constants of the
motion other than the Hamiltonian and this is
extraneous to our present purpose. To avoid this
we therefore choose a system which is actually
mathematically similar to the two-dimensional
oscillator but which we shall in fact introduce
as a non-Hamiltonian system.

We thus return to the notion of a flow in
phase space and choose a two-dimensional phase
space with motion described by the equations

$$\dot{x} = 1 \quad ; \quad \dot{y} = \alpha \qquad (1.14)$$

where α is a parameter. Furthermore, we take
as a subset of phase space of finite measure the
unit square

$$0 \leqslant x \leqslant 1 \quad ; \quad 0 \leqslant y \leqslant 1 \qquad (1.15)$$

and we impose periodic boundary conditions. This,
in fact, is equivalent to choosing the phase space
to be a torus.

The equations of motion have the solution

$$x = x_0 + t \; ; \; y = y_0 + \alpha t, \quad (\text{mod } 1) \tag{1.16}$$

and the phase trajectory is

$$y = y_0 + \alpha(x - x_0) . \tag{1.17}$$

Should α be rational, $\alpha \equiv p/q$ where p and q $\neq$ 0 are integers, the trajectory is periodic, being retraced after a time q. Thus for $\alpha = {}^3/2$, say, a typical trajectory is as shown:

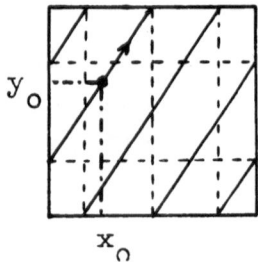

The time average of a phasefunction, taken over the closed trajectory, is not in general equal to the phase average of the phasefunction taken over the entire phase space, that is, the unit square.

Should α be irrational, however, we obtain not a closed trajectory but one that is <u>dense</u> in the unit square:

In pictorial language the trajectory passes arbitrarily close to all points in the unit square. It does **not**, however, pass through all points in the unit square, since nowhere does it intersect itself and being thus one-dimensional it cannot fill a two-dimensional surface. These remarks have relevance to the so-called ergodic hypothesis, the original form in which ergodicity was expressed.

Consider any integrable function f on the unit square. In view of the periodic boundary conditions we expand f in a Fourier series

$$f = \sum_{\ell,m=-\infty}^{\infty} A_{\ell m}\, e^{2\pi i(\ell x + my)} \qquad (1.18)$$

and so obtain the time average

$$\langle f \rangle_J = \lim_{T\to\infty} \frac{1}{T} \int_{t_0}^{t_0+T} \sum_{\ell,m} A_{\ell m} \times e^{2\pi i[\ell(x_0+t)+m(y_0+\alpha t)]}\, dt$$

$$= A_{00} + \lim_{T\to\infty} \frac{1}{T} \sum_{\ell,m\neq 0} A_{\ell m} \times e^{2\pi i[\ell(x_0+t_0)+m(y_0+\alpha t_0)]} \times \frac{\left[e^{2\pi i(\ell+\alpha m)T} - 1 \right]}{2\pi i(\ell+\alpha m)}$$

Since α is irrational, $(\ell + \alpha m) = 0$ only if $\ell = 0 = m$. It follows that

$$\langle f \rangle_J \;=\; A_{oo} \;, \tag{1.19}$$

whereas

$$\langle f \rangle_\Gamma \;=\; \int_0^1 \int_0^1 f(x,\,y)\,dx\,dy$$

$$=\; A_{oo} \;. \tag{1.20}$$

Thus this system is ergodic when α is irrational.

Again we inquire as to what is irreversible in this ergodic flow. And again the answer would seem to be 'nothing at all'.

1.6 It will no doubt have been realised by now that the theme of this lecture is that ergodicity does not imply irreversibility. Aside from its interest to mathematicians, what then, one may ask, has ergodic theory to contribute to physics? My own view is that the relevance of ergodic theory to physics lies in what it may provide towards justifying the basic procedure of equilibrium statistical mechanics, — that extraordinary procedure whereby macroscopic properties of a single system, the microscopic state of which evolves in time, are obtained as averages taken over a time-dependent ensemble of systems.

It is well known that from the postulate

that the microcanonical ensemble is appropriate
to the description of an isolated system one may
deduce that the canonical ensemble is appropriate
to the description of a closed system in a thermostat, the grand canonical ensemble to the description of an open system in a thermostat, and
so on.   In setting up the framework of equilibrium statistical mechanics it is the first step,
the assertion that the microcanonical ensemble
describes an isolated system, that is difficult
to substantiate.

It is this assertion that ergodic theory
seeks to justify.

For this purpose we interpret the time average
that occurs in ergodic theory in the following
manner.   Measurement of a thermodynamic obser-
vable is not carried out instantaneously but requires a non-zero interval of time.   Thus a
measured value is to be represented theoretically
by a time average

$$\tau^{-1} \int_{t_o}^{t_o + \tau} f(T_t x)dt.$$

Although the time interval τ may be comparatively short on a macroscopic time scale, it
may yet be very long on the microscopic time
scale that is appropriate to the motion of individual particles.   Indeed, it may well be
appreciably longer than macroscopic relaxation
times.   In such circumstances we may suppose
that the time interval τ can be extended without bound on the microscopic time scale and the

measured value thereby identified with the time
average over an infinite period that appears in
ergodic theory. The ergodic theorem, if valid,
then permits macroscopic properties of a single
system to be computed as averages over a
statistical ensemble, the ensemble for a conservative Hamiltonian system being the microcanonical ensemble on the appropriate energy
surface.

This interpretation of the relevance of
ergodic theory to physics not merely fails to
account for irreversibility but actually presupposes equilibrium insofar as the extension to
infinity of the time interval τ is involved.
For it seems impermissible to allow this extension unless the system is in equilibrium
throughout the duration of the measurement or
relaxes to equilibrium during a small part of
the time interval. In this context equilibrium
is defined in terms of certain quantities being
constant in time. Ergodic theory, if valid and
if interpreted thus, indicates that for an
ergodic system in equilibrium such quantities
may be computed as averages over a statistical
ensemble; this statistical ensemble in turn
may be utilised to provide a definition of
equilibrium in other contexts.

It may be noted moreover that ergodic theory
makes no prediction regarding the magnitude of
relaxation times. This is so even if we accept
- which I am unwilling to do - the claim of

some that ergodic theory does indeed describe
the approach of a system to equilibrium. And a
prediction of the rate of approach to equilibrium
is a demand we should legitimately make of any
adequate theory of irreversibility. Thus, even
if this alternative interpretation be allowed,
in it the ergodic theorem has but the character
of a mathematical existence theorem and in no
way accounts for the processes by which equilibrium may be attained.

1.7   To sum up: we have defined the concept of
ergodicity and claimed it to be inadequate to
account for irreversibility, indeed irrelevant
to irreversibility.

Here would end a course on 'ergodicity and
irreversibility'. We should therefore disband
at this stage had I not taken the precaution of
adding to the title given to me the phrase 'and
related topics'.

In the remainder of this course, then, I
propose to discuss concepts which, like ergodicity
derive from the behaviour of general dynamical
systems but which, unlike ergodicity, seem much
more relevant to the problem of irreversibility
than merely to the problem of justifying the
use of ensembles.

2.   Abstract Dynamical Systems

2.1.   **The concepts we are** to discuss here
have been introduced and developed by mathematicians in the study of abstract dynamical
systems, generalisations of the systems of
classical mechanics. Although early work on
the ergodicity of such systems gave results that
proved not to be of utility within physics, yet
it is noteworthy that from this field there has
but recently emerged what is essentially the
only significant contribution to the ergodic
problem in physics in the course of about one
hundred years. It is to be hoped that results
of like significance for the wider problem of
irreversibility may also appear from this field
of study.

In describing these concepts we shall make
extensive use of the excellent book: V.I. Arnold
and A. Avez, Ergodic Problems of Classical
Mechanics (W.A. Benjamin, Inc., New York,
Amsterdam, 1968), of which there is an earlier
edition in French, Problèmes Ergodiques de la
Mécanique Classique (Gauthier-Villars, Paris,
1967.)

In that book can be found proofs of several
of the theorems that we shall merely state and
references to lengthy proofs of others. (It
may be noted that the phrase 'ergodic problems'
is used therein in a much broader sense than
that in which we employ it - in fact, in a
sense broad enough to include many of our

(related topics'.)

Let us first recall our characterisation of
a system in classical mechanics by means of
(i) a phase space Γ (ii) a group of automorphisms in Γ (iii) an invariant measure on
Γ.

As an abstraction from this physical concept
an <u>abstract dynamical system</u> may be defined[1] as
a triplet (Γ, μ, T_t), where (Γ, μ) is a
<u>measure space</u> in which there is a measure-
preserving one-parameter group T_t of automorphisms[2] of (Γ, μ).

It is clear that concepts of measure theory
are here of prime importance. It may therefore
be profitable to mention briefly some of the
basic ideas in measure theory.

With the eventual aim of obtaining a more
general and more versatile concept of integral
than that given by the Riemann theory, the
notion of an open interval of a certain length
is generalised to that of a measurable set of a
certain measure and the notion of a continuous
function to that of a measurable function. There
are different ways in which this may be done. We
choose a method[3] whereby the concepts
<u>measurable space</u>, <u>measurable set</u> and <u>measurable
function</u> are introduced in a manner clearly
analogous to that by which can be obtained the
similarly related concepts of <u>topological space</u>,
<u>open set</u> and <u>continuous function</u>.

Although it is a measure space that appears

in the definition of an abstract dynamical system,
we first require a measurable space. As with so
many constructs in mathematics this is defined
by way of a class of subsets that have certain
properties. In this instance the subsets form
a so-called **σ-algebra** a , this being a class
of subsets of a set Γ ,   the class containing
Γ and being closed with respect to the operations of forming (i) complements relative to
Γ and (ii) countable[4] unions of subsets.
If a is a σ-algebra in Γ , then Γ is a
measurable space and the subsets forming a are
the measurable sets in Γ .

A topological space and open sets are de-
finable similarly with a topology as intermediary,
a class of subsets of a set Y, this class containing Y and the empty set[5] and being closed
with respect to the operations of forming (i)
the intersection of a finite number of subsets
and (ii) the union of any number of subsets.

If X and Y are topological spaces and
if f is a mapping of X into Y, we may
define f to be continuous provided $f^{-1}(V)$ is
an open set in X for every open set V in Y.
Similarly, if Γ is a measurable space and Y
a topological space and if f is a mapping of
Γ into Y, we may define f to be measurable
provided that $f^{-1}(V)$ is a measurable set in
Γ for every open set V in Y.

It is conventional to denote a measurable
space Γ by a single symbol rather than by a

pair $(\Gamma, \mathcal{a})$; the existence of the σ-algebra
$\mathcal{a}$, all important though it be, is not given
explicit mention.

Having acquired a measurable space Γ we
now need to introduce a measure μ in Γ and
thereby obtain a <u>measure space</u> (Γ, μ).

A <u>measure</u> is a countably additive function
μ of range $[0, \infty]$ and defined on a σ-
algebra $\mathcal{a}$. Countable additivity means that if
$\{A_i\}$ is a <u>disjoint</u> countable class of members
of $\mathcal{a}$ then

$$\mu(\bigcup_{i=1}^{\infty} A_i) = \sum_{i=1}^{\infty} \mu(A_i) \quad . \quad (2.1)$$

A measure space (Γ, μ) is a measurable space
Γ which has a measure μ defined on the σ-
algebra $\mathcal{a}$ of its measurable sets. In what
follows it is convenient to assume that
$\mu(\Gamma) = 1$.

The remaining constituent of an abstract
dynamical system is the measure-preserving
group T_t of automorphisms such that for any
measurable sets A and B in Γ
(i) $\mu(T_t A \cap B)$ is a measurable function of t
and (ii) $\mu(T_t A) = \mu(A)$ for all t. Thus the
group T_t defines a measurable <u>flow</u> on (Γ, μ).
Although the notation used here suggests that
T_t is a continuous group with parameter t, the
group may be a discrete group generated by an
automorphism T.

The advantage of this very general concept

is indeed its generality. Thus the concept
embraces, for instance, not only Hamiltonian
systems and non-Hamiltonian flows such as the
ones we have encountered already, linear harmonic oscillator and translation of a torus, but
also processes of a stochastic nature that have
either a continuous or a discrete time variable.
(As an example of the last-mentioned class we
cite the so-called Bernoulli trial[6] that describes the tossing of a coin.)   Furthermore,
properties of abstract dynamical systems that
depend only on μ and T_t are the same for any
class of <u>isomorphic</u> systems, those for which the
respective measure spaces are isomorphic to one
another and the groups of automorphisms likewise.

The disadvantage of the general formulation,
as far as physics is concerned, is that the behaviour of a Hamiltonian system of interest in
physics may not be typical of the behaviour of
abstract dynamical systems in general. It is
customary to confine attention to generic properties of such systems, effectively to introduce a measure on the set of systems and to
obtain properties that are valid for all systems
except those forming a subset of measure zero.
This procedure is perfectly acceptable, of course,
provided one is indeed interested in abstract
dynamical systems as such. It is less profitable,
however, if the systems in which we are interested
happen all to fall into the subset of measure
zero. A subset of measure zero is not automatically

of no interest. The exception in a generic frame
of reference may well be the exemplar in a
specific frame of reference.

   With this caution, let us turn to a considera-
tion of properties of abstract dynamical systems
that to a greater or lesser extent may indicate
irreversible behaviour.

2.2   The concept of <u>ergodicity</u> may be introduced
into the theory of abstract dynamical systems just
as for classical-mechanical systems, in terms,
that is, of the equality of time averages of
integrable[7] functions on the measure space and
of averages of these functions with respect to
the measure over the space. The two systems we
considered are examples of abstract dynamical
systems, the linear harmonic oscillator having
a Hamiltonian flow that is non-ergodic in the
entire phase space but is ergodic on an energy
surface and the translational flow on the torus
being ergodic when the parameter characterising
the flow is irrational. As we have already discussed, the property of ergodicity is irrelevant
for irreversibility.

2.3   The next concept we look at is that of
<u>mixing</u> and here the prospects for irreversibility
look decidedly more favourable. Let us study
first a particular system, again a two-dimensional

system but on this occasion not with a translational flow defined on it. Indeed we shall
consider not a flow at all but rather a single
automorphism of the torus which we can, of
course, apply repeatedly. Specifically, we consider

$$\begin{pmatrix} x' \\ y' \end{pmatrix} = \begin{pmatrix} 1 & 1 \\ 1 & 2 \end{pmatrix} \begin{pmatrix} x \\ y \end{pmatrix} . \text{(mod 1)}$$

$$(2.2)$$

The measure we choose is the Lebesgue measure
representing area, $d\mu = dx\, dy$, and we note
that the mapping preserves this measure since
the transformation matrix (or Jacobian) has
determinant of value unity.

This is the precise system that Arnold and
Avez discussed[8] as an example and the reason we
choose the same system is that we must not lose
the opportunity of reproducing their celebrated
feline illustration.

A second application of the mapping gives something of the form

It is a curious circumstance that in physics
cats seem to be singled out for ill treatment.
In quantum mechanics we spend much time inquiring
whether Schrödinger's cat[9)] is alive and well
or whether it has been poisoned. There seems to
be little doubt as to the state of health of the
Arnold and Avez cat.

We may contrast the latter's fate with the
life history of an ergodic cat. To exhibit the
contrast we extract a single mapping

$$x \longrightarrow x + \beta \quad ; \quad y \longrightarrow y + \gamma \quad (\text{mod } 1)$$
$$(2.3)$$

from the flow

$$x = x_0 + t \quad ; \quad y = y_0 + \alpha t \quad (\text{mod } 1)$$
$$(1.16)$$

considered previously.

Provided that both β and γ are irrational, repeated application of this mapping gives
ergodicity. A single application of the mapping
transforms

and an iteration gives

Remembering that the cat is actually on a torus
and so is not in fact chopped in pieces and stuck
together again, we see that, unlike the cat subjected to mixing, the ergodic cat's only danger
lies in the distortions it suffers when I attempt
to draw it.

Note in passing that it would not be sufficient to take, say

$$x \rightarrow x+1 \quad ; \quad y \rightarrow y + \alpha \quad (\text{mod } 1)$$
$$(2.4)$$

as the mapping, even with α irrational. This
mapping repeated indefinitely would translate the

cat parallel to the y-axis but the x-coordinate
of any point remains unaltered; such a mapping
would not give rise to ergodicity.

The illustration conveys an immediate
impression that mixing, unlike ergodicity, may
be significant for irreversibility. Indeed, the
mathematical description of mixing by way of
measure theory is derivable as an abstraction
from the description of the stirring together
of two incompressible fluids, a process which,
whether the ingredients be ink and water as discussed by Gibbs or more potable concoctions as
favoured by other authors, is often held to be
a paradigm of an irreversible process.

The group of automorphisms in a measure
space (Γ, μ) such that $\mu(\Gamma) = 1$ is <u>mixing</u>
provided

$$\lim_{t \to \infty} \mu\left[T_t A \cap B\right] = \mu(A)\,\mu(B) \quad (2.5)$$

for all measurable subsets A and B of Γ .
This definition is an abstraction from the
notion that after a sufficiently long time the
proportion of, say, ink in <u>any</u> region of a container in which ink and water are stirred together is approximately equal to the proportion
of ink placed in the container before the stirring was begun.

Mixing is a stronger property than ergodicity
in the sense that mixing implies ergodicity
whereas the converse is not true.

To see this, let A be an <u>invariant</u>
measurable subset of Γ . Then for any t
we have

$$T_t A \cap A \quad = \quad A . \qquad (2.6)$$

The mixing property (2.5) with B taken equal
to A now yields

$$\mu(T_t A \cap A) \quad = \quad \mu(A)\mu(A) \qquad (2.7)$$

It follows that

$$\mu(A) \quad = \quad 0 \ \text{or} \ 1 . \qquad (2.8)$$

This property, which we have shown a mixing
system to possess, that any invariant measurable
subset has measure either zero or unity is known
as the <u>metrical indecomposability</u> of the system
(or as the metrical transitivity of the group of
automorphisms), a concept of some notoriety in
the application of ergodic theory to physics.

In fact, the metrical indecomposability of
an abstract dynamical system is a necessary and
sufficient condition for the ergodicity of the
system.   That this is so is a consequence of
the theorem[10] due to Birkhoff which establishes
(i)   the existence almost everywhere in Γ of
the time average

$$\langle f(x) \rangle_{\mathcal{J}} \quad \equiv \quad \lim_{\tau \to \infty} \frac{1}{\tau} \int_{t_0}^{t_0 + \tau} f(T_t x) dt, \quad x \in \Gamma , \qquad (2.9)$$

of an integrable function f on Γ ,

(ii) the almost everywhere invariance of $\langle f \rangle_{\mathcal{J}}$ in the sense that

$$\langle f(T_t x) \rangle_{\mathcal{J}} = \langle f(x) \rangle_{\mathcal{J}} \qquad \text{a.e. in } \Gamma, \text{ all } t, \tag{2.10}$$

(iii) the integrability of $\langle f \rangle_{\mathcal{J}}$ and the result

$$\int_{\Gamma} \langle f \rangle_{\mathcal{J}} \, d\mu = \int_{\Gamma} f \, d\mu \tag{2.11}$$

that the mean of the time average is equal to the mean of the function itself. From (2.11) it follows immediately that ergodicity is equivalent to the almost everywhere constancy of $\langle f \rangle_{\mathcal{J}}$ for all measurable f.

(a) <u>Necessity</u>: Let (Γ, μ, T_t) be metrically decomposable. Then

$$\Gamma = \Gamma_1 \cup \Gamma_2, \qquad \Gamma_1 \cap \Gamma_2 = \emptyset$$
$$T_t \Gamma_1 = \Gamma_1, \qquad T_t \Gamma_2 = \Gamma_2, \text{ all } t;$$
$$1 > \mu(\Gamma_1) > 0, \qquad 1 > \mu(\Gamma_2) > 0. \tag{2.12}$$

Choose as an integrable function the so-called characteristic function $\chi_1(x)$ of Γ_1, defined by

$$\chi_1(x) = \begin{cases} 1 & x \in \Gamma_1 \\ 0 & x \notin \Gamma_1. \end{cases} \tag{2.13}$$

The invariance of Γ_1 and Γ_2 gives

$$\langle \chi_1(x) \rangle_{\mathcal{J}} = 1 \qquad x \in \Gamma_1 \tag{2.14}$$

and

$$\langle \chi_1(x) \rangle_{\mathcal{J}} = 0 \qquad x \in \Gamma_2 \tag{2.15}$$

Thus $\langle \chi_1(x) \rangle_{\mathcal{J}}$ is not constant a.e. in Γ and so the system is not ergodic.

(b) <u>Sufficiency</u>: Let (Γ, μ, T_t) be non-ergodic. Then there exists an integrable function $f(x)$ such that $\langle f(x) \rangle_{\mathcal{J}}$ is not constant a.e. in Γ. Thus there exists a number c such that

$$\langle f(x) \rangle_{\mathcal{J}} > c \qquad\qquad x \in \Gamma_1 \qquad (2.16)$$

and

$$\langle f(x) \rangle_{\mathcal{J}} \leqslant c \qquad\qquad x \in \Gamma_2 \qquad (2.17)$$

where

$$1 > \mu(\Gamma_1) > 0, \qquad 1 > \mu(\Gamma_2) > 0. \qquad (2.18)$$

From the almost everywhere invariance (2.10) of the time mean it follows that Γ_1 and Γ_2 are invariant. Thus the system is not metrically indecomposable.

We have thus demonstrated that a mixing system, being metrically indecomposable, is ergodic. The converse is disproved by the counterexample[11] of ergodic translations of a torus. A subset of the torus may easily be found such that the intersection of the cat with this subset is sometimes empty and sometimes of positive measure; the mixing property (2.5) is not then satisfied.

The reason for this apparent digression on metrical indecomposability is that the concept can serve as a criterion of whether a system has a mixing property. The property in question is not quite that of mixing as defined in (2.5) but that

of <u>weak mixing</u>, which is expressed by the requirement

$$\lim_{\tau \to \infty} \frac{1}{\tau} \int_0^\tau \left| \mu(T_t A \cap B) - \mu(A)\mu(B) \right| dt = 0 \quad (2.19)$$

for all measurable subsets A and B in Γ .
This property is intermediate between those of
ergodicity and of mixing (otherwise known, for
purposes of distinction, as strong mixing) but
as regards its relevance to irreversibility it
lies much closer to mixing than to ergodicity.
Interpreted in terms of the stirring of ink and
water it indicates that after a sufficiently long
time the proportion of ink in any region of the
container is approximately equal to the proportion of ink placed initially in the container
except, at most, at a set of isolated instants
in time when the proportion may differ.

To Hopf[12)] is due the theorem that a
system is weakly mixing iff the symmetric product
space of Γ with itself is metrically indecomposable. This product space is formed from
pairs of points of Γ taken in either order; on
the σ -algebra that can be constructed in the
product space from that in Γ is defined a
measure derived from that in Γ .

Unfortunately, as we shall see when we turn
shortly to consider the application to physics
of the theory of abstract dynamical systems,
metrical indecomposability does not seem to be
a practical criterion.

2.4   An alternative criterion for weak mixing
can be obtained in terms of the spectral theory
of operators in a Hilbert space. This formalism,
as applied to Hamiltonian systems, was introduced
by Koopman[13]. The set of measurable single-
valued functions f on Γ that are not only
integrable but also square-integrable,

$$\int_{\Gamma} |f| \, d\mu < \infty, \quad \int_{\Gamma} |f|^2 \, d\mu < \infty \quad (2.20)$$

is a Hilbert space $\mathcal{H}$ with scalar product

$$(f, g) \equiv \int_{\Gamma} f^* g \, d\mu, \quad (2.21)$$

where f^* is the complex conjugate of f.

A transformation T_t in Γ induces in $\mathcal{H}$
an operator U_t that maps $\mathcal{H}$ onto itself such
that

$$U_t \, f(x) = f(T_t x) \qquad f \in \mathcal{H} \qquad (2.22)$$

The invariance under T_t of the measure μ in
Γ implies that

$$
\begin{aligned}
(U_t f, U_t g) &= (f(T_t x), g(T_t x)) \\
&= \int_{\Gamma} f^*(T_t x) g(T_t x) d\mu(x) \\
&= \int_{\Gamma} f^*(T_t x) g(T_t x) d\mu(T_t x) \\
&= \int_{\Gamma} f^*(x) g(x) d\mu(x)
\end{aligned}
$$

i.e. $(U_t f, U_t g) = (f, g) \qquad (2.23)$

which together with the linearity property

$$U_t(af + bg) = a \, U_t f + b \, U_t g \quad a \in \mathbb{C}, \, b \in \mathbb{C}$$
$$(2.24)$$

indicates that U_t is unitary. Furthermore, the

group property of the T_t $(-\infty < t < \infty)$ in turn
implies that the $U_t(-\infty < t < \infty)$ form a one-
parameter group of automorphisms of $\mathcal{H}$.

　　If rather than consider arbitrary abstract
dynamical systems we restrict attention to those
such that $(f, U_t g)$ is continuous in t for
all $f, g \in \mathcal{H}$, - and this condition is satisfied for the Hamiltonian systems of interest
within physics - we can set

$$U_t = \exp(iLt), \qquad (2.25)$$

where L is a self-adjoint operator, a linear
operator with domain $\mathcal{D}$ dense in $\mathcal{H}$ and such that

$$(Lf, g) = (f, Lg) \qquad (2.26)$$

We have used here the symbol L because in
physics this self-adjoint operator is generally
known as the <u>Liouville operator</u>

$$L \equiv i \sum_{k=1}^{n} (\frac{\partial H}{\partial q_k} \frac{\partial}{\partial p_k} - \frac{\partial H}{\partial p_k} \frac{\partial}{\partial q_k}) , (2.27)$$

where $H = H(q_1, \ldots, q_n; p_1, \ldots, p_n)$ is the
Hamiltonian of the system. The Liouville operator plays a fundamental role in the description
of irreversibility that has been developed[14] by
Prigogine and his colleagues in Brussels; this
is dealt with in the lecture course given at this
school by Professor Balescu, a prominent member
of the Brussels group.

　　In the present context a criterion for weak
mixing and a further criterion for (strong) mixing itself are to be sought in terms of properties

of the spectrum of the operator L. In general,
the spectrum of this operator possesses both a
discrete part and a continuous part. The spectrum may be defined[15] by means of the resolvent
operator, a concept which has been utilised extensively by the Brussels school.

Consider the equation

$$(L - z I)f = g, \qquad\qquad (2.27)$$

where $z \in \mathbb{C}$, I is the unit operator and
$f \in \mathcal{D}$. This establishes a correspondence between
the elements of the domain $\mathcal{D}$ of L and the
elements of the range $\Delta(z)$ of $(L - z I)$; in
particular, $\Delta(0)$ is the range of L. If this
correspondence is one-to-one the operator $(L - zI)$
has an inverse $(L - zI)^{-1}$, known as the resolvent of L, for which the domain is $\Delta(z)$ and
the range is $\mathcal{D}$. Those values of z for which
$\Delta(z) = \mathcal{H}$ are termed regular values of L; all
other values of z constitute the spectrum of L.

An invariant function

$$f = U_t f = e^{iLt} f \qquad\qquad (2.28)$$

corresponds to the Liouville operator's having
the eigenvalue zero. Now the arguments whereby
we demonstrated that metrical indecomposability
is a necessary and sufficient condition for
ergodicity also yield the result that ergodicity
is equivalent to all invariant functions being
constant a.e. in Γ . Since constant functions
are scalar multiples of one another, it follows
that a necessary and sufficient condition for

ergodicity is that zero be a simple (or non-degenerate) eigenvalue of L.

Similarly, a system has the property (2.19) of weak mixing iff[16] zero is the only member of the discrete part of the spectrum and moreover is non-degenerate; apart from this value the **spectrum for a weakly mixing system is continuous.**

Another property provides a criterion for (strong) mixing. A system with Lebesgue spectrum is mixing[17]. The statement that an abstract dynamical system has Lebesgue spectrum L^I means that for each U_t $(t \neq 0)$ there exists an orthonormal basis of $\mathcal{H}$ consisting of the function 1 and of functions f_{ij} $(i \in J, j \in Z)$ such that

$$U_t f_{i,j} \qquad f_{i,j+1} \qquad \text{for every } i, j. (2.29)$$

The index I is termed the multiplicity of the Lebesgue spectrum.

Example: The automorphism

$$\begin{pmatrix} x' \\ y' \end{pmatrix} = \begin{pmatrix} 1 & 1 \\ 1 & 2 \end{pmatrix} \begin{pmatrix} x \\ y \end{pmatrix} \quad (\text{mod } 1) \quad (2.2)$$

that mixes cats has Lebesgue spectrum[18]. To see this we first note that the functions

$$g_{p,q}(x,y) \equiv e^{2\pi i (px + qy)} \quad p,q \in Z \ (2.30)$$

form an orthonormal basis in the Hilbert space of square-integrable functions. Now the automorphism U induced in $\mathcal{H}$ by the automorphism T given in (2.2) is such that

$$U\, g_{p,q}(x,y) = g_{p,q}(T(x,y))$$

$$= e^{2\pi i\left[p(x+y) + q(x+2y)\right]}$$

$$= g_{p+q,\ p+2q}\ (x,y)\ . \qquad (2.31)$$

For $p = 0 = q$ we have $g_{oo}(x,y) = 1$. Other functions $g_{p,q}$ in the set may be separated into classes characterised by the index i such that if $f_{i,o}$ denotes a particular function in the i-th class then the other members of the class are $U^n f_{i,o}$, $n \in Z$ -- e.g. g_{10}, g_{11}, $g_{23},\ldots,$ g_{01}, g_{12}, $g_{35},\ldots$. Hence this automorphism has (denumerably infinite) Lebesgue spectrum and in consequence is mixing and ergodic.

2.5   The last of these general measure-theoretical concepts that we shall look at before turning to their significance within physics is that of a K-system. The nomenclature has arisen because this class of abstract dynamical systems was first discussed, although not by this title, by Kolmogorov[19].

A K-system is defined by way of a certain type of measurable decomposition of Γ . A decomposition or partition $\alpha = \{A_i\}_{i \in I}$ of Γ is a class of disjoint, non-empty measurable subsets that cover Γ ;

$$\mu(A_i \cap A_j) = 0 \quad i \neq j \; ; \; \mu(\Gamma - \bigcup_{i \in I} A_i) = 0 . \tag{2.32}$$

For such a decomposition to be measurable it is required that there exist a denumerable class $\beta = \{B_j\}_{j \in J}$ of measurable sets such that
(i) each B_j is a union of elements of α
(ii) for each pair A_i, A_j of α there exists a B_k such that $A_i \subset B_k$, $A_j \not\subset B_k$ or $A_i \not\subset B_k$, $A_j \subset B_k$. In particular, it follows that a finite or denumerable partition is measurable.

Since subsets of measure zero play no role in a measurable decomposition they may be ignored; thus two measurable partitions α_1 and α_2 are identical if their elements are the same apart from subsets of measure zero. We may write $\alpha_1 = \alpha_2$, suppressing the (mod 0) that is involved. Should two partitions be such that every element of α_2 is a subset of some element of α_1, it is customary, although perhaps initially confusing, to write $\alpha_1 \leqslant \alpha_2$. Note particularly that this means that α_2 is a finer decomposition than α_1.

To define[20] a K-system we require a certain type of measurable decomposition α, one for which the following three properties are satisfied:
(1) $\quad T_t \alpha = \alpha_t \geqslant \alpha_o = \alpha$ for any $t > 0$
(ii) $\quad \bigvee_t T_t \alpha = \epsilon$
(iii) $\quad \bigwedge_t T_t \alpha = \nu .$ $\qquad$ (2.33)

Here ϵ is the decomposition of Γ into the
point elements of Γ and ν is the decomposition
of Γ that has as its only element Γ itself.
The meaning of the symbols in (ii) and (iii) is
such that these conditions express the requirements that ϵ be the smallest partition that
contains all α_t and that ν be the largest
partition that is contained in all α_t. Remembering that sets of measure zero have been ignored,
we note that these conditions imply that the
elements of an invariant partition, for which
$T_t\alpha = \alpha$, have measure zero or unity. Thus a
K-system is ergodic.

In fact a K-system has much stronger properties
than ergodicity; it has denumerably infinite
Lebesgue spectrum[21] and in consequence is mixing
as well as ergodic. As an example[22] of a K-
system we may cite again the automorphism (2.2)
of the two-dimensional torus.

2.6   Introducing a decomposition of Γ is a
procedure by no means unfamiliar to physicists,
although it is generally referred to by them as
'dividing phase space into phase cells'. Such
a procedure is standard practice in defining the
entropy of a classical system. In the theory of
abstract dynamical systems it turns out that a
generalised version of the concept of entropy,
introduced by **Kolmogorov**, developed by Sinai
and known **simply as** <u>**entropy**</u>[23], plays an important

role, particularly in relation to K-systems. The
concept is more closely allied to that known as
entropy in information theory than to the entropy
of equilibrium thermodynamics.

    To arrive at a definition of the entropy of
an automorphism we introduce first the entropy
of a finite measurable decomposition
$\alpha = \{A_i\}_{i \in I}$,   I finite, as

$$h(\alpha) \;=\; -\sum_{i \in I} \mu(A_i) \log \mu(A_i). \qquad (2.34)$$

    The next step is to define the entropy of
one finite partition $\alpha = \{A_i\}_{i \in I}$ relative to
another finite partition $\beta = \{B_j\}_{j \in I}$ as

$$h(\alpha/\beta) \;=\; -\sum_j \sum_i \mu(A_i \cap B_j) \log\!\left[\mu(A_i \cap B_j)/\right.$$
$$\left./\mu(B_j)\right] \quad (2.35)$$

This is needed merely as an auxiliary. From a
study of the properties of $h(\alpha/\beta)$ it is
possible to prove[24] the existence of the
entropy of a finite partition α relative to an
automorphism T when this is defined as

$$h(\alpha,T) = \lim_{n \to \infty} n^{-1} h(\alpha \vee T\alpha \vee \ldots \vee T^{n-1}\alpha) \qquad (2.36)$$

Here $\alpha \vee \beta$ means the smallest partition that
contains both α and β ; its members are
$A_i \cap B_j$ where $A_i \in \alpha$, $B_j \in \beta$.

    Finally we obtain the entropy of the auto-
morphism T as

$$h(T) = \sup \ h(\alpha, T), \qquad\qquad (2.37)$$

the supremum   (or least upper bound) being
taken over all finite measurable decompositions
α.   The entropy of a group T_t ($-\infty < t < \infty$)
of automorphisms is taken to be the entropy of
the automorphism T_1.

   This property of entropy provides a means
of categorising what we may describe loosely as
the 'degree of mixing' in a system. From its
definition $h(T)$ is such that $0 \leqslant h(T) \leqslant \infty$.
The extremity $h = \infty$ corresponds to stochastic
systems. For present purposes the most interesting results[25] concerning entropy are
(i) K-systems have positive entropy ($h > 0$)
(ii) classical systems (those defined by differential equations, not necessarily Hamiltonian in
form) have finite entropy ($h < \infty$).
It should not be thought that systems with
entropy zero are those without mixing properties;
examples have been constructed[26] both of classical and of non-classical systems that have entropy
zero and so are not K-systems although they possess
denumerably infinite Lebesgue spectrum and so are
mixing.

3.   Application of Abstract Dynamical Theory to
                    Physical Systems

3.1   Having mentioned briefly some of the
measure-theoretical ideas of abstract general

dynamics, we now turn to consider the application
of these to systems of interest within physics.

Our first criterion for mixing is that of
the metrical indecomposability of the product
space $\Gamma \times \Gamma$; correspondingly, a criterion
for ergodicity is the metrical indecomposability
of Γ itself. Should the system be a conservative Hamiltonian system, Γ in the foregoing
criteria is to be replaced, as discussed previously, by an energy surface S_E such that
$\mu_E(S_E) = 1.$

These criteria were established about forty
years ago and so there has been ample opportunity
for the study of the metrical indecomposability
of conservative Hamiltonian systems. This study,
however, has not proved fruitful in that we do
not know what features of a Hamiltonian ensure
metrical indecomposability.

The problem is related to the existence of
a certain type of constant of the motion.
An <u>integral of the motion</u> is a function
$f(T_t x; t)$ that   (i) is constant along each trajectory $x = x(t)$ but (ii) is not constant
everywhere in Γ .   A <u>constant of the motion</u> is
a conservative integral of the motion, one, that
is, with no explicit dependence upon time. Often
the adjective 'conservative' is omitted and the
terms 'integral of the motion' and 'constant of
the motion' are used indiscriminately.

A Hamiltonian system with n degrees of
freedom is specified by 2n first order differen-

tial equations and thereby possesses no more than
2n <u>independent</u> integrals of the motion. It is
often stated that it always does possess as many
as 2n independent integrals of the motion. This
is generally true only in a local sense, that is,
if the initial conditions are confined to some
bounded and closed domain and if the time parameter is restricted to a certain interval dependent upon this domain. In these circumstances
suitable elimination of time from among the 2n
local integrals of motion yields (2n - 1) constants of the motion which likewise, of course,
are of purely local significance.

        If solutions are to be considered for all
instants of time $(-\infty < t < \infty)$ and the initial
conditions are not restricted, such local constants of the motion may have singularities -
they do not then exist as <u>global</u> constants of
the motion. The definition of a conservative
Hamiltonian system implies the existence of
energy as a global constant of the motion. Likewise a system in uniform translational or
rotational motion has linear or angular momentum
respectively as a global constant of the motion;
these are usually excluded in statistical
mechanics by the device of supposing the system
of interest to be enclosed within a massive
rigid container at rest (and with rough walls if
the walls have the shape of a surface of
revolution).

Global constants of the motion such as energy that confine trajectories to a hypersurface in phase space and thereby isolate points on trajectories from other points of the phase space are termed <u>isolating</u>[27] constants of the motion. It is these that are of concern as regards the metrical indecomposability of Γ or, for a conservative system, of S_E. For, if there exists an isolating constant of the motion additional to, but independent of, the energy, then possible trajectories are confined to the intersection of the energy surface with the surface defined by this additional isolating constant of the motion. The set of possible trajectories is of measure zero on the energy surface S_E and the time mean $\langle f \rangle_\mathcal{T}$ of a phasefunction f is thus not constant almost everywhere on S_E. From the argument given previously it follows that the energy surface is not metrically indecomposable and that in consequence the system is non-ergodic and thereby non-mixing.

If the existence of a set of isolating constants of the motion is known, it is possible to confine attention to the surface of reduced dimension that is formed by the intersection of all these isolating constants and to consider questions of ergodicity or of mixing with respect to this surface only - just as, in fact, is done with respect to an energy surface when the system is known to be conservative. But this, of course,

presupposes a knowledge of all the isolating
constants of motion that exist. And therein lies
the problem, since it is not known in general how
many isolating constants of the motion a Hamiltonian system possesses. In applications of
statistical mechanics it is customary to proceed
as if the energy were the only isolating constant
of the motion; the fact that this procedure usually gives results consistent with those of experiment may suggest that the energy is in general
the only isolating constant. But no proof of
this has yet appeared.

Accordingly, the criterion of metrical indecomposability does not serve as ·a practical
means of determining whether or not a Hamiltonian
system is mixing.

Before passing on to consider what other
criteria yield we shall illustrate the concept
of a non-isolating constant of the motion by
means of an example. We choose a system that
we have not yet dealt with but that nevertheless
turns out to have familiar properties, the
system described by the Hamiltonian[28]

$$H(x,y,p_x,p_y) \;=\; \tfrac{1}{2}(p_x{}^2 + p_y{}^2) \qquad (3.1)$$

and by the requirement that x and y are
periodic with period unity. The equations of
motion are thus

$$\dot{x} = p_x \;;\; \dot{y} = p_y \;;\; \dot{p}_x = 0 \;;\; \dot{p}_y = 0. \quad (3.2)$$

There exist global constants of the motion p_x, p_y and H, only two of which, say p_y and H, are independent. There exists also the constant of the motion $(xp_y - yp_x)$. Let us set $p_x = 1$ and take for the parametric value of p_y the parameter α. We then have as a constant of the motion the function $(\alpha x - y)$.

The isolating constants of the motion $H = E$ and $p_y = \alpha$ reduce the available phase space to a two-dimensional region specified by the coordinates x and y (mod 1). We recognise this region as representing the two-dimensional torus. From equation (1.17) we further recognise the relation

$$(\alpha x - y) = c \qquad (3.3)$$

where c is a parametric constant, say $(\alpha x_0 - y_0)$, as defining trajectories on this torus. As was discussed previously, such trajectories pass arbitrarily close to any point in the unit square when α is irrational. In this circumstance the constant of the motion is non-isolating; it does not serve to delimit any region of the unit square and its existence is of no significance as regards ergodicity or mixing.

Note that in this example the Hamiltonian system is non-ergodic with respect to the four-dimensional phase space and with respect to the three-dimensional energy surface but, as was seen previously, it is ergodic (but not mixing) with respect to the two-dimensional unit square when

α is irrational.

3.2   Metrical indecomposability having proved
an impractical tool, we turn to spectral properties. These, however, tend to be subsumed
within K-systems - as mentioned already, a K-
system is the current paragon, having Lebesgue
spectrum and being mixing. And the easiest way
to obtain the latter properties for a system
having a differentiable flow is, in general, to
prove that the system is a K-system.
    The reason this approach is feasible is
that there exists a certain wide class of differentiable systems, called C-systems by Arnold
and Avez, that possess certain local properties
giving rise to asymptotic behaviour of a highly
stochastic nature, whereby trajectories diverge
from each other at an exponential rate. Although
each refers to a strongly stochastic system, the
notions of K-system and of C-system are actually
quite different; the former is a measure-
theoretical construct pertaining to a system with
a measure-preserving flow, the latter a topological construct pertaining to a system with differentiable flow. A flow that is both differentiable and measure-preserving, as is a Hamiltonian
flow, may well prove to be simultaneously a C-
flow and a K-flow.
    About forty years ago Hedlund, Hopf and
others studied geodesic flows on a surface of

constant negative curvature, the motion that is, of a particle constrained to move on such a surface but otherwise free from applied forces. Such a flow they showed to be ergodic[29] and mixing[30], a result that derives from the dispersive effect of the negative curvature in causing phase points to separate at an exponential rate. These geodesic flows are now known[31] to be C-systems; indeed the concept of a C-flow is a generalisation from these geodesic flows that have highly stochastic behaviour.

A C-system[32] is either a C-diffeomorphism or a C-flow, the former a single mapping and the latter a one-parameter group T_t of differmorphisms of a smooth manifold M such that

(i) the velocity vector $v_o \equiv \frac{d}{dt}(T_t m)_{t=0}$ is non-zero, $m \in M$,

(ii) the tangent space to the manifold at any point m may be decomposed into the one-dimensional space of the velocity vector at m and the so-called dilating and contracting subspaces X_m and Y_m respectively, that have the properties

$$\| T_t\, v_o \| \geqslant a\, e^{\lambda t}\, \| v_o \|; \quad \| T_{-t}\, v_o \| \leqslant b e^{-\lambda t}\, \| v_o \|,$$

$$v_o \in X_m$$

and

$$\| T_t\, v_o \| \leqslant b e^{-\lambda t} \| v_o \|; \quad \| T_{-t}\, v_o \| \geqslant a e^{\lambda t}\, \| v_o \|,$$

$$v_o \in Y_m \qquad\qquad (3.4)$$

where a, b and λ are positive constants and
where t > 0.

A C-diffeomorphism requires only the decomposition of the tangent space into X_m and Y_m,
where a positive integer n here plays the role
of time t in the C-flow. The mapping

$$\begin{pmatrix} x' \\ y' \end{pmatrix} = \begin{pmatrix} 1 & 1 \\ 1 & 2 \end{pmatrix} \begin{pmatrix} x \\ y \end{pmatrix}, \qquad (3.5)$$

which is that of (2.2) but no longer restricted
to the unit square, provides a simple example of
a C-diffeomorphism. It has two eigenvalues
$\lambda_1 > 1$ and $0 < \lambda_2 < 1$, where $(1-\lambda)(2-\lambda) = 1$
for $\lambda = \lambda_1, \lambda_2$, and the corresponding eigenfunctions define two directions X and Y such
that $dX' = \lambda_1 dX$ and $dY' = \lambda_2 dY$. These
satisfy the conditions analogous to (3.4) with
a = 1 = b, $e^{\lambda} = \lambda_1$, $e^{-\lambda} = \lambda_2$. Pictorially
there is elongation in the X-direction and
shrinking in the Y-direction:

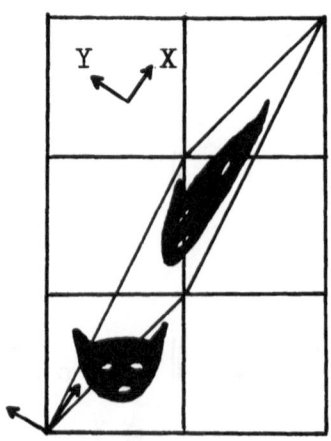

This is, of course, the same diagram, that of Arnold and Avez, as has been presented earlier, except that there is here no translation back into the unit square.

It may also be proved[33] that this particular system (2.2) is a K-system. This is merely a particular instance of a theorem due to Sinai[34] that asserts every C-diffeomorphism to be a K-system. Although every C-system, whether a differmorphism or a flow, is known[34] to be ergodic, it is not true[34] that every C-flow is a K-system; there is the possibility that the C-flow has eigenfunctions that are not constant.

Now this is all very well from the point of view of general dynamics, even if restricted to Hamiltonian systems. But the question of interest to physicists is what type of Hamiltonian system possesses tne properties of being a C-system and a K-system. A particle on a surface of negative curvature has both these properties[35], but this is far from being the kind of system that is of interest within statistical mechanics. A system of particles interacting via a two-body potential that has a strongly repulsive core and a weak long-range attraction - such is the system we should like to know about.

3.3 All the previous history of the contributions made by general dynamics to statistical mechanics might well lead one to suppose that

what the abstract theory can usefully provide
about our system of particles is precisely
nothing at all. It is perhaps surprising but
certainly gratifying to find this expectation to
be in part confounded.

To Krylov[36] is due the suggestion that a
system of hard spheres enclosed in a box and
making elastic collisions with one another might
exhibit similar behaviour to that of geodesic
flows on surfaces of negative curvature; the
dispersive role of the negative curvature would
be taken by the elastic collisions.

To Sinai[37] is due the proof that this
system of hard spheres in a box is a K-system,
and hence is both mixing and ergodic. The collisions are utilised in constructing dilating
and contracting subspaces but since these have
discontinuities a generalised version of C-system
is developed.

When news of this result first permeated to
the physics community some eight years or so ago,
the hopes of some physicists were high. Here
perhaps was the vindication for statistical
mechanics of so many years of, again for
statistical mechanics, ineffectual endeavour in
the field of general dynamics. The proof would
rapidly be extended to a system with any form of
short-range repulsive potential and thence to a
potential having a repulsive core and a weak
long-range attraction. But the expectations were
vain. No such extensions have been found. Again

from the viewpoint of statistical mechanics,
Sinai's result stands alone in its splendour.

4.   Irreversibility and Infinite Systems

4.1   What should be noted in particular about the
results of abstract general dynamics and especially about Sinai's result for the system of hard
spheres in a box is that they refer to <u>finite</u>
systems. We restricted attention to measure
spaces or subspaces of finite total measure such
that we could set $\mu(\Gamma) = 1$. For phase spaces
or energy surfaces relating to Hamiltonian systems the implication is that the volume of the
system is finite. And Sinai's proof for a system of N hard spheres in a box of finite
volume V is valid for any finite number of particles greater than unity.   There is certainly
no question of taking the thermodynamic limit
$N \rightarrow \infty$, $V \rightarrow \infty$, N/V finite, so as to obtain
the result; the proof is valid even for N = 2.
        This, of course, is at variance with what
we are accustomed to in equilibrium statistical
mechanics. Not only is N required to be large
so as to reduce fluctuations and permit ensembles
to be used interchangeably but the limit $N \rightarrow \infty$
is needed to obtain thermodynamic properties and,
in particular, phase transitions. Likewise in
non-equilibrium statistical mechanics it is often
supposed that the limit $N \rightarrow \infty$ is required if

irreversible behaviour is to appear. Indeed, the
title of this school is not 'Irreversibility' but
'Irreversibility in the Many Body Problem'. It
is of considerable interest, therefore, to determine what aspects of irreversibility appear in
the finite system of hard spheres.

To do this we first reinterpret the concept
of mixing in terms that are perhaps more familiar
to physicists. We have defined mixing by the
relation

$$\lim_{t \to \infty} \mu\left[T_t \, A \cap B\right] \quad = \quad \mu(A)\mu(B) \qquad (2.5)$$

for all measurable sets A and B. An equivalent[38] definition may be given that utilises
square-integrable functions in place of sets;
this is the relation

$$\lim_{t \to \infty} (U_t f, \, g) \quad = \quad (f,1)(1,g) \qquad (4.1)$$

which has to hold for all pairs of square-
integrable functions f and g. As before, the
scalar product is defined as

$$(f,g) \quad = \quad \int_\Gamma f^* \, g \, d\mu \, , \qquad (2.21)$$

and, as we shall be concerned only with real
functions, the complete conjugate sign may be
omitted.

Now it may be seen from (2.21) that the
mixing relation (4.1) may be written as

$$\lim_{t \to \infty} \langle f(t)g \rangle \quad = \quad \langle f \rangle \langle g \rangle \quad , \qquad (4.2)$$

where the angular brackets are used to denote
the <u>correlation function</u>

$$\langle f(t) \; g \rangle \quad = \quad \int_{\Gamma} f(t) \; g \; d\mu \qquad (4.3)$$

of the pair of functions f and g. In particular, it follows that if a system is mixing
the autocorrelation function of any square-
integrable phasefunction satisfies the relation

$$\lim_{t \to \infty} \langle f(t)f \rangle \quad = \quad \langle f \rangle^2 . \qquad (4.4)$$

For the system of N hard spheres in a
finite box the momentum of any one of the
particles is a square integrable function, the
phase average of which is zero. Thus Sinai's
result indicates, among other things, that

$$\lim_{t \to \infty} \langle p_i(t) \; p_i(0) \rangle \quad = \quad 0, \qquad (4.5)$$

the autocorrelation function of the single-
particle momentum vanishes in the limit of
infinite time.

Now this result for the autocorrelation
function itself is to be compared with the behaviour of the integral of the autocorrelation
function taken over an interval of time. Consider a function f such that

$$f \quad = \quad iLF \quad = \quad \frac{dF}{dt} \quad , \qquad (4.6)$$

where both f and F are square-integrable.
For a function of this particular form we have

$$\int_0^T \langle f(t)f \rangle \, dt \quad = \quad \int_0^T \langle \tfrac{dF}{dt}f \rangle \, dt$$

$$= \quad \langle F(T)f \rangle - \langle Ff \rangle . \quad (4.7)$$

But

$$\langle Ff \rangle = \langle F\tfrac{dF}{dt} \rangle = \quad \tfrac{1}{2}\tfrac{d}{dt}\langle F^2 \rangle = 0 \quad . \quad (4.8)$$

Hence for such a function f we have

$$\int_0^T \langle f(t)f \rangle \quad = \quad \langle F(T)f \rangle \quad . \qquad (4.9)$$

If the system is mixing, it follows from
(4.2) that

$$\lim_{T \to \infty} \int_0^T \langle f(t)f \rangle \quad = \quad \langle F \rangle \langle f \rangle . \qquad (4.10)$$

But

$$\langle f \rangle = \langle \tfrac{dF}{dt} \rangle = \tfrac{d}{dt}\langle F \rangle \quad = \quad 0 \quad . \qquad (4.11)$$

Hence we obtain

$$\lim_{T \to \infty} \int_0^T \langle f(t)f \rangle = \quad 0 \quad . \qquad (4.12)$$

As a particular instance of this we again
consider a single particle in a system of N
hard spheres in a finite box and choose

$$f = \frac{p_i}{m} ; \quad F = q_i ; \quad f = \frac{dF}{dt} . \quad (4.13)$$

Then we obtain

$$\lim_{T \to \infty} \int_0^T \langle p_i(t) \, p_i(0) \rangle \, dt \;\; = \;\; 0. \qquad (4.14)$$

Lebowitz[39], who presented these results, has emphasised the significance for irreversibility of the conclusion that in a mixing system not only autocorrelation functions but also the integral over time of autocorrelation functions may vanish in the limit of infinite time. For the Kubo formulae[40] for transport coefficients all involve the integral over an infinite time of the autocorrelation function of a function of the type (4.6); accordingly, these transport coefficients have here the value zero. In particular, (4.14) indicates that the diffusion coefficient is zero.

Thus the finite hard sphere system, being mixing, exhibits irreversibility in the sense that correlations may decay to zero; yet it does not exhibit irreversibility in the sense of possessing transport coefficients. It is this that indicates what may be the role of the thermodynamic limit in producing irreversibility. In the limit $N \to \infty$ not only may correlations decay but also we may obtain non-zero values for transport coefficients and so a hydrodynamical description.

4.2   This, of course, leads on to a study of the
temporal behaviour of infinite dynamical systems.
Although the infinite-system approach has been
developed extensively with regard to equilibrium
statistical mechanics, much less has been done
with reference to non-equilibrium situations.
Indeed it is only with considerable effort that
the evolution in time of an infinite dynamical
system can even be defined. The difficulty in
obtaining a proof of the existence of a unique
temporal evolution is that the infinite number
of degrees of freedom allows the local density
of particles to be infinite, a situation that
cannot be tolerated in an appropriate physical
theory. Thus constraints must be set on particle velocities and initial configurations that
permit an acceptable time evolution may thereby
be delineated. Although Ginibre[41] obtained a
local existence theorem, this required unduly
restrictive conditions that, for instance, exclude
a Maxwellian distribution of velocities. Thus,
until very recently at any rate, certain one-
dimensional classical systems[42] interacting by
finite-range two-body forces and, secondly,
quantum spin systems[43] were the only infinite
systems possessing interactions for which a
satisfactory definition of time evolution had
been given.

   With infinite systems that do not possess
interactions much more can be done; in fact,
Sinai and Volkoviskij[44] have proved that an

infinite one-dimensional ideal gas is not merely
ergodic but is actually a K-system. By noting
that a similar argument holds for a semi-infinite
one-dimensional ideal gas and that this system is
isomorphic to, and hence has the same dynamical
properties as, a semi-infinite one-dimensional
system of hard rods, de Pazzis[45] has shown the
latter system to be a K-system, and has reported
that Sinai has solved the problem of an infinite
system of hard rods when these have a Maxwellian
velocity distribution. As with the finite system
of hard spheres mentioned earlier, the infinite
systems with no interparticle interactions do not
exhibit hydrodynamical features - there is no
establishment of local equilibrium followed by
the eventual accession to global equilibrium.

It will be clear that in this brief mention
of infinite systems I have already encroached
upon the preserve of Professor Miracle who is
lecturing at this school on the time evolution of
infinite dynamical systems; further detailed
discussion of this field may safely be left to him.

4.3   Instead of dealing directly with an infinite
system one may study the thermodynamic limit of
a finite system.

Although again I must be careful not to
impinge upon the content of other lecture courses
given here, I wish to point out the relation

of the work to be described by Professor Mazur[*]
to that which we have looked at earlier.

Recall that the two-dimensional oscillator
of (3.1), or equivalently a system of two non-
interacting linear harmonic oscillators, is non-
ergodic and so non-mixing in that it possesses
isolating constants of the motion other than the
energy. Likewise a system of N non-interacting
linear harmonic oscillators is non-ergodic and
non-mixing; it is an example of a so-called
integrable system. (This is a term that has no
precise definition; it refers to the type of
system studied in elementary classical mechanics
whose behaviour is determinable generally because
it involves periodic motions.) The existence of
such systems that are known to be non-ergodic and
non-mixing, and the near non-existence of systems
of physical interest that are known to be ergodic
and mixing, suggested that attention might profit-
ably be focussed not on mixing systems but rather
on mixing properties of systems. Whereas a system
is mixing if

$$t \xrightarrow[\to\ \infty]{lim} \langle f(t)g \rangle \ = \ \langle f \rangle \langle g \rangle \qquad\qquad (4.2)$$

for all square-integrable f and g, the system
has a mixing property if (4.2) holds for some

---

[*]Editor's note: the following part of Dr.
Farquhar's lectures were designed to link up
with an expected course by Prof. Mazur which,
unfortunately, could not be delivered at the
school.

square-integrable f and g. More particularly,
the square-integrable phase function f is mixing if

$$\lim_{t \to \infty} \langle f(t)f \rangle = \langle f \rangle^2 . \qquad (4.4)$$

Likewise the square-integrable phasefunction f
is ergodic if its time mean is equal to its
phase mean, whereas for a system to be ergodic
this property must be satisfied for all phasefunctions f .

Professor Mazur and collaborators have calculated the asymptotic time behaviour of the autocorrelation function of local properties of harmonic systems - for instance, the momentum of
an impurity particle in a chain of particles with
harmonic interactions, the impurity being dis-
tinguished by its different mass. Others have
studied spin systems similarly. To obtain mixing
properties it is necessary in these studies to
take the thermodynamic limit. One of the interesting features that emerges is that the existence of
mixing or ergodicity in the thermodynamic limit is
associated with the absence of isolated local modes
of oscillation. This we mention not for the purpose of forestalling Professor Mazur's own exposition of the topic, but so as to point out the
analogy with the result of abstract general
dynamics that the Liouville operator of a weakly
mixing system has no discrete spectrum other than
the value zero. This correspondence is very suggestive:, nevertheless it should be noted that

general dynamics deals with a finite system
whereas the absence of isolated localised modes
is needed in the thermodynamic limit of a finite
system.

5.   Integrable Systems

5.1   We began this course with a topic,
ergodicity, that we claimed to have no significance for irreversibility. We conclude by
looking at systems that in one sense at least
may be regarded as being even farther removed
from the theme of this school.

A system of N uncoupled harmonic oscillators is the most noted example of an integrable
system. As has been mentioned already, this
system has mixing properties as $N \longrightarrow \infty$. Here,
however, we shall be concerned with a finite
system.

Action-angle variables are peculiarly suited
to a description of a system with periodic
motions[46) and hence especially to a system of
harmonic oscillators. To introduce these in the
Hamiltonian

$$H = \sum_{i=1}^{N} H_i = \sum_{i=1}^{N} \frac{1}{2}\left(\frac{p_i^2}{m_i} + m_i \omega_i^2 q_i^2\right) \quad (5.1)$$

we set

$$q_i = (2J_i/m_i \omega_i)^{1/2} \sin \alpha_i$$

$$p_i = (2m_i \omega_i J_i)^{1/2} \cos \alpha_i \qquad (5.2)$$

and thereby obtain

$$H = \sum_{i=1}^{N} H_i = \sum_{i=1}^{N} \omega_i J_i \qquad (5.3)$$

Thus the Hamiltonian has a particularly simple form, depending only on the so-called action variables J_i ; the corresponding angle variables α_i are cyclic.

The equations of motion

$$\dot{J}_i = -\frac{\partial H}{\partial \alpha_i} \quad ; \quad \dot{\alpha}_i = \frac{\partial H}{\partial J_i} \qquad (5.4)$$

are

$$\dot{J}_i = 0 \quad ; \quad \dot{\alpha}_i = \omega_i \qquad (5.5)$$

and these are immediately integrable to give

$$J_i = J_i^{\,0} \quad ; \quad \alpha_i = \omega_i t + \alpha_i^{\,0}. \qquad (5.6)$$

The set of action variables $J_i = J_i^{\,0}$ provides N constants of the motion and, since $H_i = \omega_i J_i$, the physical interpretation of this is that the **energy in each mode of oscillation is** constant. These are isolating constants of the motion and their existence confines trajectories to an N-dimensional hypersurface in the 2N-dimensional phase space, a hypersurface on which the motion is describable in terms of the angle variables α_i alone. In fact, the hypersurface is an N-dimensional torus whose radii are specified by $J_i = J_i^{\,0}$.

We have already come across this in the

instance $N = 2$, as this is the system specified
by the Hamiltonian (3.1). There we had motion
restricted to the unit square with periodic
bounaary conditions and this is topologically
equivalent to the 2-dimensional torus (often
referred to merely as 'the torus'.)

Returning to the system of N oscillators
we note that the entire phase space Γ is decomposable into invariant hypersurfaces,

$$\Gamma = T^N \times R^{+N} , \qquad (5.7)$$

where T^N is the N-dimensional torus and where
$R^{+N} = R^+ \times R^+ \ldots \times R^+$ is the N-space of the
action variables J_i. The initial conditions
specify the constant values of these J_i and
hence determine on which of the invariant N-
tori the trajectory lies.

Now this general structure that we have seen
to exist for a system of N oscillators is reproduced[47] in any Hamiltonian system that
possesses N isolating constants of the motion.
What characterises the system of oscillators
is that the frequencies ω_i of the constituent
periodic motions are independent of the action
variables J_i. In an arbitrary integrable sys-
tem these frequencies $\omega_i = \partial H/\partial J_i$ depend on
the action variables J_i and so vary from one to
another of the continuous family of invariant N-
tori.

Our previous considerations of motion on a
two-dimensional torus, the invariant surface for

a system of two uncoupled harmonic oscillators,
indicate that trajectories pass arbitrarily close
to any point on the torus if the ratio of the
frequencies ω_1 and ω_2 is irrational. This
is the situation in which we have ergodicity on
the torus, although, of course, the system of two
oscillators is itself non-ergodic because of the
existence of two independent isolating constants
of motion. When the ratio ω_1/ω_2 is rational,
closed trajectories are found. For an arbitrary
integrable system the behaviour is similar but
not completely analogous to this. If the frequencies are incommensurable, if, that is, the
relation

$$\sum_{i=1}^{N} n_i \omega_i = 0 , \qquad n_i \in Z \qquad (5.8)$$

holds only when all $n_i = 0$, trajectories are
everywhere dense on the N-torus; likewise, if
the frequencies are commensurable closed trajectories are obtained - the complete motion is
periodic rather than merely multiply or conditionally periodic. However, the difference
now is that the type of trajectory is specific
to the particular N-torus; this is a consequence of the frequencies' dependence on the
action variables J_i, which in turn specify the
particular N-torus. Thus a given N-torus bears
either a commensurable or an incommensurable frequency. Those N-tori with commensurable frequencies are exceptional and form a set of measure
zero in the phase space.

Now, whereas integrable systems may perhaps
form a set of measure unity within elementary
classical mechanics, they may be regarded as of
measure zero in the wider context of classical
mechanics. What are of great interest are systems that deviate slightly from being integrable
systems, integrable systems that are subjected to
some non-integrable perturbation. The major problem in celestial mechanics has been that of the
stability of the solar system.

A most significant advance in the theory of
the stability of dynamical systems is afforded
by what is generally referred to as the KAM
theorem[48]. (The acronym derives from Kolmogorov,
who formulated the theorem, and from Arnold and
Moser, who later and independently provided proof
of the theorem.) This asserts that if the perturbation is sufficiently small[49] there exist
invariant N-tori for the perturbed system that
are close to invariant N-tori of the unperturbed
system; this is so for all sets of frequencies
ω_i except a set of measure zero. Furthermore
the set of invariant N-tori for the perturbed
system is of positive measure as is the complement of the set of invariant N-tori; the
measure of this complementary set tends to zero
as the amplitude of the perturbation approaches
zero.

This complementary set that appears in the
perturbed system is of considerable interest.
For $N = 2$ the invariant tori confine trajectories

in the complementary set to toric annular regions
between the invariant tori. For $N > 2$, however,
the invariant N-tori do not divide the (2N-1)-
dimensional energy surface and so trajectories
not on N-tori may wander throughout the entire
energy surface. Not much seems to be known in
general about the behaviour of such trajectories
and it has been conjectured that trajectories may
pass arbitrarily close to all points of the complementary set.

It is then tempting to suppose that as the
amplitude of the perturbation is increased and
the measure of the set of invariant N-tori decreases so does the system pass towards ergodic
or mixing behaviour.

5.2   A system of two coupled oscillators has been
discussed[50)] by Walker and Ford as a simple illustration of the way in which a perturbation can
break down those unperturbed N-tori that have
either closed trajectories or multiply periodic
motion with frequencies that, although incommensurable, can yet be approximated closely by
frequencies such that $\sum_i n_i \omega_i = 0$ where all
the n_i are small.

The Hamiltonian for this system is

$$H = H_o(J_1, J_2) + V(J_1, J_2, \alpha_1, \alpha_2), \qquad (5.9)$$

and the perturbation V may be expanded in a

Fourier series to give

$$H = H_0(J_1, J_2) + V_{n_1 n_2}(J_1, J_2)\cos(n_1\alpha_1 + n_2\alpha_2) + \ldots , \tag{5.10}$$

where for simplicity of argument we have followed Walker and **Ford** in displaying only one term of the Fourier series.

In the spirit of the KAM approach we should now search for a convergent sequence of canonical transformations that will yield a Hamiltonian free from angle variables; then, as for integrable systems, there exist constants of the motion here determined by the new action variables. Should each member of the **sequence** of canonical transformations be close to the identity transformation, then tori for the perturbed system are close to tori for the unperturbed system. However, to exemplify the procedure it may be sufficient to consider a single canonical transformation to new action and angle variables such that the part of the Hamiltonian that appears explicitly in (5.10) is cyclic in the new angle variables.

To this end may be introduced the generating function[51)

$$F = G_1\alpha_1 + G_2\alpha_2 + A_{n_1 n_2}(G_1, G_2)\sin(n_1\alpha_1 + n_2\alpha_2), \tag{5.11}$$

where the function $A_{n_1 n_2}(G_1, G_2)$ is to be determined in such a way as to eliminate the

specific angle-dependent term displayed in (5.10).
We then have

$$
J_i = \frac{\partial F}{\partial \alpha_i} = G_i + A_{n_1 n_2} \, n_i \cos(n_1 \alpha_1 + n_2 \alpha_2)
$$
$$
(5.12)
$$

and

$$
\Theta_i = \frac{\partial F}{\partial G_i} = \alpha_i + \frac{\partial A_{n_1 n_2}}{\partial G_i} \sin(n_1 \alpha_1 + n_2 \alpha_2),
$$
$$
(5.13)
$$

where the Θ_i, $i = 1, 2$, are the new angle
variables conjugate to the new action variables
G_i. Note that if $A_{n_1 n_2}$ is zero the identity
transformation is obtained.

Substitution of (5.12) and (5.13) in (5.10),
followed by Taylor expansion and explicit reten-
tion only of lowest terms yields

$$
H = H_0(G_1, G_2) + \left[V_{n_1 n_2}(G_1, G_2) \right.
$$

$$
\left. + A_{n_1 n_2} \left\{ n_1 \frac{\partial H_0}{\partial G_1} + n_2 \frac{\partial H_0}{\partial G_2} \right\} \right] \times
$$

$$
\times \cos(n_1 \Theta_1 + n_2 \Theta_2) + \ldots \ldots \qquad (5.14)
$$

We now set

$$
\omega_i(G_1, G_2) \equiv \frac{\partial H_0}{\partial G_i} \, . \qquad i = 1,2. \qquad (5.15)
$$

To eliminate the angle-dependent term appearing explicitly in (5.14) we choose $A_{n_1 n_2}$ to be

$$A_{n_1 n_2} = - \frac{V_{n_1 n_2}(G_1, G_2)}{n_1 \omega_1 (G_1, G_2) + n_2 \omega_2 (G_1, G_2)} . \quad (5.16)$$

For this to be possible it is necessary, of course, that $\sum_{i=1}^{2} n_i \omega_i \neq 0$. Furthermore, we see that, should the denominator in (5.16) be small compared to $V_{n_1 n_2}$, the canonical transformation is not close to the identity transformation. There is thus considerable distortion of unperturbed tori for which the frequencies satisfy the inequality

$$\left| n_1 \omega_1 (J_1, J_2) + n_2 \omega_2 (J_1, J_2) \right| \ll \left| V_{n_1 n_2}(J_1, J_2) \right| . \quad (5.17)$$

The larger is the perturbation the more readily may this inequality be satisfied.

Walker and Ford suggested that, if there are many angle-dependent terms of the form $\cos(n_1' \alpha_1 + n_2' \alpha_2)$ such that $n_1'/n_2' \approx n_1/n_2$, the resonant coupling of the oscillators, in the instance that their frequencies satisfy (5.17) and similar relations, may lead to such overlapping of grossly distorted tori that the system then exhibits some type of mixing property.

5.3   In the absence of analytical predictions concerning this supposition one can turn to computer studies, of which there have been several[52)]

The earliest computer work related to
irreversibility is probably that of Fermi, Pasta
and Ulam[53], who integrated numerically the
equations of motion of a harmonic chain subjected
to an anharmonic perturbation. Their intention
was to study the approach to equilibrium of the
system when the initial conditions were such that
the energy resided in only a few of the lower
energy modes. Their result, that the expected
equipartition of energy among all modes did not
materialise, occasioned considerable surprise at
the time. Since the advent of the KAM theorem
this model has been reanalysed[54] and has been
found to exhibit considerable, but not complete,
equipartition of energy among the different modes
when the amplitude of the perturbation, and so
the total energy, is appreciably larger than that
introduced by Fermi, Pasta and Ulam.

Numerical work designed to test the existence
of a constant of the motion additional to the
energy, work which again was later reinterpreted[56]
in the light of the KAM theorem, was carried out
by Hénon and Heiles; this has formed the basis
for much subsequent investigation. The motivation for this lay in astronomical observations
of stellar motions; the mathematical model became the motion of a particle of constant energy
constrained to a plane and subject to the potential

$$U(x,y) = \frac{1}{2}(x^2+y^2+2x^2y - \frac{2}{3}y^3), \qquad (5.18)$$

this potential chosen as being sufficiently simple
to facilitate computation but sufficiently complicated to yield non-trivial trajectories. The
phase space is four-dimensional but the motion
is confined to a three-dimensional energy surface and so may be described in terms of the
coordinates (x, y, p_y). A graphical representation in two dimensions, those of (y, p_y), may
be obtained by the device of considering successive intersections of the trajectory (in the
positive x-direction) with the plane $x = 0$. If
there exists an isolating constant of the motion
additional to the energy these successive intersections should lie on a curve in the (y, p_y)
plane; in the absence of such an isolating constant of the motion the points of intersection
may be expected to be found anywhere within the
area on the (y, p_y) plane that is determined by
the energy surface.

A sequence of diagrams for increasing values
of the total energy shows a dramatic change in
the behaviour of the trajectories obtained by
numerical integration of the Hamiltonian equations
of motion:

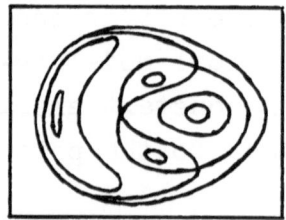

Each curve represents
one trajectory on
$E = 1/12$

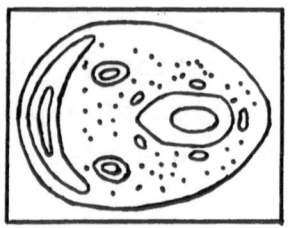

Some curves belong to
the same $E = 1/8$ tra-
jectory

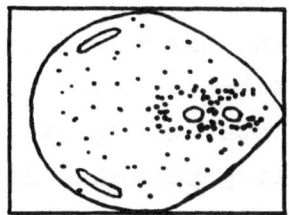

All isolated points lie on
one trajectory on E = 1/6

Later computer studies undertaken with the
specific purpose of investigating the so-called
KAM amplitude instability have given widespread
evidence of a change from stable to unstable behaviour of the phase trajectories. The degree
of instability is uncertain in that the isolated
points of an unstable trajectory exhibit a certain slight order.

The magnitude of the energy at which instability becomes prominent is of considerable
interest. In the many models studied the
potentials chosen have generally indicated onset
of instability only at comparatively high energies.
Walker and Ford[50] have conjectured that instability may arise at much lower energies for
systems with a nearly hard-core repulsive part
in the pair potential - a repulsive part, that
is, that conforms to the generally accepted form
for physical systems. This conjecture was made
in consequence of the results of Northcote and
Potts[57], who, in following up the work of Fermi,

Pasta and Ulam, introduced hard-cores as a
source of nonlinearity in an otherwise harmonic
chain of particles and thereby found a high
degree of equipartition of energy among different
modes, even at low total energies.

A rather startling proposal is that made[58]
very recently by Cercignani, Galgani and Scotti,
that the energy at which KAM instability occurs
in a classical system may be related to the
quantum zero-point energy. This emerges from an
analysis of the recent computer studies[59] of
Bocchieri, Scotti, Bearzi and Loinger on a one-
dimensional system with nearest-neighbour
Lennard-Jones interactions. The general problem,
that statistical mechanics appears to work quite
satisfactorily without any regard having been
paid to the existence of KAM invariant N-tori,
is neatly overcome; motion on such N-tori is
relegated to zero temperature and statistical
mechanics is concerned only with behaviour at
temperatures greater than zero, at which the ordered motions have disappeared.

Fascinating though it may be, the proposal
obviously requires further investigation before
it can be accepted.

5.4   It will have been realised that, although
in the KAM theory we start with an integrable
system and add a perturbation, what is being conjectured in several of the computer investigations

is that the ideas are applicable more generally.
Let us take the standard type of interparticle
potential of common interest in statistical
mechanics. At low energy the potential near its
minimum may be regarded as approximately harmonic; may we then suppose that it corresponds
to a perturbed version of an integrable system
for which KAM stability obtains? At very high
energy does the system likewise correspond closely to a hard-sphere gas and exhibit mixing behaviour?

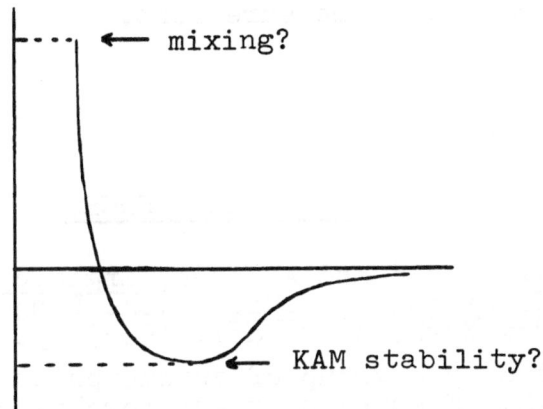

We have as yet no answers to these questions.

Clearly, however, the subject is very much
alive and we may look forward to further knowledge of what aspects of irreversibility such
studies as we have mentioned in this course may
show a finite system to possess. To obtain an
adequate theory of irreversibility we may then
perhaps adjoin, as necessary, concepts relating
to a reduced description of the system and concepts relating to an infinite system.

As has been seen, I have attempted in this course merely to provide a sort of tourist's guide to the field, pointing out the major sights of interest and their locations relative to one another. To get to know a country it is necessary to live in it. The following references to concepts and proofs of theorems may enable one to begin the process of becoming familiar with that country of which I have merely brought back travellers' tales. It is to be hoped that those of you who start out on the journey will form a set of more than measure zero.

NOTES AND REFERENCES

1) V.I. Arnold and A. Avez, Ergodic Problems of Classical Mechanics (W.A. Benjamin, Inc., New York, Amsterdam, 1968), p. 7.

2) Actually a group of automorphisms (mod 0), this meaning that automorphisms that differ only on a set of measure zero are regarded as being equivalent.

3) W. Rudin, Real and Complex Analysis (McGraw-Hill, Inc., New York, 1966).

4) It is to this that the symbol 'σ' refers; were (ii) to refer to finite unions only the class a would be an algebra.

5) The definition given of a σ-algebra automatically ensures that the empty set $\emptyset$ belongs to a, since $\emptyset$ is the complement of Γ relative to Γ.

6) For the Bernoulli trial or Bernoulli scheme $B(\frac{1}{2}, \frac{1}{2})$ the space Γ is the set of sequences ... $a_{-1} a_0 a_1$ ... of tosses wherein a_i = 0, 1 (heads or tails). The σ-algebra $\mathcal{Q}$ is formed by subsets A_i of sequences in which, say, a head appears at the i-th toss. The measure is such that $\mu(A_i) = \frac{1}{2}$ for all i, and the group of automorphisms is a discrete group of shifts or translations which map ..., a_i, ... into ..., a_i', ... where for all i we have $a_i' = a_{i-1}$.

7) An integrable function is a measurable function f such that $\int_\Gamma |f| d\mu < \infty$.

8) Ref. 1, p. 5.

9) E. Schrödinger, Naturwiss. <u>48</u>, 52 (1935).

10) G.D. Birkhoff, Proc. Natl. Acad. Sci. U.S. <u>17</u>, 656 (1931).

11) Ref. 1, p. 21.

12) E. Hopf, Proc. Natl. Acad. Sci. U.S. <u>18</u>, 333 (1932).

13) B.O. Koopman, Proc. Natl. Acad. Sci. U.S. <u>17</u>, 315 (1931).

14) See, for instance, I. Prigogine, Non-Equilibrium Statistical Mechanics, (Wiley-Interscience, New York, 1962.)

15) N.I. Akhiezer and I.M. Glazman, Theory of Linear Operators in Hilbert Space, (Frederick Ungar Publishing Co., New York, (1966) Vol. 1, p. 91.

16) P.R. Halmos, Ergodic Theory (Chelsea Publishing Co., New York, 1956) p. 39.

17) Ref. 1, p. 30.

18) Ref. 1, p. 28.

19) A.N. Kolmogorov, Dokl. Akad. Nauk. 119, 861 (1958).

20) This definition in terms of a measurable partition is essentially that given by Ja.G. Sinai in Statistical Mechanics: Foundations and Applications, T.A. Bak ed. (Benjamin, New York 1967), p. 564.

21) Ref. 1, pp. 34, 154.

22) Ref. 1, p. 70.

23) Ref. 1, p. 35.

24) Ref. 1, p. 38.

25) Due respectively to Kolmogorov and to Kouchnirenko; see Ref. 1, pp. 43, 46.

26) See Ref. 1, p. 45.

27) A. Wintner, The Analytical Foundations of Celestial Mechanics (Princeton University Press, Princeton 1941) p. 96.

28) A.I. Khinchin, Mathematical Foundations of Statistical Mechanics (Dover Publications, New York, 1949) p. 59.

29) G.A. Hedlund, Ann. Math. 35, 787 (1934).

30) E. Hopf, Ergodentheorie, (J. Springer, Berlin, 1937).

31) Ref. 1, p. 60.

32) Ref. 1, p. 55.

33) Ref. 1, p. 70.

34) See Ref. 1, p. 75.

35) Ref. 1, pp. 60, 76.

36) N.S. Krylov, Papers about the foundations of statistical physics, Moscow 1950. (In Russian).

37) A brief outline of the proof is presented by J.G. Sinai in Statistical Mechanics: Foundations and Applications, T.A. Bak ed. (Benjamin, New York 1967), pp. 559-573. The proof itself is of inordinate length; see Ref. 1, pp. 76-79.

38) Ref. 1, p. 25.

39) J.L. Lebowitz, Hamiltonian Flows and Rigorous Results in Non-Equilibrium Statistical Mechanics, lecture given at IUPAP Conference on Statistical Mechanics, Chicago, March 1971.

40) R. Kubo, J. Phys. Soc. Japan $\underline{12}$, 570 (1957).

41) J. Ginibre, unpublished.

42) O.E. Lanford III, Commun. math. Phys. $\underline{9}$, 176 (1968); $\underline{11}$, 257 (1969).

43) D.W. Robinson, Commun. math. Phys. $\underline{7}$, 337 (1968).

44) J.G. Sinai and K.L. Volkoviskij, unpublished.

45) O. de Pazzis, Commun. math. Phys. $\underline{22}$, 121 (1971).

46) See H. Goldstein, <u>Classical Mechanics</u> (Addison-Wesley, Cambridge, 1953) p. 288.

47) Ref. 1, p. 210.

48) Ref. 1, p. 95.

49) A further condition is that the Jacobian $\partial(\omega_1,\ldots,\omega_N)/\partial(J_1,\ldots,J_N)$ for the unperturbed system is non-zero. Concerning the removal of this restriction see Ref. 1, p. 97, and N. Saitô, N. Ooyama, Y. Aizawa and H. Hirooka, Prog. Theor. Phys. Suppl. $\underline{45}$, 209 (1970) p. 213.

50) G.H. Walker and J. Ford, Phys. Rev. $\underline{188}$, 416 (1969).

51) Ref. 46, p. 242. In Goldstein's notation the generating function is of the form $F_2(q,P,t)$.

52) There are at least two forthcoming articles by participants in such computer studies, in which not only are these methods and results discussed and many references provided but also a lengthy review is given of the analytical work in general dynamics that provides the motivation for these studies. The articles referred to are:

(i) L. Galgani and A. Scotti, to be
published in La Rivista del Nuovo Cimento
(I am grateful to these authors for a
draft copy of this article), and
(ii) J. Ford, to be published in Advances
in Chemical Physics.

53) E. Fermi, J. Pasta and S. Ulam, Los Alamos
Scientific Laboratory Report LA-1940 (1955);
see Enrico Fermi: Collected Papers, Vol. II
(University of Chicago Press, Chicago,
1965) p. 978.

54) N.J. Zabusky and G.J. Deem, J. Comput. Phys.
2, 126 (1967).

55) W.H. Jefferys, Astron. J. 71, 306 (1966).
G.H. Walker and J. Ford, reference 50.

56) M. Hénon and C. Heiles, Astron. J. 69, 73
(1964).

57) R.A. Northcote and R.B. Potts, J. Math. Phys.
5, 383 (1964).

58) C. Cercignani, L. Galgani and A. Scotti,
Physics Letters 38A, 403 (1972). See
also L. Galgani and A. Scotti, Phys.
Rev. Letters 28, 1173 (1972).

GENERAL REFERENCES

1) V.I. Arnold and A. Avez, Ergodic Problems
of Classical Mechanics (Benjamin, New
York, 1968).

2) R. Abraham with J.E. Marsden, Foundations
of Mechanics (Benjamin, New York, 1967).
This contains in an appendix an English
translation of an address entitled The
General Theory of Dynamical Systems and
Classical Mechanics that was given by
A.N. Kolmogorov to the 1954 International
Congress of Mathematicians.

3) S. Sternberg, Celestial Mechanics, Vols. I
and II (Benjamin, New York, 1969).

4) S. Smale, Bull. Amer. Math. Soc. 73, 747
(1967).

IRREVERSIBILITY IN MANY-BODY SYSTEMS

H. Wergeland

University of Trondheim
Trondheim, Norway

I. MOTION OF SYSTEMS OVER LONG PERIODS

Introduction. Part of the time(viz. after
Gibbs) during which the question of irreversibility has been on the agenda of Theoretical Physics,
Liouville's theorem on the motion of volume elements in Phase Space has been the basis from which
this discussion takes off: The points,

$$\mathfrak{X}(t) = \left\{ q_1(t) \ldots p_s(t) \right\} ;$$

—each of which may represent a separate mechanical system— move around like an incompressible
flow in the 2s-dimensional Cartesian space spanned
by the coordinates (q) and the momenta (p).

This may, of course, be stated in a variety
of other ways and we shall recall those which are
useful in Statistical Mechanics. In its purely
mathematical form, the theorem appears as a mere
corrollary to the general characteristics of Hamiltonian systems [1]. Let us for the moment suspend
an optimum of compactness —which, by the way,
would have to be paid for by more careful preliminaries.

a) Ensemble theory is concerned with a density $\varphi(q,p,t)$ distributed over phase space. By
analogy to hydrodynamics this concept provides a
nice picturesque version of Liouville's theorem:
Since representative points (of an ensemble of
systems) can neither arise nor vanish in the course
of motion, their flow in phase space must be conservative:

$$\frac{\partial \varphi}{\partial t} = - \text{div} \left(\varphi \vec{v} \right) \qquad (I.1)$$

where velocity and gradient are 2s-dimensional
Cartesian vectors

$$\vec{v} = \dot{\vec{x}} \qquad grad \quad = \left\{ \frac{\partial}{\partial q} , \quad \frac{\partial}{\partial p} \right\} .$$

By Hamilton's equations the velocity field is
divergence free.

$$div \; \dot{\vec{x}} \; = 0. \tag{I.2}$$

This shows that in eq. (I.1)

$$div \; (\varrho \, \vec{v}) = (\vec{v} \; grad) \varrho + 0,$$

and accordingly that the <u>substantial derivative</u>

$$\frac{d}{dt} = \frac{\partial}{\partial t} + \sum_{1}^{s} (\dot{q} \frac{\partial}{\partial q} + \dot{p} \frac{\partial}{\partial p}) , \tag{I.3}$$

when applied to the phase density — gives zero:

$$\frac{d \varrho}{dt} = 0 . \tag{I.4}$$

<u>Thus the density stays constant around a moving
point.</u>

This is one form of Liouville's theorem of
which we shall soon avail ourselves. But it should
be admitted of course that from a more mathematical
point of view the phase density ϱ is only an orna-
ment in this context. Just as in Hydrodynamics,
incompressibility of the flow is already contained
in the equation <u>div $\vec{v}$ = 0.</u>

b) Looking at the motion of ϱ, solely from
the point of view of Hamiltonian dynamics, we
would at once write down

$$\frac{d \varrho}{dt} = \frac{\partial \varrho}{\partial t} + \left\{ \varrho , \mathcal{H} \right\} . \tag{I.5}$$

Combining this with the conservation equations
(I.1) and (I.2), which can be written

$$\frac{\partial \varrho}{\partial t} = \{ \mathcal{H}, \varrho \} \tag{I.6}$$

we may thus express Liouville's theorem in the form (I.4) by saying that $\underline{\varrho(q,p,t)}$ is an integral of the motion: it makes the r.h.s. of eq. (I.5) equal to zero.

   c) We shall make use of Liouville's theorem in still another form:

   An element of the phase fluid does not change in volume during its natural motion.

   This is implied in the preceding equations. Being independent of special properties of the density, it can, of course, also be proved directly. Consider the two integrals

$$I_o = \int_{\Omega(t_o)} (dq_o \, dp_o)^s \quad \text{and} \quad I = \int_{\Omega(t)} (dq \, dp)^s$$

Here $\Omega(t_o)$ denotes the set of points $q,p \in \Omega(t_o)$ at $t=t_o$, and $\Omega(t)$ encompasses the same set of representation points at a subsequent stage t of their motion. In these two integrals the volume elements $(dq \, dp)^s$ may be conceived as instantaneous subdivisions of phase space. But we can transform them to the same time by the equations of motion:

$$\int_{\Omega(t)} (dq \, dp)^s \longrightarrow \int_{\Omega(t_o)} (dq_o dp_o)^s \; D(t,t_o)$$

where

$$D(t,t_o) = \frac{\partial \, (qp)}{\partial (q_o p_o)}$$

is the Jacobian connecting $\{q(t) \, p(t)\}$ and $\{q_o p_o\}$. It is a well-known theorem in Mechanics that this

Jacobian is equal to unity

$$D(t, t_0) = 1 \qquad (I.7)$$

Accordingly

$$\int_{\Omega(t_0)} d\Omega_0 = \int_{\Omega(t)} d\Omega . \qquad (I.8)$$

In particular if Ω comprises the whole available phase space

$$\Omega(t_0) = \Omega(t) ,$$

every point moves into the hole left by another one. The most compact form of Liouville's theorem is perhaps just eq.(I.7). The proof follows from Hamilton's equations, or else as one of the properties of canonical transformations.

<u>Gibbs' Mixing</u> [2]

Let us now look at Gibbs' definition of entropy in a state —not necessarily equilibrium— and its evolution in time. Initially we should have (with Boltzmanns constant = 1)

$$S(o) = - \int_{\Omega(o)} \varrho(o) \log \varrho(o) \, d\Omega. \qquad (I.9)$$

$S(t) = ?$

Following the orbit $\mathbf{x}(t)$ of a single system:

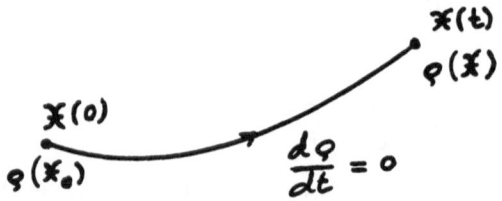

we have $\varrho(\mathbf{X}_0) = \varrho(\mathbf{X}_t)$ constant. Furthermore,
as we have seen, $\Omega(t) = \Omega(0)$. It is true that
in the figure we consider only one single orbit,
while the entropy as defined by Gibbs is a functional $S[\varrho]$ over the entire phase space. Nevertheless —as long as nothing prevents us from
using volume elements $d\Omega(t)$ in the integral
(I.9) co-moving with the flow— the conclusion
is inevitable that the phase average

$\overline{\log \varrho}$ = constant in time

as if the Entropy were constant, although we no-
where presumed a <u>stationary density</u> $\partial\varrho/\partial t = 0$,
as is necessary for equilibrium!

Before proceeding to Gibbs' own answer to this
we may perhaps remark that it is really a result
which could be anticipated. The entropy, as defined
here, is a measure of the latitude in the mechanical definition of the system. At time zero this is
given by a distribution $\varrho(0)$ over the phases. In
the interval $[0,t]$ we let the system evolve in a
determinate way prescribed by its Hamiltonian:

$$\mathbf{X}(0) \longrightarrow \mathbf{X}(t;\mathbf{X}(0))$$

The whole indeterminacy therefore rests in the
<u>initial conditions</u> which have a probability distri-
bution $\varrho(\mathbf{X}(0))$. And this indeterminacy is precisely the same at time t as at time zero. Equivalently
one may consider the ensemble as a bundle of motions emerging with a certain initial latitude
$\Delta\Omega(0)$. In Boltzmann's terminology the entropy
corresponding to (I.9) is

$$S = \log \Delta\Omega, \qquad\qquad (I.10)$$

and again by Liouville's theorem $\Delta\Omega$ stays constant (*).

Gibbs riposted to this apparent constancy of
$S[\varrho]$ by the following imaginary experiment: Let
some dye be distributed with a density $\varrho(x,y,z)$
in an incompressible liquid, which does not change
its hydrodynamical properties, and let us neglect
<u>diffusion</u> of the dye.

If ϱ initially were constant over the whole
volume it would stay constant during all motions
we could produce, e.g. by stirring the liquid.
Accordingly, this corresponds to an equilibrium
state and an extremum in the average $\overline{\log\varrho}$.

If on the other hand the distribution $\varrho(o)$
had not been uniform, the density of the dye would
still be constant along every streamline. Choosing
volume elements moving with the liquid and making
them small enough we should also in this case have

$$\overline{\log\varrho(t)} \;=\; \overline{\log\varrho(o)}$$

Yet it seems from our most immediate experien-
ces that sufficient stirring will produce a uniform
distribution of the dye.

How far this (under these circumstances) is an
illusion can be seen from the next of Gibbs'
<u>examples</u>: A cylinder of liquid, one sector of

(*)   The situation would be quite different if the
Hamiltonian of the system were not determinate,
but time dependent in a stochastic way. This seems
to be a natural description of a thermodynamical
system [3] but we shall not take it up here.

which is black, the rest white, is set into rotation with an angular velocity which is a function
of the distance from the axis:

Rotating liquid $\omega = \omega(r)$ according to Gibbs[2]

In the course of time the black part of the
liquid will be extended into laminae which will be
so thin, and so densely spaced, that the entire
liquid will appear uniformly grey to the eye.

Nevertheless, the black liquid will occupy
precisely the same volume as before, and integrating
over the black and white parts of the liquid separately one would always find just the same value
of $\overline{\log \varrho}$.

The presumption is that in the Riemann sum
corresponding to the integral, one can make the
volume elements small in a way depending upon
how far the mixing has proceeded.

Quite another result ensues if, instead, one
subdivides the space into <u>fixed</u> cells $\Delta\Omega$ of finite magnitude, and considers —instead of the
<u>microscopic</u> phase density— the <u>coarse grained</u>
density defined by

$$P_i = \frac{1}{\Delta\Omega_i} \int_{\Delta\Omega_i} \varrho \, d\Omega \qquad (I.11)$$

This concept, invented by the Ehrenfests [4],
actually accounts for the <u>loss of information</u> which

is indicated by Gibbs in his allusion to the
limited accuracy of observation. As we emphasized,
the purely dynamical evolution in an ensemble can
not, of itself, entail any loss of information.
According to Gibbs and the Ehrenfests however,
it may still over long times bring about a mixing
which cannot be resolved by an observer. The size
of the cells $\Delta\Omega_i$ used for coarse graining is therefore something relative to the experimental situation

In this way one can easily prove a monotonic
increase of the entropy defined as

$$S = - \sum_\lambda P_\lambda \, \log P_\lambda \, \Delta\Omega_\lambda \quad (I.12)$$

$t = 0$

$P(0) = \wp(0)$

Since the fixed cells $\Delta\Omega$ are somewhat arbitrary,
this entropy is no longer an absolute concept.
But that is perhaps as it should be. The corresponding developments for Quantum Mechanics can be
carried out by means of the von Neumann-Dirac
density operator (see e.g. [5]). We will not go
into this now but shall briefly return to it
later.

Proof that:

$$\int \wp \log \wp \, d\Omega - \sum_\lambda P_\lambda \, \log P_\lambda \, \Delta\Omega_\lambda \geq 0 \qquad (I.13)$$

Whether in the first term we set $\wp(t)$ or $\wp(t_0)$ is
immaterial for the value of the integral. Locally
however, $\wp(t) \neq \wp(t_0)$ unless, incidentally, $\wp(t_0)$
had been chosen stationary, $\wp = \wp(E)$.

According to the definition of the coarse
grained density (I.11) we may also write (I.13)

as

$$\Delta S = \sum_{\lambda} \int_{\Delta\Omega_\lambda} d\Omega \left\{ \varrho \log \varrho - \varrho \log P_\lambda \right\}$$

because P is constant over a cell. Adding to this

$$\sum_{\lambda} \int_{\Delta\Omega_\lambda} d\Omega \, (P_\lambda - \varrho) = 0$$

one may write

$$S = \sum_{\lambda} P_\lambda \int_{\Delta\Omega_\lambda} d\Omega \left\{ \frac{\varrho}{P_\lambda} \log \frac{\varrho}{P_\lambda} - \frac{\varrho}{P_\lambda} + 1 \right\}$$

Since the $\{\}$ can never be negative the proposition
follows.

The H-theorem

Much earlier and rather different was
Boltzmann's famous approach to the problem of
irreversibility. His functional

$$H = \int f \log f \, d^3\vec{v} \, d^3\vec{r} \tag{I.14}$$

$$f(\vec{r}, \vec{v}, t) = \text{number density of gas mole-cules}$$

extends only over the 6-dimensional μ-space. The
evolution of H was comprised in the collision
equation for f

$$\frac{\partial f}{\partial t} + \vec{v} \cdot \frac{\partial f}{\partial \vec{r}} + \vec{a} \cdot \frac{\partial f}{\partial \vec{v}} = \int \left[f' f_1' - f f_1 \right] gb \, db \, d\epsilon \, d^3\vec{v}. \tag{I.15}$$

By using this and the symmetries in the expression
for dH/dt one may write

$$\frac{dH}{dt} = \frac{1}{4} \int\!\!\int \left\{ \left[f' f_1' - f f_1 \right] \log \frac{f f_1}{f' f_1'} \right\} gb \, db \, d\epsilon \, d^3\vec{v}_1 d^3\vec{v} d^3\vec{r}$$

$$\leq 0$$

hence Boltzmann's $H = -S$ is monotonically decreasing, and the collision equation embodies irreversibility. We shall not repeat the well-known

objections to this result and their refutation
by Boltzmann himself. The essential thing is that
the Boltzmann equation (I.15) in contrast to the
Liouville eq. (I.6) —to which it can be related—
displays irreversibility, as demonstrated by
Boltzmann. Somehow —with the "Stosszahl Ansatz"—
it has left the solid ground of General Dynamics.
The B.E. is certainly no longer a purely dynamic
equation.

II.   RECURRENCES AND IRREVERSIBILITY
Poincaré's Recurrence Theorem

Concomitant to Liouville's theorem, though
not identical with it, we have the other dictum of
General Dynamics that all bounded motions have
a repetitive character. Poincaré noted this en
passant in his great memoir on the Three Body
Problem [6]. Like Liouville's theorem it belongs
to the absolutely firm foundations which Statistical Mechanics must carefully observe.

As an example of more learned terminology
we may state it as follows:

$$\forall t: \quad \mathbf{x}(t) \in \Delta\Omega$$
$$\exists T: \quad \mathbf{x}(t+T) \in \Delta\Omega$$

Proof:

$\alpha)$ $\underline{\omega \subset \Omega}$: $\omega(\mathbf{x} \in \Delta\Omega; t=t_0)$, $\mathcal{M}(\omega) = \Delta\Omega$

$\beta)$ $\underline{\mathcal{T}(\tau)}$: $\mathcal{T}\mathbf{x}(t) = \mathbf{x}(t+\tau), \mathcal{T}^{-1}\mathbf{x}(t) = \mathbf{x}(t-\tau)$

$$\mathcal{M}(\mathcal{T}\omega) = \mathcal{M}(\omega)$$

Then almost every point of the set ω is recurrent.

(that is: $\mathcal{T}^n \mathcal{X} \in \Delta\Omega$, for some n). For consider those points of ω which do <u>not</u> reenter into $\Delta\Omega$ after the

first epoch τ: $\qquad \omega \cap \mathcal{T}^{-1}\bar{\omega} \qquad (\bar{\omega} = \Omega\backslash\omega)$

2nd epoch τ: $\qquad \omega \cap \mathcal{T}^{-2}\bar{\omega}$

.   .
.   .
.   .

Their intersection is

$$\delta = \omega \cap \mathcal{T}^{-1}\bar{\omega} \cap \mathcal{T}^{-2}\bar{\omega} \cap \ldots \cap \mathcal{T}^{-n}\bar{\omega}$$

If now $\mathcal{X} \in \delta$ then $\mathcal{T}\mathcal{X} \notin \delta$, $\mathcal{T}^2\mathcal{X} \notin \delta$ ... and <u>no</u> two of the sets

$$\delta, \mathcal{T}\delta, \mathcal{T}^2\delta \ldots \text{ ad infinitum,}$$

will overlap. Since $\mathcal{T}$ is measure preserving and the total measure $\mathcal{M}(\omega + \bar{\omega}) = \Omega$ is finite, we must therefore conclude that

$$\underline{\mathcal{M}(\delta) = 0 \qquad \text{q.e.d.}}$$

Although we knew quite well that:

$$\mathcal{T} = \exp\left(-\tau\{\mathcal{H}, \}\right), \mathcal{M}(\omega) = \int_{\Delta\Omega}(dqdp)^s$$

or, if Ω is the energy surface Σ,

$$\mathcal{M}(\omega) = \int_{\Delta\Sigma} \frac{d\Sigma}{\text{grad } E}$$

we have here stripped the argument of all this physical imagery. Each of us may judge for himself what can be gained (or lost) by reducing it to the logical skeleton essential in the proof. Perhaps it is somewhat like the situation of the painter who —by means of only a few white and black pebbles— was to make a picture of Mona Lisa. He succeded —but for a certain smile on that enigmatic countenance$^{(*)}$.

With regard to the Physics we have so far used the language of Classical Mechanics. We shall, of course, in many cases need Quantum Mechanics. But the paralellism between classical and quantal equations is close in the context of irreversibility. It is conveniently expressed by the Statistical operator ϱ [5] whereby both Liouville's and Poincaré's theorems can be retrieved in analogous form.

Recurrence in Quantum Mechanics

It is well emphasized that, for Harmonic Oscillators, wave packets stick nicely together. Unlike those for free particles, they do not diffuse in the course of time. They only show some periodic expansions and contractions if the initial state has not been carefully chosen.

It seems to be widely believed, however, that, in general, wave packets must diffuse even for bounded systems, and in fact, an argument to this effect can apparently be drawn from the following remark of Heisenberg:

The time after which the outer parts of a wave packet of band with $\Delta\nu$ will begin to interfere destructively is $T \sim 1/\Delta\nu$. Here,

$$\Delta\nu = \frac{\partial^2 \mathcal{E}}{\partial J^2} \Delta J = \frac{1}{h} \frac{\partial^2 \mathcal{E}}{\partial n^2} \Delta n$$

The Harmonic Oscillator however is a singular case in as much as

---

$^{(*)}$ Example by Henning Sinding Larsen, Norwegian journalist.

$$\frac{\partial^2 \mathcal{E}}{\partial n^2} = 0 \quad \therefore \quad T = \infty$$

At first glance this looks convincing. Considering, for instance, free particles in a box, the general wave packet can be expressed by theta functions and, using their nice transformation to inverse time arguments [7], it is tempting to conclude that an initially peaked probability density will spread out indefinitely over long times.

But this is fallacions: A general wave packet on the line [o,a] can be written

$$\psi(x,t) = \sum_n C_n \sin \frac{n \pi x}{a} \exp\left\{\frac{n^2 \pi^2 \hbar}{2ma^2} it\right\}$$

Whatever its initial shape and however fast it will fall apart in the initial stage, it will be completely resurrected once in every epoch

$$\frac{8 \, ma^2}{h} \times \text{integer}$$

Now, this example is also very special, but a general recurrence theorem for Quantum Mechanics can be stated in the form [8]

$$\forall t, \, \Psi(t)$$
$$\exists T: \quad \| \Psi(t+T) - \Psi(t) \| < \delta$$

Proof: $\quad \Psi(t) = \sum_n C_n \, \psi_n(o) \, e^{-(i/\hbar)E_n t}$

$$\| \quad \| = (\sum_1^N + \sum_{N+1}^\infty) \, |C_n|^2 \cdot 2 \left[1 - \cos(E_n T/\hbar) \right]$$

Firstly, N can be chosen such that

$$\sum_{N+1}^\infty < \delta/2$$

Secondly, the sum

$$\sum_1^N < 2 \sum_1^N \left[1-\cos\,(E_n T/\hbar)\right]$$

is an almost periodic function; i.e. a time T can
always be found such that also the sum

$$\sum_1^N < \delta/2 \quad,$$

whereby the proposition is proved.

In a more qualitative manner we could say
that the density matrix of a pure state

$$\langle x \mid \rho \mid x' \rangle = \sum C_n C_m^* \, e^{\frac{i}{\hbar}(E_m - E_n)t} \, \psi_n(x)\,\psi_m^*(x')$$

will have an almost periodic dependence upon time
unless, incidentally, all frequencies are commensurable. That is, denoting its diagonal element
by R(t) we shall have

$$\left| R(t+T) - R(t) \right| < \epsilon$$

when T(ϵ) is one of the translation numbers of
R(t) belonging to ϵ.

Thus the phenomenon of recurrence —despite
the different definition of a state— subsists
also in Quantum Mechanics.

Recurrence time

Recurrence times care —relatively speaking—
enormously long, even for quite simple systems.
Evidently the duration of a Poincaré Cycle must
increase with the number (s) of degrees of freedom and with the sharpness ($\Delta\Omega$) with which the
recurrent state is specified.

In the cases known, the dependence is roughly of the form $T_{rec} \sim \Omega/\Delta\Omega = (\omega/\Delta\omega)^S$. It can be evaluated exactly for a certain class of systems, namely those which are <u>conditionally periodic</u>:

$$q_K = \sum_n C_n^{(k)}(J)\, e^{i(n\cdot\omega)t}$$

$$n = \{n_1, n_2, ..., n_s\} \quad \text{(integers)}$$

$$\omega = \{\omega_1, \omega_2, ..., \omega_s\}$$

We shall be interested only in the case when all frequencies are <u>rationally independent</u>,

i.e. $\quad n_1\omega_1 + n_2\omega_2 + \cdots n_s\omega_s = 0$

insoluble in integers not all zero.

The action variables J_j are constants, while the angle variables ω_j move on an (s-1)-dimensional torus i.e. the Cartesian product of s-1 circles $z_j(t)$:

$$0 \leq \arg z_j = \omega_j t \leq 2\pi \quad j=1,2,...,s-1$$

<u>Motion in the angle variables</u>

Following an idea of H. Tornehave [*] we depict the motion of the angle variables by a set

---

[*] Colloquium of the Swedish Math. Soc. 1954

of $s-1$ clocks the hands of which rotate with an-
gular velocities $\omega_1, \omega_2, ..., \omega_s$. Given such a set
of rotating vectors $z_1(t), ..., z_{s-1}(t)$ it is always
possible —by a suitable choice of the time parameter— to restore any situation

$$z_j \in \Delta\varphi_j \, , \text{ all } j$$

This is a celebrated theorem in Number Theory due
to Kronecker. Since re-entry into a configuration
$z_j \in \Delta\varphi_j$ evidently means recurrence of a mechanical state once occupied, Kronecker's theorem asserts
that there will be recurrences. So far, it appears
here as a mathematical exemplification of Poincaré's
theorem. In this particular case however, we are
also able to say something about how long one must
expect to wait between recurrences [9].

One needs two lemmata (Hemmer 1956):

α) The probability of finding a special situation
$z_j \in \Delta\varphi_j$ at an arbitrary time is:

$$\text{Prob} (...z_j \in \Delta\varphi_j ...) = \prod_{j=1}^{s-1} \left(\frac{\Delta\varphi_j}{2\pi} \right)$$

β) The average duration of a configuration where
every vector is within its sector $\Delta\varphi_j$ is

$$T_{coinc.} = 1 \bigg/ \sum_{j=1}^{s-1} \frac{\omega_j}{\Delta\varphi_j}$$

Once these are proved one can easily find the average recurrence time defined by Hemmer as

$$T_{Rec.} = \lim_{r \to \infty} (t_r/r)$$

where

t_r = time of r'th recurrence.

The probability in the first lemma α) can be given the alternative form

$$\prod \left(\frac{\Delta \varphi_j}{2 \pi}\right) = \lim_{t \to \infty} \left(\frac{\text{time in coincidence}}{\text{total time}}\right)$$

Dividing numerator and denominator by the number of recurrences up to time t, one obtains

$$\prod \left(\frac{\Delta \varphi_j}{2 \pi}\right) = \frac{T_{\text{coinc}}}{T_{\text{Rec}}}$$

from which follows

$$T_{\text{Rec}} = \prod_1^{s-1} \left(\frac{2 \pi}{\Delta \varphi_j}\right) \Big/ \sum_1^{s-1} \left(\frac{\omega_j}{\Delta \varphi_j}\right)$$

The most important of Hemmer's many ingenious tricks was probably to go first for T_{coinc} which is easier to find than T_{Rec}

Coupled harmonic oscillators, that is, all kinds of linear systems, (e.g. Hooke's Law crystal lattices) belong to the class here considered. For a linear chain of s=10 particles, a frequency spectrum of width $2 \omega_o$ = 10/sec, and a latitude of $\Delta \varphi = \pi/100$ in the angle variables, one finds a recurrence time exceeding the age of the universe ($T_{\text{Rec}} \sim 10^{10}$ years).

This is, to my knowledge, the most precise information so far available on recurrence times. Whether the limit of (t_r/r) as $r \to \infty$ converges pointwise or in some other sense is not known. It would also have been of considerable interest to know a distribution of recurrence times, not only an average.

III.   REMARKS ON THE ERGODIC PROBLEM

Ergodic theory seems still to be an unfinished chapter of Mathematics although its inception in Physics goes back to Maxwell and Boltzmann. Specifically it was the assumption they made in order to justify the equivalence of time and phase average. By "ergodic theorem" we shall just mean the statement (when true): "Time average equals phase average", and not the several different propositions from which this conclusion can be drawn.

Let the mechanical state of the system be given again by a point

$$\mathbf{X}(t) = \{q_1 \cdots p_s\} \tag{III.1}$$

or, eventually, be delimited by a Cartesian volume element

$$d\,\Omega = dq_1 \cdots dp_s \tag{III.2}$$

A real system will be restricted by external conditions to a finite part of the phase space. The total phase volume (Ω) available may, for instance, be the energy shell,

$$E \leq \mathcal{H}(q,p) \leq E + \Delta E \tag{III.3}$$

A thermodynamical variable should primarily be identified with a time average:

$$\widetilde{f} = \frac{1}{t} \int_0^t f(\mathbf{X}(t))\, dt \tag{III.4}$$

of some phase function $f(\mathbf{X})$. Let us denote the limit for infinite time by

$$f^* = \lim_{t \to \infty} \widetilde{f} \tag{III.5}$$

assuming that it exists, in a sense to be specified later.

As a rule, however, we are only able to compute phase averages

$$\bar{f} = \frac{1}{\Omega} \int f(\mathbf{X}) \, d\Omega . \qquad\qquad (III.6)$$

But it can be shown (on certain premises) that

$$f^{\ast} = \bar{f} \qquad\qquad (III.7)$$

which is Birkhoff's ergodic theorem [10].

As will be discussed at the end of this lecture, <u>Birkhoff's result (7) is not quite sufficient for the foundation of statistical mechanics</u>. But it is <u>necessary</u>, and we shall first try to deduce eq. (7), albeit only in an intuitive way. More complete mathematical expositions are found in the books of Khinchin and Farquhar and in Rosenfeld's lecture notes.

For the moment I have taken the integrals (4) and (6) to be ordinary ones and the limit (7) to exist point wise. But we shall return later to the reasons for considering (4) and (6) as Lebesgue integrals and (7) to hold only for almost all orbits $\mathbf{X}(t)$. As a precaution we might therefore have written $\mathcal{M}(\Omega)$ and $d\mathcal{M}(\Omega)$ instead of Ω and $d\Omega$ to emphasize a problem of measure here.

Transitivity.

An essential aid for the whole discussion is Birkhoff's idea of transitivity. In order to explain this concept it is convenient to recall some other notions of general dynamics:

α) A region Ω of phase space is said to be invariant if for any point $\mathbf{X}$ in Ω the whole orbit through $\mathbf{X}$ belongs to Ω .

β) Such a region Ω is said to be <u>indecomposable</u> if it cannot be divided into parts $\Omega' + \Omega''$ which are separately invariant; the corresponding motion is said to be transitive.

As before we consider closed systems only, and the energy shell Ω as the region to be inspected for transitivity. Ergodicity of systems with stochastic Hamiltonian is probably trivial.

A further insight into the significance of transitivity comes from a certain classification of the integrals of the equations of motion

$$\frac{dq_1}{\partial \mathcal{H}/\partial p.} = \ldots\ldots \frac{dp_s}{-\partial \mathcal{H}/\partial q_s} = dt \ . \qquad \text{(III.8)}$$

2s-1 of them can be given in the time-independent form,

$$g_i (\mathcal{X}) = c_i \qquad (i=1,2,\ldots 2s-1) \qquad \text{(III.9)}$$

Here a characteristic alternative shows up: <u>Such an integral may or may not define a surface in Ω</u>. In the former case (9) considered as an equation for any one single variable has only isolated solutions (Poincaré's "Integrales uniformes").

The alternative is infinitely many roots in any finite interval. Integrals defining surfaces we may denote as <u>univalued</u> (finitely multivalued).

The existence of a univalued integral will lower the <u>dimensionality</u> of Ω by one. In the extreme case where there are 2s-1 of them Ω will be reduced to the one-dimensional continuum of the orbit which now appears as the section between the surfaces $c_1, \ldots c_{2s-1}$. This means a closed orbit and exact periodicity.

When the number of univalued integrals is less than $2s-1$, a phase space of more than one dimension remains, and fixing the "non uniform" integrals will not further reduce the dimensionality.

It is of course true that specification of all the constants of integration $c_1, c_2 \ldots$ will determine the orbit completely. But it is <u>only the constants fixing the univalued integrals which affect the long time averages</u>. Whatever values we choose for the others —because of the infinite multivaluedness of any variable— the orbit must interweave the remaining phase space densely.

We can also say that the univalued integrals decompose the phase space into invariant parts (m in number):

$$g_1(\mathbf{x}) = c_1, \ldots, g_m(\mathbf{x}) = c_m$$

Their intersection —of dimension $2s-1-m$— is indecomposable.

Simple examples are afforded by separable problems, for example those satisfying Stäckels conditions or the conditionally periodic systems. For these the action variables form a set of $m \leq s-1$ univalued integrals —in addition to the energy which is always univalued (or two-valued). The remaining phase space can be mapped on an m-dimensional torus (Poincaré) which in general will be invariant.

For illustration one can think of the

Lissajous figure traced by two incommensurable
harmonic oscillations in a plane:

$$X_2 = \cos \left(\frac{\omega_2}{\omega_1} \arccos x_1 + c \right) \qquad (III.11)$$

To a given value of $x_1 \in [-1, +1]$ corresponds a
multiplicity of values $x_2^{(n)} \in [-1, +1]$:

$$x_2^{(n)} = \cos \left(\frac{\omega_2}{\omega_1} \text{Arc} \cos x_1 + c + 2\pi n \frac{\omega_2}{\omega_1} \right) (III.12)$$

$$n = 0, 1, \ldots$$

Arc = principal branch of arc cos

If ω_2 / ω_1 is rational the multiplicity will be
finite and the orbit $x_2(x_1)$ will be closed. But
if ω_2 / ω_1 is irrational every finite segment
$\Delta x_2 \in [-1, +1]$ on the line $x_1 = \underline{\text{const}}$ will contain an infinite number of points $x_2^{(n)}$ because
the numbers

$$\varepsilon(n) = n \frac{\omega_2}{\omega_1} - n \left[\frac{\omega_2}{\omega_1} \right] \quad n = 1, 2, \ldots \infty$$

will be uniformly distributed in $(o, 1)$. (Weyl)

The number of univalued integrals (Levi
Civita: "Degree of Imprimitivity") is of course
of great interest and there are a few general
results [11] (Poisson, Bruns, Poincaré, Hadamard
and Levi Civita) which pertain to it. But the
general theory of orbits is one of the uncompleted
chapters in analytical mechanics and we shall gain
by deferring this question to the context of specific systems.

Returning to the proposition (7) we shall
now assume that $\mathcal{H}(\mathfrak{X}) = E$ is the only univalued

integral. (For extension to higher order of imprimitivity see H. Grad [10]). Apart from 1) the hypothesis of invariance of the region Ω, we shall
need only well established theorems of Dynamics
in the proof of eq. (7) viz. 2) Liouville's theorem on the incompressibility of phase volumes, and
3) Poincaré's on the recurrence of any state $x \in \Delta\Omega$.
These three premises will be used repeatedly.

We shall need two lemmata:

1º Life time of $x \in \Delta\Omega_i$

When the representation point $x(t)$ passes
through a particular cell $\Delta\Omega_i$ in phase space the
duration of the passage depends upon how it hits
the cell, whether centrally or grazingly. The mean
duration or 'lifetime' τ_i must be equal to the
mean segment $\bar{l}_i$ of the orbit inside $\Delta\Omega_i$ —divided
by the velocity v_i at this point in phase space:

$$\tau_i = \frac{\bar{l}_i}{v(x_i)}$$

(III.13)

$$v = \left[\dot{q}_1^2 + \dot{q}_2^2 + \dots \dot{p}_s^2\right]^{\frac{1}{2}}_{x = x_i}$$

To find $\bar{l}$ one may argue as follows: In the course
of long times the orbit will intersect $\Delta\Omega$ with a
dense array of parallel chords. Introducing by an
obvious geometrical analogy the "projected area"
σ of $\Delta\Omega$ along $\vec{v}$ we have $\bar{l}\sigma = \Delta\Omega$ and consequently

$$\tau = \frac{\Delta\Omega}{\sigma v}$$

(III.14)

2º Recurrence time of $x(t) \in \Delta\Omega$

Consider now an area σ orthogonal to the

velocity field in phase space. In a time interval
t a phase volume $\sigma v t$ will have passed through the
cross section σ . At most after a time $\Omega/\sigma v$ recurrences must have taken place. It is a natural guess
that

$$T = \frac{\Omega}{\sigma v} \qquad\qquad\qquad (III.15)$$

is just the mean recurrence time, and this can be
corroborated [13].

   We are now ready for the demonstration of
Birkhoff's result (7). In the integral (4) we
understand the representation point $\mathfrak{X}(t)$ to move
according to a determinate Hamiltonian, transitive
in Ω . The first step is to replace the integral
(4) by its Riemann sum:

$$\int_0^t f(t)\, dt = \lim \sum_n f(t_n)\, \Delta t_n \qquad\qquad (III.16)$$

When this exists, the subdivision of the time interval

$$t = \sum \Delta t_n$$

is arbitrary. In particular we may choose the time
differences Δt equal to the transit times of
 $\mathfrak{X}(t)$ through cells $\Delta\Omega_i$ of a fixed subdivision
of Ω .

   By Poincaré's theorem every cell will be tra-
versed many times (for large t). Let Δt_{ir} be the
time spent by $\mathfrak{X}(t)$ inside $\Delta\Omega_i$ at the r-th passage
and let $N_i(t)$ be the number of recurrences to this
cell in the interval (o,t). Then one can write:

$$\sum f(t_n)\,\Delta t_n = \sum_{i=1}^{\Omega/\Delta\Omega} \sum_{i=0}^{N_i} f(\mathbf{X}_i)\,\Delta t_{ir} = \sum_i f_i \sum_{r=0}^{N_i} \Delta t_{ir} \qquad (*)$$

$$\text{(III.16)}$$

In the second sum we may write

$$\sum_{r=0}^{N_i} \Delta t_{ir} = N_i \quad \overline{\Delta t_i} \qquad\qquad \text{(III.17)}$$

where $\overline{\Delta t_i}$ is the average transit time over N_i passages. After these preparations we can write the time average

$$\frac{1}{t}\int_0^t f(\mathbf{X}(t))\,dt = \lim_{\Delta\Omega\to 0} \sum_{i=1}^{\Omega/\Delta\Omega} f_i \frac{\overline{\Delta t_i}}{t/N_i}$$

In eq. (7) we have still to consider the limit $t\to\infty$ and this must be carried out before making the cells $\Delta\Omega$ infinitesimal. Assuming that $\overline{\Delta t_i}$ and $t/N_i(t)$ approach limits for $t\to\infty$:

lim $\Delta t_i = \tau_i$ mean time of sojourn $\mathbf{X}(t) \in \Delta\Omega_i$.

lim $t/N_i = T_i$ "   recurrence time for " "   ,

we have for an infinite period

$$f^* = \lim_{\Delta\Omega\to 0} \sum_i f_i \frac{\tau_i}{T_i} \qquad\qquad \text{(III.18)}$$

Now, according to the lemmata (13) and (14)

$$\frac{\tau_i}{T_i} = \frac{\Delta\Omega_i}{\Omega}$$

$$\therefore \sum f_i \frac{\tau_i}{T_i} = \sum_i f(\mathbf{X}_i) \frac{\Delta\Omega_i}{\Omega}$$

(*) From the last reshuffling of the time subsets (16) one sees that it is natural to assume only measurability of $f(\mathbf{X})$ in Ω. Then the limit (15)=(16) exists as Lebesgue integral. That the subsets $\Delta t_{ir} \subset (0,t)$ are σ- additive is evident.

and passing to the second limit

$$\lim_{t \to \infty} \int_0^t f\left(\mathbf{X}(t)\right) dt \;=\; \frac{1}{\Omega} \int_\Omega f\left(\mathbf{X}\right) d\Omega \qquad (\text{III}.19)$$

as was to be shown.

The ´eggregial´ part of Birkhoff´s proof, however, is not eq. (7) but eq. (5) that <u>the time average</u>

$$f^{*} = f^{*}\left(\mathbf{X}(t=o)\right)$$

<u>exists and is independent of the initial point for a.e.</u> $\mathbf{X}_o$. The equality of f^{*} with the phase average —which is the main concern of Statistical Mechanics— Birkhoff concludes off hand from transitivity. Therefore, eq. (7) or (19) is often denoted as the "Corollary to Birkhoff´s ergodic theorem" the dignity of a theorem being reserved only for the <u>existence</u> of f^{*}.

Anyway —theorem or not, eq. (19) is still too weak as a foundation for Statistical Mechanics: One must remember that the statistical interpretation of thermodynamics presumes time averages over periods which are extremely short on a labo-ratory scale, while the averages considered above must involve at least several Poincaré cycles!

To my knowledge, the first who saw a problem here was Einstein in his independent development of Statistical Mechanics which was published in the "Annalen der Physik" in the years 1904-1912. Einstein´s point of view seems also to contain the solution. Essentially his answer is this: what matters for the averages is recurrence of the <u>physical state</u>, not a complete mechanical recurrence

in Poincaré's sense. Since all phases emerging
from a given one by a mere permutation of identical particles are equivalent, the physical, or in
Gibbs' terminology the "generic" recurrences must
be vastly more frequent than the "specific" ones:

While the durations of Poincaré cycles are
beyond all experience, the interval between two
generic recurrences of a macroscopic system

$$T_{Einstein} \sim T_{Poincaré} / N! \; ; N=\text{number of particles}$$

$$(III.20)$$

may well be short enough to secure a <u>stronger</u> equivalence of time and phase average than Birkhoff's
eq. (7), namely in sufficient approximation

$$\widetilde{f} = \overline{f} \qquad\qquad (III.21)$$

which is what actually is requested.

This is the point where an assumption about
the degree of coarse graining naturally enters:
It must not be so coarse that we average out experimentally discernible inhomogeneitis. On the other
hand we are not interested in subdivisions $\Delta\Omega$ finer
than corresponding to a few times the average phase
volume per particle in μ-space ($\Omega = \omega^N$).

On the basis of past experience we would then
guess:

$$\frac{T_{Rec}}{T_{coinc}} = \frac{\Omega}{\Delta\Omega} = \left(\frac{\omega}{\Delta\omega}\right)^N$$

$$T_{coinc} \sim \frac{1}{N} \cdot \tau \quad , \quad \tau = \text{molecular time of transit}$$

Taking somewhat arbitrarily

$$\Delta\omega = e \frac{\omega}{N}$$

we have

$$T_{Einstein} \sim \frac{1}{N!} \left(\frac{N}{e}\right)^N \cdot \frac{1}{N} \; \chi \; (molecular \; time)$$

Thus, when supplemented with Einstein's remark, Birkhoff's theorems may be sufficient.

IV. ENERGY TRANSPORT IN LINEAR AND NON-LINEAR CHAINS
Energy Transport in one dimensional lattices

It seems intuitively clear that the most probable distribution of energy over a homogeneous many particle system is spatially uniform, and that energy gradients, once set up, for the majority of initial conditions will start by levelling out. Looking not only at a single orbit but at a whole pencil corresponding to a distribution $\varphi(\chi)$ at t = o we must expect to find the statistical mechanics of heat conduction.

Transport theory, e.g. the kinetic equations of Maxwell and Boltzmann —not to speak of the so-called master equations— proceeded, of course, quite differently. And the programme just sketched is certainly inefficient for applications. But it would appear to be the most direct approach to time dependent statistical mechanics from the theory of equilibrium as it was conceived by Gibbs. The problem in the foundations of transport theory is to know where the probabilistic assumptions are added to the dynamics. In Gibbsian statistical mechanics this was always clear: it is in the initial conditions.

When feasible, the Gibbsian method throws some interesting sidelights upon current transport theory, but its implementation depends, of course, on the availability of mechanical models of many-body systems which can really be analysed. Until recently this was limited to the class of linear systems, i.e. to coupled harmonic oscillators. But a great advance was made in 1967 when Toda discovered that the equations of a certain type of non-linear chain could be solved exactly.

Linear Chains

Let us first consider the linear chains. The simplest of them is the homogeneous one

$$m\ddot{x}_n = \alpha (x_{n-1} + x_{n+1} - 2x_n), \quad n=0, \pm 1, \pm 2 \ldots \quad (IV.1)$$

It was first considered, with respect to heat conduction, in a famous paper of Schrödinger [15] and later by many authors [16]. About the same time (1914) however, Debye had rejected all linear models in the context of heat conduction. The reason was that in a harmonic lattice, energy is propagated by completely independent elastic waves. Thus the propagation of energy —so was Debye´s contention— can not be diffusive as befits heat conduction. Having successfully related heat expansion of solids to anharmonicity of the lattice forces, Debye associated heat conduction also with the same cause: anharmonicities scatter the elastic waves and thus conjure the propagation of energy to be diffusive instead of wave like - - .

Debye´s doctrine of heat conduction was translated to quantum mechanics by Peierls. In the quantum theory of solids the lattice energy

is carried by phonons. Using the elementary formula for the heat conductivity of gases:

$\varkappa = \frac{1}{3}$ (specific heat)(velocity)(mean free path)

Debye´s argument means that the free path of
phonons will be infinite (or $\sim V^{\frac{1}{2}}$) unless anharmonicities are there to bring about their mutual
scattering.

I think that this criticism of Schrödinger´s
model is partly valid. Yet it is so instructive
that I would like to expound it a little. Schrödinger wrote the solution of (IV.1) as

$$\xi_n(\tau) = \sum_k \xi_k^0 \; J_{n-k}(\tau) \tag{IV.2}$$

$$\xi_{2n} = \sqrt{m} \; \dot{x}_n \; , \quad \xi_{qn+1} = \sqrt{\alpha}\,(x_n - x_{n-1})$$

$$\xi^0 = \xi(t=o) \quad \text{initial values}$$

$$\tau = \omega_o t \; , \quad \omega_o = 2\sqrt{\alpha/m}$$

J = Bessel functions of 1st kind

First example, point source at n=o. This case was
considered by Schrödinger and by Hemmer [9]

$$\xi_k(o) = \xi_k^0 = \delta_{ko}$$

$$^v\xi_{2n} = \text{velocity of n'th particle } \sqrt{m}$$

$$\epsilon_n = J_{2n}^2(\tau) \sim \text{kinetic energy of n'th particle}$$

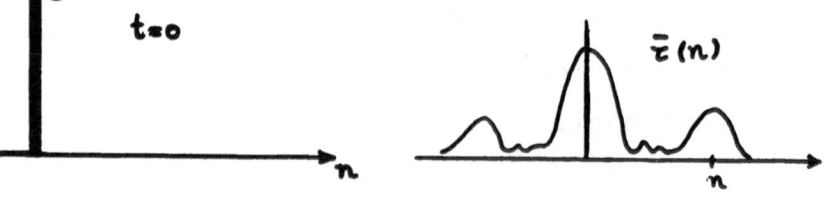

First max. of J_{2n} = first zero of J'_{2n} occurs
at $\bar{\tau} = 2$ nf where $1 < f < \sqrt{2}$.

$$\frac{s(n)}{\bar{\tau}(n)} \sim \frac{1}{2} \, \omega_0 a \cdot t$$

Group velocity: $\dfrac{\omega_0 a}{2} \cos(k\,a/2)$

Phase   "   : $\dfrac{\omega_0 a}{2} \cdot \dfrac{\sin(k\,a/2)}{(k\,a/2)}$

so we have a wave-like precursor as stressed by
Hemmer [9], and also admitted by Schrödinger [15].
The latter, however, maintained that the central
peak is always higher than the wings of the temperature distribution, while Hemmer's sketched
curve seems to indicate the opposite. Actually
after a long time both maxima are about equal
viz. $\sim n^{-\frac{1}{2}}$. For moderate values of τ however,
the central peak is still dominant. The front
passing the point n=4 ($\bar{\tau} = 4.49$) for instance
is less than one tenth of the central peak
($J_8(4.9)/J_0 (4.9) \approx 3 \cdot 10^{-2}$). The dispersion is
very weak in this lattice therefore smooth wave
forms will not be strongly distorted. Important
for the question of heat conduction is the following theorem of Hemmer [9]:
Contrary to Fourier's Law, the Heat flow in the
linear chain is not proportional to the gradient
of energy density.
Specifically: Starting with two halves of a chain
each at its own constant temperature, the heat
flow across the junction will approach a constant
value while the gradient goes to zero. This can be

seen from the solution 2 as follows:

Initial state:

$$\langle \xi_k^0 \rangle = 0; \langle \xi_k^0 \; \xi_\ell^0 \rangle = \begin{cases} \delta_{k\ell} & k \text{ and } \ell < o \\ o & k \text{ and } \ell > o \end{cases}$$

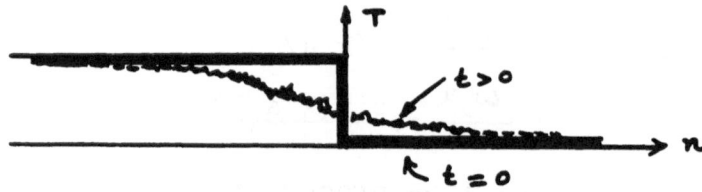

Temperature curves

Heat passed into $n > o$ $(E = \tfrac{1}{2} \sum \xi_n^2)$:

$$Q = \sum_{n \geqslant o} \sum_{k \leqslant o} J_{n-k}^2 \; \varepsilon_{n-k} \quad (\; \varepsilon_m = 1 - \tfrac{1}{2} \, \delta_{mo})$$

$$= \sum_{n \geqslant o} \sum_{k \geqslant o} J_{n+k}^2 \; \varepsilon_{n+k}$$

$$= \sum_{m \geqslant o} \varepsilon_m \; (1+m) \; J_m^2$$

Here $\sum_0^\infty \varepsilon_m J_m^2(\tau) = 1$ by a well-know addition theorem, therefore

$$Q (\tau) = \text{const.} + \sum_0^\infty m \; J_m^2 (\tau)$$

$$\frac{dQ}{d\tau} = 2 \sum m \; J_m \; J_m'$$

$$= \sum_0^\infty \frac{\tau}{2} \; (J_{m-1}^2 - J_{m+1}^2) = \frac{\tau}{2} \; (J_0^2 + J_1^2)$$

Now for $\tau \gg 1$:

$$J_n (\tau) \sim (2/\pi\tau)^{\frac{1}{2}} \cos (\tau - [n+\tfrac{1}{2}] \tfrac{\pi}{2})$$

$$J_0^2 + J_1^2 = \frac{2}{\pi\tau} \; (\sin^2 + \cos^2)$$

$$\therefore \quad \frac{d\,Q}{d\,\tau} = 0 \quad q.e.d.$$

Hence, although the 'temperature curves' sketched have a certain similarity to the Error integrals of phenomenological theory, the model does by no means follow Fourier's equation.

The presence of wave like precursors to a heat front in example 1 should however not be held against Schrödinger model. A complete kinetic theory must in this respect always differ from a linear parabolic equation. The latter contradicts the "Einstein Causality" which must show up as soon as a finite velocity of waves emerges from the mechanics. This leads us to expect that the partial differential equations of the continuum limit, by which we endeavour to link lattice models to a phenomenological description of heat, should neither be hyperbolic, like a wave theory, nor parabolic like diffusion theory. The simplest case of a mixed type is the "Telegraphers Equation" which actually has solutions similar to heat propagation, yet complying with Einstein causality. Therefore the question of diffusive or wave like energy transport may be wrongly posed. The continuum limit of the homogeneous linear chain is a purely hyperbolic equation. But smoothness of the initial conditions must thus be presumed before transition to the continuum! (Schrödinger)

Anharmonic chains.

The question is now: will it really help to make the coupling anharmonic? The failure of the

linear chain was usually connected with lack of
ergodicity. Its motion can be separated in normal
coordinates, no exchange of energy between different modes can take place. Anharmonicity can
achieve such exchange, but will it secure ergodicity? Fermi, Pasta and Ulams computer experiment
of 1955 was designed to study this question.

They considered linear chains perturbed by
anharmonic coupling terms and observed how —after
initial exitation of the fundamental vibration—
the energy was transferred to other modes. Surprisingly, they found that the initial exitation
recurred with full amplitude after a short time!

The chains studied were of the type

$$m\ddot{x}_n - \alpha(x_{n+1} + x_{n-1} - 2x_n) = \mu\{(x_{n+1} - x_n)^2 + (x_{n-1} - x_n)^2\}$$

which typically belongs to the class

$$\mathcal{H} = \mathcal{H}_0 + \mu\,\mathcal{H}_1 + \ldots$$

considered by Poincaré in his theorem on imprimitivity, and also to Fermi´s "Normal Systems", which
for general values of μ possess no univalued ("uniform") integral other than the energy. Yet their
ergodicity is not certain. According to the mathematicians [17] isolating integrals ("invariant tori")
may exist for small values of μ.

Recently much more extensive computer experiments have been done by Japanese physicists [18].
The typical outcome of them is sketched in figs.
1 and 2

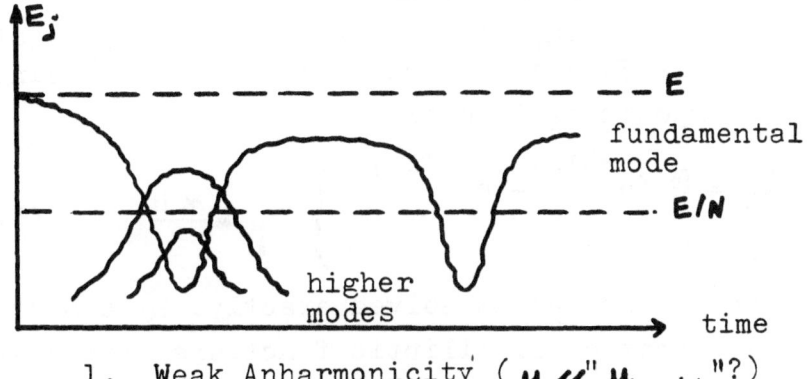

1. <u>Weak Anharmonicity</u> ($\mu \ll$" μ_{crit}."?)

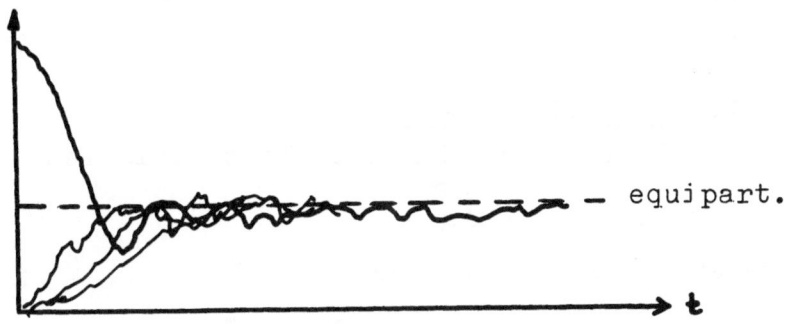

2. <u>Strong Anharmonicity</u> ($\mu \gg$" μ_{crit}.?)

The Japanese group suggests that there is a critical strength for the anharmonicity such that ergodicity obtains when the parameter μ exceeds a
certain value μ_{crit}. For $\mu < \mu_{crit}$. the Fermi-
Pasta-Ulan recurrence phenomenon is more or less
conspicuous.

Now computers cannot cover any long interval
of time, but the experiments seem on this point
to agree with the mathematical conjectures mentioned by Kolmogorov.

Toda´s Exact Solutions

The most important advance in this part of
physics is Toda´s discovery [19] that for chains
with ´exponential´ coupling:

$$m\ddot{x}_n = \frac{\alpha}{b} (e^{-br_n} - e^{-br_{n+1}})$$

$$r_n = x_n - x_{n-1}$$

$$\Bigg\} \quad " \mu " \Leftrightarrow \frac{\alpha b}{2} \qquad (IV.2)$$

the equations can be solved exactly. By a master-
ful application of elliptic functions Professor
Toda found two types of solutions:

1: Monochromatic travelling waves

2: Pulses or ´solitary´ waves

1. Travelling waves

Introducing dimensionless variables

$$y = br \qquad \tau = \sqrt{\frac{\alpha}{m}}\, t$$

these solutions can be written

$$e^{-y_{n-1}} = (2 K \nu)^2\, Z´\, (2K\left[\frac{n}{\Lambda} \mp \nu\tau\right]) \qquad (IV.3)$$

where

$$Z´(u) = (\frac{\pi}{K})^2 \sum_{m=1}^{\infty} m \cos (\frac{m\pi u}{K})/ \operatorname{Sinh} (m\pi\frac{K´}{K})$$

is the derivative of Jacobi´s Zeta function, K(k)
and K´= K(k´) the complete elliptic integrals of
1st kind

$$K \sim \begin{cases} \frac{\pi}{2} (1+ \frac{1}{4} k^2 +..) & k \ll 1 \\ \log 4/ \sqrt{1-k^2} \end{cases} \quad k´= \sqrt{1-k^2} \text{ the}$$
complementary
modulus

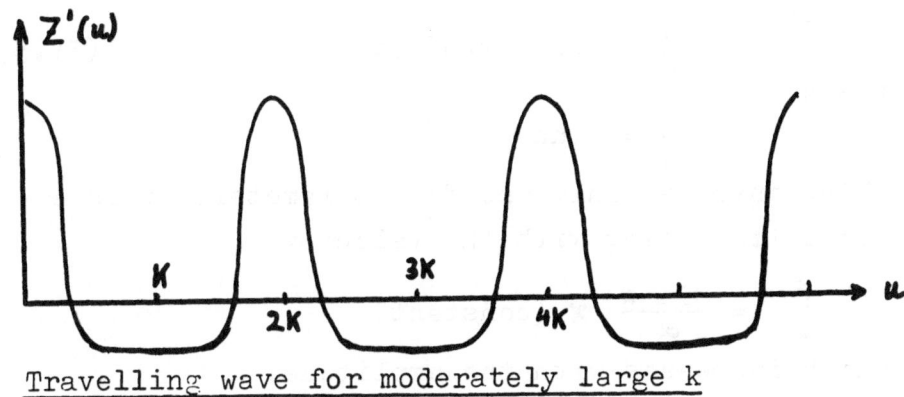

Travelling wave for moderately large k

At the maxima $r_n = x_n - x_{n-1}$ is negative, the lattice being compressed there, but as a whole the chain expands in motion. If one chooses modulus k and frequency ν as the free parameters, the 'wave length' Λ (an integer) is determined by :

$$cs^2 \left(\frac{2\,K}{\Lambda}\right) = (2\,K\,\nu)^{-2} - (E/\,K)\qquad\qquad\text{(IV.4)}$$

Here

$$E \sim \begin{cases} \frac{\pi}{2}\left(1 - \frac{k^2}{4} + ..\right), & k \ll 1 \\ \\ 1 & k \approx 1 \end{cases}$$

is the complete elliptic integral of 2nd kind and cs=cn/sn. For small k the waves approach the harmonic form: $\cos(2\pi[\nu\tau\mp\frac{n}{\Lambda}])$. For large k (k~1) the peaks are sharpened and the minima flattened.

2. Solitons[*]

In addition, Professor Toda found a sequence of highly interesting particular solutions,

---

[*] The name is evidently drawn from hydrodynamics from the 'solitary wave' of Stokes and Rayleigh.

the simplest of which has the form

$$e^{-y}{}_{n}-1 = \beta^2 \operatorname{Sech}^2 (\alpha n \mp \beta \tau) \tag{IV.5}$$

where

$$\beta = \operatorname{Sinh} \alpha$$

which involves only one free parameter. It is a pulse travelling with the velocity

$$\frac{\beta}{\alpha} = \frac{\operatorname{Sinh}\alpha}{\alpha} \times (\text{constant})$$

which increases with the amplitude.

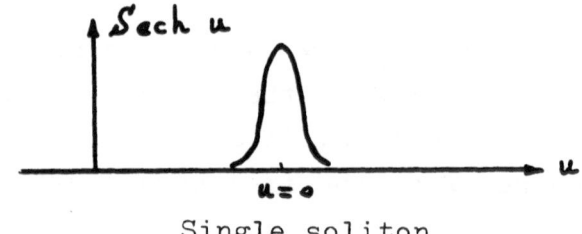

Single soliton

There are also solutions corresponding to two solitons:

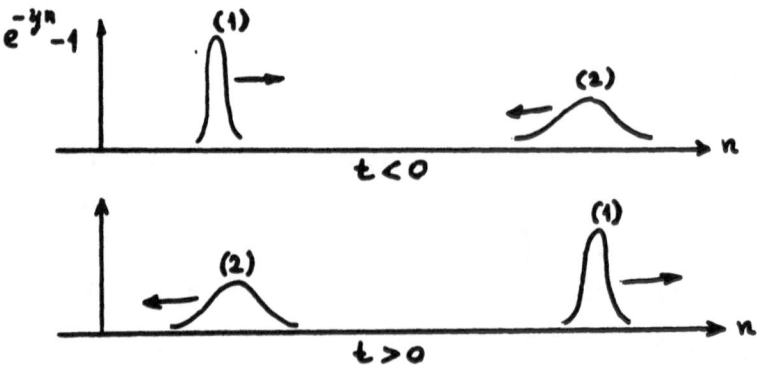

colliding at t=o and moving off to $n = \pm \infty$. They interact but regain their shape after collision. Even triple collisions can be produced, but the adjustment of parameters to find the exact solutions becomes increasingly laborious. Professor Toda's beautiful work has thus provided us ·with <u>exact solutions of a non linear many body problem</u>. There is one drawback in this scheme as compared to the linear one: we cannot stipulate arbitrary initial conditions.

But I think, that Toda's results show that Debye and Peierls' verdict on heat conduction in harmonic crystals may hit the anharmonic ones as well and I suppose that no one will consider the present theory of lattice heat conduction to be more than a partly phenomenological affair.

With regard to ergodicity, the solutions found by Toda show that at least this particular non-linear chain is not ergodic.

<u>Heat source</u>.

There is a degenerate case (unpublished) of Professor Toda's double solitons which could represent a bell shaped temperature distribution at

$\tau = 0$

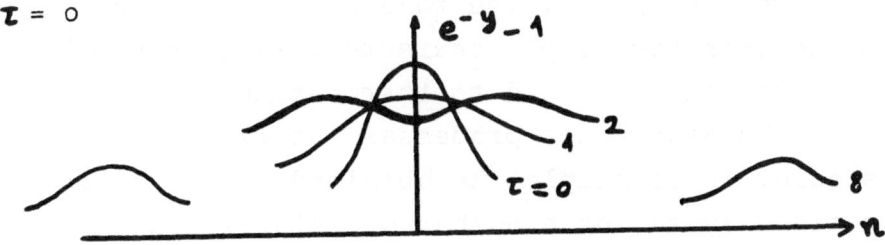

<u>Heat source in Toda model ?</u>

It resembles rather strongly the 'point source' studied by Hemmer [9] on the linear chain:

$$\xi^2_{2n}(t) = J^2_{2n}(\tau)$$

V. TIME REVERSAL SYMMETRY IN BROWNIAN MOTION
Time Reversal and Brownian Motion

The theme of this school has its root in the old thesis

"Time symmetry in mechanics" and the antithesis

"Unidirectedness in phenomena of nature" Boltzmann meant to have accomplished the synthesis, and I think all of us here will endorse his attempt. To me, Boltzmann's explanation appears perhaps the most impressive of all the many great things which happened in Physics before the advant of Quantum theory and Relativity. It is now precisely 100 years ago. Nevertheless, this same issue is taken up over and over again. Perhaps because it is so important, perhaps because the mathematical proofs are still incomplete.

Among the objects of Statistical Mechanics, the Brownian Motion offers perhaps the easiest possibility for elucidation of the paradoxes of of irreversibility. The reason is that this phenomenon has simple characteristics both relative to mechanics and to thermodynamics.

The stochastic processes met with in physics should, in principle, be obtained from dynamic motions by taking a number of integration constants (initial data) to be indeterminate. In particular the Brownian Motion must emerge in this way, if in a many-body system we consider the probable motion of one selected particle, treating all the other degrees of freedom as "hidden variables".

No explanation of this sort, however, was really attempted in the classical papers of Einstein, Smoluchowski, Langevin and the other founders of the theory of B.M.

The standard theory, as embodied for example in the Fokker-Planck equation

$$\frac{\partial P}{\partial t} = \beta \frac{\partial}{\partial v} (vP) + \beta \frac{kT}{m} \frac{\partial^2 P}{\partial v^2} \; ; \; P = P \left({}^v_t \big| {}^{v_o}_o \right)$$

is already remote from dynamics as is seen by its obvious lack of time symmetry.

The missing link of the theory was only picked up in the 1950's by Prigogine and Klein, then by Rubin, Hemmer and many others in their extensive studies of linear chains. Here we have, in fact, a many-body system which is tractable in complete detail and yet rich enough to represent the mechanical substratum of Brownian motion.

The simplest characteristic of stationary stochastic process $v(t)$ is the self correlation.

$$R(t) = E\left\{v(t) \, v(o)\right\} / E\left\{v^2(o)\right\} \qquad (V.1)$$

If v is the velocity of a Brownian particle R is a falling exponential. For a selected particle in a homogeneous linear chain

$$R(t) = J_o \left(2\sqrt{\frac{\alpha}{m}} \, t\right) , \text{ not exponential!} \qquad (V.2)$$

This model process is thus not Markoffian as it should be [20].

With embedding of a heavy Brownian particle $M \gg m$, so that

$$m_n = \begin{cases} m & n \neq o \\ M & n = o , \end{cases}$$

Hemmer [9] proved that

$$R\,(\tau) = e^{-\gamma\tau} + r(\tau) \quad, \quad |r| < 5\gamma \text{ when } \gamma < \tfrac{1}{5} \qquad (V.3)$$

Here

$$\gamma = \frac{m}{M} \text{ , and } \quad \tau = \omega_o t, \; \omega_o = 2\sqrt{\frac{\alpha}{m}} \text{ as before.}$$

Hemmer's theorem is exact but a little too weak to show that the exponential is the leading term for times $\tau \gtrsim 1/\gamma$.

A solution of the dynamical problem is required which in every step permits a full account of causality, reversibility, statistical assumptions and limit operations [21].

Schrödinger coordinates: $\xi_{2n} = \sqrt{m}\,\dot{x}_n$,

$$\xi_{2n+1} = \sqrt{\alpha}\,(x_n - x_{n+1})$$

$$\frac{d\xi_n}{dt} = \frac{1}{2}\,(\xi_{n-1} - \xi_{n+1}) \cdot \begin{cases} 1, n \neq o \\ \gamma, n = o \end{cases} \qquad (V.4)$$

save us a detour into normal coordinates and an extra limit process $N \to \infty$. The generating function:
$Z(\tau, \theta) = \sum_n \xi_n(t)\,e^{in\theta}$ obeys a Volterra type integral equation whose solution can be written

$$Z = Z_o + (\gamma - 1)\,K * \frac{1}{1-(\gamma-1)\bar{K}} * \dot{Z}_o \qquad (V.5)$$

Here

$$K = e^{i\tau\sin\theta}$$

$$K*f = \int_o^\tau K(\tau - \tau')\,d\tau'\,f\,(\tau')$$

$$\bar{f} = \frac{1}{2\pi} \int_{-\pi}^{+\pi} f\,d\theta$$

$$Z_o = Z\,(o,\theta)\,e^{i\tau\sin\theta}$$

From (V.4) every component $\xi_n(\tau)$ can be found as

$$\xi_n(\tau) = \overline{Z\,e^{-i\,n\,\theta}}$$

$$= \overline{Z_0 e^{-i\,n\,\theta}} + (\gamma-1)\,\overline{K\,e^{-i\,n\,\theta}}\,\frac{1}{1-(\gamma-1)\overline{K}}\,\dot{Z}_0$$

Using well-known properties of Bessel functions:

$$\overline{Z_0 e^{-i\,n\,\theta}} = \sum_{k=-\infty}^{\infty} J_{n-k}(\tau)\,\xi_k\,(0)$$

$$\overline{K\,e^{-i\,n\,\theta}} = J_n(\tau) \quad , \quad \dot{\overline{K}} = -J_1(\tau)$$

one can write the complete solution as an affine transformation of the initial values

$$\xi_n(\tau) = \sum_{k=-\infty}^{\infty} a_{nk}(\tau)\,\xi_k(0)$$

where

$$a_{nk} = (J_{n-k}+(\gamma-1)\,J_n * \frac{1}{1+(\gamma-1)J_1} * \dot{J}_{-k})\,(\tau)$$

$$(V.6)$$

Our stochastic process $v_t = \xi_0(\tau)/\sqrt{m}$; its autocorrelation function ($\langle\ \rangle$ = ensemble average)

$$R = \langle \xi_0(\tau)\,\xi_0(0)\rangle / \langle \xi_0(0)^2\rangle = \sum a_{ok}\langle \xi_k^0\,\xi_0^0\rangle\langle\xi_0^2\rangle$$

also satisfies

$$R = \langle \xi_0(\tau)\,;\xi_0(0)=1\rangle = \sum a_{ok}\,\langle \xi_k^0\ ;\ \xi_0(0)=1\rangle$$

the latter due to the <u>linearity</u>. Using the <u>regression function</u> we can generalize the statistical assumptions considerably.

<u>Statistical assumptions</u>:

$$\langle \xi_k(0)\rangle = \delta_{ko} = \langle\xi_k(0)\,\xi_0(0)\rangle$$

At $\tau = 0$ $\xi_0=1$ so $v=v_0$ and all $x_n x_n'$ $n\neq o$ are symmetrical around zero. There is no correlation between the initial values $\xi_{k\neq 0}(o)$ and $\xi_o(o)$. That is all. Therefore the process v_t <u>does not have to be Gaussian</u>. In any case, if there is no correlation between the initial values we obtain

$$R(t) = a_{oo}(\tau) \qquad\qquad (V.7)$$

writing out the symbolic expression (V.6) this means

$$R(\tau) = J_o(\tau) + (1-\rho)J_o * \frac{1}{1-(1-\rho)\,J_1} * J_1$$

$$= J_o(\tau) + (1-\rho)\int_0^\tau J_o(\tau-\tau_1)d\tau_1 J_1(\tau_1) + \ldots + $$

$$m'\text{th term} = (1-\rho)^m \int_0^\tau J_o(\tau-\tau_1)d\tau_1 \int_0^{\tau_1} \ldots \int_0^{\tau_{m-1}} J_1(\tau_{m-1}-\tau_m) \times$$

$$\times J_1(\tau_m)d\tau_m$$

which can be unravelled by means of <u>'Causal'</u> Bessel functions:

$$\frac{1}{2\pi}\int_c \chi(\omega)d\omega\ e^{i\tau\omega} = \begin{cases} J(\tau) & \tau > 0 \\ 0 & \tau < 0 \end{cases}$$

$$\chi_o = (1-\omega^2)^{-\frac{1}{2}}, \quad \chi_1 = 1 - i\omega\chi_o$$

$$c = \left[-\infty, +\infty\right] \text{ through neg. imaginary values.}$$

So far positive time τ has been assumed but $a_{oo}(\tau)$ (and in general $a_{o,2k}$) must be <u>symmetric in time</u>. This follows from the dynamics, as can be seen from (V.6) by reversing time and velocities.

Actually one finds the following expression for R:

$\underline{\tau > 0}$: $R(\tau) = \frac{1}{2\pi}\int_c \alpha(\omega)\, d\omega\, e^{i\omega\tau}$ and

$\underline{\tau < 0}$ "   = " $\int_{\bar{c}}$ "   "   "

where

$$\alpha(\omega) = 1/\gamma\sqrt{1-\omega^2} - i(1-\gamma)\omega \ ,$$

   c = real axis $-\ i\epsilon$,

   $\bar{c}$ =   "   " $+\ i\epsilon$

and both paths c and $\bar{c}$ lie on the sheet (I) of the Riemann surface for α —with branch cut $[-1,+1]$ — where $\sqrt{1-\omega^2}$ is positive on the lower lip of the cut. This is indeed time symmetric. An apparent difficulty is the pair of poles

$$\omega = \pm\, i\varkappa \quad , \quad \varkappa = \gamma/\sqrt{1-2\gamma}$$

of the integrand (V.8). These are, however, both located on the 'unphysical sheet' (II) of the Riemann surface. Apart from the cut, the paths c and $\bar{c}$ can thus be freely deformed: c can be closed upwards and $\bar{c}$ downwards for positive and negative times respectively $\int_c = 0$ for $\tau < 0$ and $\int_{\bar{c}} = 0$ for $\tau > 0$.

   Half the sum of $\int_c$ and $\oint_{\bar{c}}$ gives

$$R(\tau) = \int_0^\infty h(\omega)\cos\omega\tau\, d\omega \qquad\qquad (V.9)$$

$$h(\omega) = \begin{cases} \dfrac{2}{\pi}\dfrac{\gamma\,|\sqrt{1-\omega^2}|}{\gamma^2 + (1-2\gamma)\omega^2} & \omega < 1 \\[4mm] 0 & \omega > 1 \end{cases}$$

(V.9) is the Wiener-Khinchin formula for the
correlation function. In this form it is <u>always</u>
time symmetric, but in a trivial sense since there
is no dynamics in it .

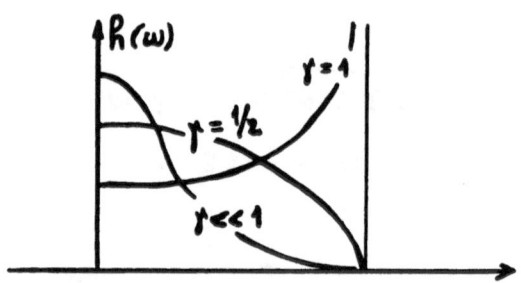

<u>Spectral function for the stochastic motion of</u>
<u>heavy particle in linear chain</u>

For a Markov process one should have

$$R(t) = e^{-\beta t} \qquad\qquad (V.9')$$

as was proved for stationary Gaussian processes
e.g. in 20). Actually it is true for a somewhat
wider class as was stated without proof in 21)
(f.c.p. 142 footnote) and proved in 22). But the
spectral function h(ω) (V.9')shows that our process v_t is <u>not</u> quite Markoffian. Neither could
this be expected, save as a limit property of the
model:

    <u>The limiting procedure</u> 1: Mass ratio m/M= $\rho \rightarrow 0$
2: Time scaling $\tau/t = \omega_0 \rightarrow \infty$ corresponds to the
experimental situation of the Brownian phenomenon:
(1) The observed particle is practically infinitely
heavy in comparison to the impinging molecules of
the surrounding medium.
(2) The frequency of its encounters with the

molecules is practically infinite on the laboratory scale

$$t \sim 1/\beta$$

By taking the limits in such a way that:

$$\Gamma \to 0 \qquad \omega_0 \to \infty \qquad \Gamma \omega_0 = const = \beta$$

we obtain a spectral function appropriate for
a Markov process

$$h \sim 1/\beta^2 + \omega^2$$

This has two imaginary poles $\omega = \pm i\beta$. How is
it possible to extract such a thing from a stable
conservative linear system? The answer: go back
to the expressions V.8 for the correlation function.
Here we have a spectral representation which is
much more closely related to the dynamics. In
contrast to the Wiener-Khinchin representation,
however, this one is univalued only in a Riemann
surface. On this surface it has indeed imaginary
poles $\pm i\chi$ which, in the limit $\Gamma \to 0$, $\omega_0 \to \infty$,
give rise to the exponential damping (V.9).

Nevertheless —as we saw in the mechanical
analysis retaining causality and reversibility in
every step— these poles are not located on the
sheet where the frequency integral has to be taken.
That is the reason why these poles do not affect
the time symmetry.

<u>Linear chains with two or more heavy particles</u>
Some work has also been done with chains containing more than one heavy particle. (O. Amble,
Z. Inowe and H. Wergeland to be published). One
finds as before

$$\dot{Z} = i \sin \theta \, Z + i(\Gamma - 1) \frac{1}{2\pi} \int_{-\pi}^{\pi} Q(\theta - \vartheta) d\vartheta \, Z(\tau, \vartheta)$$

$$(V.10)$$

For N heavy particles equally spaced with a distance 2 pa

$$Q = \frac{\sin \left([2\ N+1]\ p[\theta - \vartheta] \right)}{\sin \left(p[\theta - \vartheta] \right)}$$

Writing again the solution in the form

$$\xi_n(\tau) = \sum a_{nk} (\tau) \, \xi_k(o)$$

the coefficients a_{nk} express the propagation properties of the system. In particular, when n and k are sites of two neighbouring heavy particles, a_{nk} shows how energy is transmitted through the 'medium' of the intervening 2p-2 light particles.

Writing (with n=2rp k=2sp) as before

$$a_{rs} = \frac{1}{2\pi} \int \alpha_{rs} \, e^{i\tau \omega} \, d\omega$$

one finds:

$$\alpha_{rs} = \frac{N(\omega)}{\Gamma \sqrt{1-\omega^2} - i(\Gamma - 1)N}$$

where N is the Toeplitz matrix :

$$N_{rs} = \left(\sqrt{1-\omega^2} - i\omega \right)^{4p|r-s|}$$

A diagonalization is then necessary. In this model a transition to the continuum for the light particles:

$$a \rightarrow o \quad p \rightarrow \infty \quad pa = x \quad \omega_o a = c$$

$$\frac{zpa}{\omega_o a} = \frac{\text{abscissae}}{\text{sound velocity}}$$

gives a field description with Einstein causality:

$$a(x_1 x_2) \neq o \quad \text{only when} \quad |x_2 - x_1| < c$$

REFERENCES

1) H. Arnold, Foundations of Mechanics, p. 108, Benjamin, New York, 1967.

2) J.W. Gibbs, Elementary Principles, Ch. XII, p. 139, Longmans & Green, London 1928.

3) M. Delbrück and G. Moliere: Berl. Ber. Math. Phys. Klasse., n$^{\underline{o}}$ 1 (1936)

4) P. and T. Ehrenfest, Enc. Math. Wiss. IV^2.

5) R.C. Tolman, Statistical Mechanics, p.467, Oxford, 1938.

6) H. Poincaré, Acta Math. 13, 67 (1890).

7) W. Pauli, Vorlesungen über Wellenmechanik, p. 123 ETH Zürich 1962.

8) P. Bocchieri and A. Loinger, Phys. Rev. 107, 337 (1957).

9) P.C. Hemmer, Trondheim thesis 1956.

10) G.D. Birkhoff, Proc. Nat. Acad. 1932.

11) See e.g. R. Abraham, Foundations of Mechanics, Benjamin.

12) H. Grad, Comm. Pure and Appl. Math. 5,455(1952).

13) H. Wergeland, Acta Chem. Scand. 12, 1117,(1958).

14) E. Schrödinger, Ann. Phys. 44, 916 (1914) (infinite chain)

15) I. Prigogine and G. Klein, Physica 19, 1053 (1953) (finite chain)

16) A.N. Kolmogorov, Address to Int. Cong. Math. Amsterdam, 1954 (contained in ref. 11)

17) N. Saito, et al., Prog. Th. Phys. Supp. 45 (1970)

18)   M. Toda. a) J.Phys. Soc. Jap. $\underline{22}$, 431 (1967)

b) Prog. Th. Phys. Supp. $\underline{45}$, 174(1970)

c) One and two solitons in exponential chain: manuscript (1969)

d) Wave propagation in anharmonic lattices: manuscript (1969).

e) Interaction of solitons with e.m. waves, Physica Norvegica $\underline{5}$, 203 (1971)

19) M.C. Wang and G.E. Uhlenbeck, Rev. Mod. Phys. $\underline{17}$, 270 (1945)

20) I. Fujiwara, P.C. Hemmer and H. Wergeland, Tomonaga Homage Volume Prog. Th. Phys. Supp. $\underline{37}$, 141 (1966).

21) M. Kac and H. Wergeland, D. Kgl. N. Vid. Selsk. Forh. $\underline{41}$, 48 (1968).

INFINITE DYNAMICAL SYSTEMS AND
TIME EVOLUTION: RIGOROUS RESULTS

S. Miracle

Facultad de Ciencias
Universidad de Zaragoza
Zaragoza, Spain

1.   Introduction

In recent years much work has been devoted
to the study of the exact properties of infinitely extended systems in statistical mechanics.
Actually, the major part of the work in this field
has been concerned with equilibrium statistical
mechanics, and has played an important role in
the understanding of macroscopic equilibrium
phenomena (phase transitions, etc.). Similarly,
one would expect that the analysis of the time
development of such systems would aid the understanding of non-equilibrium phenomena, the most
famous problem which one should like to understand being the irreversibility of macroscopic
evolution. Very little is known, however, about
such problems in cases of real physical interest.
A major difficulty already is to define the time-
development of a dynamical system with an infinite
number of degrees of freedom, and in fact the
only systems for which one has a satisfactory
definition are non-interacting systems, quantum
lattice systems, and a class of one-dimensional
classical systems. We will try to illustrate,
in these lectures, some of the problems that have
been solved concerning those systems.

2.   Evolution equations in classical statistical mechanics

We consider a system of identical particles
whose mass m we take equal to one and whose phase

coordinates $(q_i, p_i) \in \mathbb{R}^\nu \times \mathbb{R}^\nu$ will be denoted by x_i. We suppose that the hamiltonian of these particles is

$$H(x_1, \ldots, x_n) = \sum_{1 \leq i \leq n} \frac{p_i^2}{2} + \sum_{1 \leq i < j \leq n} \phi(q_i - q_j)$$

where the interaction potential should satisfy some general conditions, that we shall make precise later, in order that thermodynamics exists. We denote by $S^t(x_1, \ldots, x_n) = (x_1(t), \ldots, x_n(t))$ the configuration into which $(x_1, \ldots, x_n)$ evolves in time under Hamilton's equations

$$\dot{p}_i = -\frac{\partial H}{\partial q_i} \quad , \quad \dot{q}_i = \frac{\partial H}{\partial p_i} \quad .$$

The problem of time evolution can be put as follows: given an initial state described by a set of correlation functions

$$\varphi(x_1, \ldots, x_n; 0) \qquad n = 1, 2, \ldots$$

how can one describe the state at time $t \neq 0$? One can expect the answer to be that the state at time t is described by a set of correlation functions $\varphi(x_1, \ldots, x_n; t)$ uniquely determined by the initial ones. We shall now give a formal derivation of the equations which will be satisfied by $\varphi(x_1, \ldots, x_n; t)$.

The correlation functions $\varphi(x_1, \ldots, x_n; 0)$ have the meaning of probability densities for finding n particles in a volume $\dfrac{dx_1 \ldots dx_n}{n!}$ around $x_1, \ldots, x_n$ (where $dx_i = dq_i \, dp_i$)

irrespective of the positions of the other particles.

Let us consider the system in a box Λ and suppose that the initial distribution is D $D_{\Lambda,n}(x_1,\ldots,x_n;0)$. This function gives the probability density of finding inside Λ exactly n particles and of finding them in a volume $\dfrac{dx_1,\ldots,dx_n}{n!}$ around $x_1,\ldots,x_n$. At time t the distributions will be $D_{\Lambda,n}(x_1,\ldots,x_n;t) =$ $= D_{\Lambda,n}(S^{-t}(x_1,\ldots,x_n);0)$ where S^t is the evolution operator introduced above. It follows immediately that

$$\frac{\partial D_{\Lambda,n}(x_1,\ldots,x_n;t)}{\partial t} = \left[D_{\Lambda,n}(x_1,\ldots,x_n;t), \ H\right]$$

where $[\ , \ ]$ denotes the Poisson bracket.

The correlation functions associated with a distribution $D_{\Lambda,n}(x_1,\ldots,x_n;t)$ are, according to their probabilistic meaning,

$$\varphi(x_1,\ldots,x_n;t) = \sum_{m=0}^{\infty} \int_{(\Lambda \times \mathbb{R}^\nu)^m} D_{\Lambda,n+m}(x_1,\ldots,x_n, \ y_1,\ldots,y_m) \ \frac{dy_1\ldots dy_m}{m!}$$

and by differentiating both sides with respect to time and using the preceding formulae one finds

$$\frac{\partial \varrho(x_1,\ldots,x_n;t)}{\partial t} = \left[\varrho(x_1,\ldots,x_n;t), H(x_1,\ldots,x_n)\right] +$$

$$+ \sum_{1\leq i\leq n} \int \frac{\partial \varrho(x_1,\ldots,x_n,x_0;t)}{\partial p_i} \; \frac{\partial \Phi(q_i-q_0)}{\partial q_i} \; dx_0.$$

Here we have neglected the presence of the walls and assumed that $D_{\Lambda,n}(x_1,\ldots,x_n;t)$ takes equal values at opposite sites on the boundary of Λ, and vanishes for large momenta.

Because of these assumptions these equations can only hold for infinite boxes Λ, but, of course, the above arguments do not really prove them. They are called the BBKGY (Born, Bogoliubov, Kirkwood, Green and Yvon) hierarchy of integro-differential equations; they relate the n-particle correlation functions to the (n+1)-particle ones. These equations are the starting point of the Bogoliubov kinetic theory.

Such systems, composed of an infinite sequence of equations with an infinite sequence of unknown functions, are typical in the study of dynamical systems with an infinite number of degrees of freedom, and as often occurs in these problems, there are no existence theorems for the BBGKY system for arbitrarily given initial conditions $\varrho(x_1,\ldots,x_n;0)$. Functional analysis provides an appropriate mathematical frame in which to formulate the problems related to this class of dynamical systems, and we shall begin to use this approach in the following section.

3.   Mathematical description of the states of a classical system

   We have first to explain what is meant by a
state in classical statistical mechanics. Typically, in statistical mechanics one wishes to study
"large" systems where the density per unit volume
in the space is finite. If we make the theoretical idealization that "large" means infinitely
extended in the space, then the number of particles
involved is infinite. We have then to define the
space of infinite configurations of particles;
let us call it $[\mathbf{X}]$ and suppose that $[\mathbf{X}]$ can be
given the structure of a topological measurable
space. Then it is natural to define a state of
the infinite system as a probability measure on
$[\mathbf{X}]$, i.e. a positive measure with total mass
equal to 1.

   A locally finite configuration of particles
is defined by giving a sequence (possibly finite)
$x = (x_1, x_2, \dots)$ of positions and momenta
(q_i, p_i) such that only finitely many q_i are
contained in bounded sets of the space or,
equivalently,

$$\lim_{i \to \infty} |q_i| = \infty$$

We will let $\mathbf{X}$ denote the set of all such configurations which are labelled configurations of particles and, since the particles are supposed to

be identical, we identify configurations which
differ only by the labelling.

   We shall denote by $[\mathbf{X}]$ the set of <u>locally</u>
<u>finite unlabelled configurations</u>, or the set
of equivalence classes of $\mathbf{X}$, two configurations
being equivalent if they differ only by a permutation of the index set.

   Now consider the functions g on $[\mathbf{X}]$ such
that $g([x]) = g([x] \cap \Lambda)$ for all $[x] \in [\mathbf{X}]$, the
configuration $[x] \cap \Lambda$ being obtained by deleting
from $[x]$ all the particles x_i such that
$q_i \notin \Lambda$. There is a simple way to construct such
functions. Let f be a function $\mathbb{R}^\nu \times \mathbb{R}^\nu$ such
that $f(x_1) = f(q_1, p_1) = 0$ for $q_1 \notin \Lambda$. Then
define

$$(Sf)([x]) \;=\; \sum_{i \geqslant 1} f(x_i) \;,$$

where $x = (x_1, x_2, \ldots)$ is a representative of
$[x]$. If Λ is bounded there are only finitely
many non-zero terms in this sum and the function
Sf has the above property of the function g.
Let us denote by $\mathcal{K}_1$ the set of all continuous
functions $f(x_1)$ in $\mathbb{R}^\nu \times \mathbb{R}^\nu$ whose support in
q_1 is bounded.

   <u>We give $[\mathbf{X}]$ the weakest topology such that</u>
<u>Sf is continuous for all $f \in \mathcal{K}_1$.</u>

   <u>The states of classical statistical mechanics</u>
<u>should be identified with Borel probability</u>
<u>measures on $[\mathbf{X}]$.</u> In fact, if Λ is a bounded open
set of $\mathbb{R}^\nu$, the mapping $[x] \to [x] \cap \Lambda$ is a
Borel mapping from $[\mathbf{X}]$ into $[\mathbf{X}] \cap \Lambda$. A Borel

measure ρ on $[\mathcal{X}]$ therefore defines a measure μ_Λ on $[\mathcal{X}] \cap \Lambda$, i.e. a sequence $\mu_{\Lambda,n}$ of symmetric Borel measures on $(\Lambda \times \mathbb{R}^\nu)^n$, $n = 1,2\ldots$. A state of classical statistical mechanics can then be equivalently described by a <u>family</u> $\{\mu_{\Lambda,n}\}$ <u>of probability distributions</u> of particles in each bounded open region Λ of the space. If each $\mu_{\Lambda,n}$ is absolutely continuous with respect to Lebesgue measure, we define the <u>density distributions</u> $D_{\Lambda,n}(x_1,\ldots,x_n)$ by

$$\mu_{\Lambda,n}(dx_1,\ldots,dx_n) = D_{\Lambda,n}(x_1,\ldots,x_n) \frac{dx_1\ldots dx_n}{n!}$$

The following remark will be useful in showing the analogy between the mathematical description of states given above for classical systems and the one that will be given for quantum systems.

The set of Borel probability measures on a topological space $[\mathcal{X}]$ coincides with the set of normalized positive functionals on the C^*-algebra $\mathcal{O}\mathcal{l}$ of continuous functions on $\mathcal{X}$. Then the states of classical statistical mechanics coincide with <u>mathematical states over the C^*-algebra $\mathcal{O}\mathcal{l}$.</u>

If $f_1,\ldots,f_n$ are functions in $\mathcal{H}_1$ we define a continuous function $S(f_1\ldots f_n)$ on $[\mathcal{X}]$ by

$$S(f_1\ldots f_n)([x]) = \sum_{i_1 < i_2 < \ldots < i_n} f_1(x_1)\ldots f_n(x_n)$$

where $x = (x_1,x_2,\ldots)$ is a representative of $[x]$. Since the functions $S(f_1\ldots f_n)$ generate, by taking linear combinations and limits, the whole algebra $\mathcal{O}$ one can also characterize the state by the following sequence of expectation values

$$\varrho(f_1,\ldots,f_n) = \int \varrho(d \, x \,) \, S(f_1\ldots f_n)([x]) \; .$$

These are related to the <u>correlation functions</u> (when they exist) by the following expression

$$\varrho(f_1,\ldots,f_n) = \int \varrho(x_1,\ldots,x_n) f_1(x_1)\ldots \, f_n(x_n) \, \times$$
$$\times \, dx_1\ldots dx_n \; .$$

Finally, let us remark how the <u>evolution in time of states of the infinite system</u> can be defined. Suppose that we have an existence and uniqueness theorem for the solutions of the equations of motion with initial data in a subset $\mathcal{Y}$ of $\mathcal{X}$, and suppose that we have a measure ϱ on $[\mathcal{X}]$ which is concentrated on $[\mathcal{Y}]$, i.e. such that

$$\varrho([\mathcal{X}]\setminus[\mathcal{Y}]\,) \; = \; 0$$

(where $[\mathcal{X}]\setminus[\mathcal{Y}]$ means the complementary set of $[\mathcal{Y}]$ on $[\mathcal{X}]$, then we can define a time evolved measure ϱ_t by

$$\varrho_t([\mathcal{X}]\setminus[\mathcal{Y}]) \; = \; 0$$
$$\varrho_t(A) \; = \; \varrho(S^t A) \qquad \text{if } A \subset [\mathcal{X}]$$

where S^t is the evolution operator assumed to exist in $\mathcal{Y}$.

Having made this remark by way of motivation, we will now devote our attention to the problem of solving the differential equations of motion.

4. <u>Preliminary remarks on the motion of a</u>
    <u>mechanical system of infinitely many particles</u>

We consider now locally finite configurations
$x \in \mathcal{X}$ and we want to solve the differential equations

$$\frac{dq_i(t)}{dt} = p_i(t) \quad , \qquad \frac{dp_i(t)}{dt} = \sum_{j \neq i} F(q_i(t) - q_j(t))$$

with the initial conditions

$$q_i(0) = q_i, \qquad p_i(0) = p_i \quad .$$

The interparticle force F will be assumed to be
bounded and to have compact support, i.e.

$$|F(r)| \leq F_{max}$$
$$F(r) = 0 \text{ if } |r| > R \quad .$$

As long as each bounded region Λ contains only
a finite number of $q_j(t)$'s, the sum on the
right of the second equation has only a finite
number of non-zero terms for each i and the
equations therefore make sense. It is clear,
however, that for some initial conditions we must
expect these equations to lead, in a finite time,
to a catastrophic situation with infinitely many
particles in some bounded interval. To take a
trivial example, if $F = 0$ and if $p_i = -q_i$ for
each i, then all the particles are at the
origin at time $t = 1$. For interacting systems,

however, another kind of catastrophe can happen;
even if the initial velocities are very reasonable, the interparticle forces can build up very
large velocities in a short period of time. The
main point in proving the existence theorem, is
to find a set of initial configurations for which
such catastrophes can be shown not to happen.

We can hope to play two sets of estimates
off against each other. Given a bound on the
local densities, we can bound the forces on the
particles and hence the accelerations; then, with
some assumed bounds on the initial velocities, we
can get bounds on the velocity at later times.
Using these velocity bounds we can estimate how
far particles can travel in a given interval of
time, and get a bound on the new local density.
This might allow us to eliminate the possible
catastrophes.

A first result in this direction was obtained
by J. Ginibre (unpublished), who proved that the
equations of motion have a solution if the initial
configuration admits an upper bound on the absolute
values of the velocities and on the local densities.
This solution is valid for $t < t_o$ in the ν-dimen-
sional case, and is valid for all values of t if
$\nu = 1$. In order to show how such estimates work,
we shall consider briefly this problem.

Given a configuration $x \in X$ we denote by
$N_\Lambda(x)$ the number of particles in $x \cap \Lambda$. Let
us suppose that the configuration
$x(t) = (x_1(y), x_2(t),...)$ at time t satisfies

$$\sup_{i} |p_i(t)| \quad < \quad P_{max}(t)$$

$$N_{S_R}(x(t)) \quad < \quad N_{max}(t)$$

for every sphere S_R of radius R. The equations of motion lead to

$$p_i(t) = p_i(0) + \int_0^t d\tau \sum_{j \neq i} F(q_i(t) - q_j(t))$$

and thence to the estimate (remember that R is the range of the interparticle force)

$$P_{max}(t) \leq P_{max}(0) + \int_0^t d\tau \, N_{max}(\tau).$$

On the other hand, if a particle is in a sphere of radius R centered at q_0 at time t, it must have been at a distance not greater than $R + \int_0^t d\tau \, P_{max}(\tau)$ from q_0 at time $t = 0$. Hence

$$N_{max}(t) \leq const. \left[R + \int_0^t d\tau \, P_{max}(\tau) \right]^{\nu} \times$$
$$\times N_{max}(0)$$

where the constant is chosen so that for any $a > 0$ a sphere of radius $R+a$ can be covered by $const \times [R+a]^{\nu}$ spheres of radius R. Combining the two estimates we obtain

$$P_{max}(t) \leq P_{max}(0) \int_0^t d\tau \, F_{max} \times$$
$$\times const. \left[R + \int_0^{\tau} d\theta \, P_{max}(\theta) \right]^{\nu} N_{max}(0).$$

Let $\mu(t)$ be a solution of the integral equation

$$\mu(t) = P_{max}(0) + C \int_0^t d\tau \left(R + \int_0^{\tau} d\theta \, \mu(\theta) \right)$$

where $C = const. \, F_{max} \cdot N_{max}(0)$. Then we have

$$P_{max}(t) \leq \mu(t) \ .$$

In order to solve the integral equation, we introduce

$$f(t) = \int_0^t d\tau \ \mu(\tau) \ .$$

Then we have $f' = \mu$ and $f'' = c(R+f)^{\nu}$. Integrating this last equation we get

$$f' = \sqrt{f'_0 + \frac{c}{\nu+1} (R+f)^{\nu+1}}$$

and hence the implicit solution

$$t = \int_0^f \frac{ds}{\sqrt{f'_0 + \frac{c}{\nu+1} (R+f)^{\nu+1}}}$$

showing that the bound becomes infinite, i.e.
$f = \infty$, when $t = \infty$ if $\nu = 1$ and for finite t
if $\nu \geq 2$. We have shown, therefore, that the
above inequalities imply bounds on $P_{max}(t)$ and
$N_{max}(t)$ in terms of $N_{max}(0)$ and $P_{max}(0)$; these
bounds remain finite at least for a finite interval of time, and for all times when $\nu = 1$.

However the restriction to initial configurations with bounded velocities is unfortunately
much too limiting for applications to statistical
mechanics in which states usually have a Maxwellian
velocity distribution, i.e. the velocity of a given
particle is independent of its position and of
the positions and velocities of the other particles,
and is distributed with the probability density
$\sqrt{\frac{\beta}{2\pi}} \exp(-\beta \frac{p^2}{2})$. It is easy to see that no such
state can be concentrated on the set of

configurations with bounded velocities as we are
always sure, by going far enough from the origin,
to find some particle with an arbitrarily high
velocity.

We must therefore consider more general configurations. What one can do is to allow a reasonable rate of growth for the velocities and for
the local densities of particles as a function of
their distance to the origin. We will choose some
slowly increasing function h(r) and consider
configurations in which the velocity increases at
most like a constant time h(r), r being the distance from the origin, i.e. configurations with

$$\sup_i \frac{|p_i|}{h(|q_i|)} < +\infty \ .$$

For the local density, the obvious thing to try,
i.e. allowing it to grow as h(r), does not work.
There is, however, a variant which is more satisfactory. This is to put

$$\frac{N_{S(q,h(|q|)^{1/\nu})}(x)}{h(|q|)} < +\infty$$

where $S(q,h(|q|)^{1/\nu})$ is the sphere centered at
q and with radius equal to $h(|q|)^{1/\nu}$. Note that,
whenever |q| is large enough so that $h(|q|) > R$
this implies that the number of particles in the
sphere S(q,R) is less than a constant times
h(|q|). Hence, if the local density condition
is preserved for some interval of time, the force
on the particle located at q increases with q
at most like h(|q|), so the velocity increases

again at most like h($|q|$). We are arguing very
roughly here and, in particular, ignoring the fact
that the location of the particle changes with
time and there is not a well-defined value of
h($|q|$). This should not be important if h is a
slowly increasing function. How then are the density estimates preserved? If a particle has to
be in a sphere S(q, h($|q|$)$^{1/\nu}$) at time t, it must
have been in a sphere S(q,h($|q|$)$^{1/\nu}$+const. h($|q|$))
at time zero by the velocity estimates. But the
number of particles satisfying this condition is
for large q, of the order of h($|q|$)$^{\nu}$ rather
than h($|q|$), so we cannot hope to have density
estimates preserved unless ν = 1. It may be hoped
that some simple change in the sort of restriction
imposed, would allow us to handle the case of
dimension ν>1; no such modification has yet
been found. We will therefore restrict our considerations to the case ν = 1, i.e. to one dimensional systems.

There remains to choose the function h(r).
Let us first remark that a typical configuration
of non-interacting particles has velocity fluctuations which increase like $\sqrt{\log(d)}$, and density
fluctuations which increase like log(d)/log(log(d))
To make this statement precise, we define the
following functions on $[X]$

$$\chi_1(x) \;=\; \sup_i \; \frac{|q_i|}{\sqrt{\log(|q_i|)}}$$

$$\chi_2(x) = \lim_{n \to \infty} \sup \frac{N_{(n,n+1)}(x)}{\log n / \log(\log n)} \quad .$$

An elementary calculation shows that, for the
state obtained by taking the infinite volume limit
of the grand-canonical ensemble for non-interacting
particles, with any temperature and any chemical
potential, (state for which the correlation functions are

$$\varrho(x_1, \ldots, x_n) = \text{const. } e^{\beta \mu n} \exp(-\beta \sum_{1 \le i \le n} \frac{p_i^2}{2})$$

we obtain that the function $\chi_1(x)$ is bounded,
and the function $\chi_2(x) = 1$, in subsets of $[\mathcal{X}]$
whose complements have zero measure with respect
to ϱ. Hence, the logarithmic rate of increase
is almost the slowest one that we can admit for
$h(r)$, while, in order to prove the existence
theorem for the equations of motion we cannot allow
$h(r)$ to have a high rate of increasing. Happily,
as we shall see, one can fulfil these two opposite
requirements by taking $h(r)$ increasing as $\log(r)$.

5.   Existence of classical motion in one-dimensional systems

Let us now give a precise definition of the
set of initial configurations for which one can
solve the equations of motion. We are considering
locally finite configurations of particles
$x = (x_1, x_2, \ldots)$ in the one-dimensional space,
i.e. $x_i = (q_i, p_i) \in \mathbb{R} \times \mathbb{R}$. We introduce the
definition

$$\log_+(r) = \log(r) \vee 1 ,$$

where $\vee$ means "supremum", to cut off the logarithmic function for small values of its argument; and, as we have discussed, the initial configurations which we want to consider will be those satisfying

1. $\quad K_1(x) = \sup_i \dfrac{|p_i|}{\log_+(|q_i|)} < +\infty$

2. $\quad K_2(x) = \sup \left\{ \dfrac{N_{[a,b)}(x)}{b-a} : \right.$

$\qquad\qquad \left. : b - a > \log \dfrac{b+a}{2} \right\} < +\infty$

Condition 1) is a condition on the rate of increase of the velocities of the particles, and condition 2) may be reformulated by saying that there is an upper bound for the mean density of particles in any interval of length greater than one and also greater than the logarithm of the distance from its centre to the origin. Condition 2) implies, in particular, that the number of particles in any interval of unit length is bounded by a constant times the logarithm of the distance from its centre to the origin.

We will denote by $\hat{\mathcal{X}}$ the set of labelled configurations satisfying conditions 1) and 2) and by $[\hat{\mathcal{X}}]$ the corresponding set of unlabelled configurations. The set $\hat{\mathcal{X}}$, although not defined in a manifestly translation- invariant way, is easily seen to be invariant under translations.

For any $x \in \hat{\mathcal{X}}$ we will put

$$\| x \| = K_1(x) \vee K_2(x) .$$

We can now formulate the existence theorem for
the equations of motion.

Lanford's theorem. Let F be a real-valued function vanishing outside the interval $(-R, R)$ and
satisfying a Lipshitz condition

$$|F(q_1) - F(q_2)| \leq K|q_1 - q_2|$$

and let $x = (x_1, x_2, \ldots)$ belong to $\hat{\mathfrak{X}}$. There
exists one and only one function $x(t) = (x_1(t),$
$x_2(t), \ldots)$, defined for $-\infty < t < +\infty$, with values
on $\mathfrak{X}$, satisfying

1. the equations of motion

    $$\frac{dq_i(t)}{dt} = p_i(t), \qquad \frac{dp_i(t)}{dt} = \sum_{j \neq i} F(q_i - q_j)$$

2. the initial conditions

    $$q_i(0) = q_i, \qquad p_i(0) = p_i$$

3. $\|x(t)\|$ is a locally bounded function of
    t, i.e. bounded on any bounded interval

   Note that the uniqueness is not as strong as
one might hope, since it does not rule out the
possibility of solutions not satisfying the
regularity condition 3).

   We shall not give here the detailed proof of
this theorem but only try to indicate the essential points of it. The underlying idea has been
described in the preceding section, but instead
of trying to make rigorous the arguments presented there, it has been found helpful in order to
get a bound for $\|x(t)\|$ in terms of $\|x(0)\|$, to

convert the problem into one which is formally
more familiar, that of an ordinary differential
equation in a Banach space. The estimates we
have described then reappear in the form of
estimates on the "infinitesimal generator".

One notices first the following fact.

Let $x(t) \in \hat{\mathcal{X}}$ defined for $|t| < T$ be such
that $\| x(t) \|$ is bounded and $\dfrac{dq_i(t)}{dt} = p_i(t)$. Then

$$\sup_i \left\{ \frac{|q_i(t) - q_i(0)| \vee |p_i(t) \vee p_i(0)|}{\log_+(|q_i(0)|)} \right\} < +\infty$$

Given a configuration $x \in \mathcal{X}$ we will then introduce a space of "neighbouring configurations"
in which the evolution takes place. We will call
this $\mathcal{Y}_x$, and according to the preceding remark
it is defined as the Banach space of $\zeta = (\zeta_1, \zeta_2 \dots)$
where $\zeta_i = (\xi_i, \eta_i) \in \mathbb{R} \times \mathbb{R}$, such that

$$\| \zeta \|_x = \sup_i \left\{ \frac{|\xi_i| \vee |\eta_i|}{\log_+(|q_i|)} \right\} < +\infty$$

the norm on $\mathcal{Y}_x$ being $\| \zeta \|_x$.

Given $\zeta \in \mathcal{Y}_x$, the configuration $x + \zeta$ is
defined as the set of pairs $(q_i + \xi_i, p_i + \eta_i)$. It
is easy to see that if $x \in \hat{\mathcal{X}}$ and $\zeta \in \mathcal{Y}_x$ then
$x + \zeta$ belongs to $\hat{\mathcal{X}}$, and we can rewrite the equations of motion in the new variable ζ in the form

$$\frac{d\xi_i(t)}{dt} = p_i + \eta_i(t),$$

$$\frac{d\eta_i(t)}{dt} = \sum_{j \neq i} F(q_j + \xi_i(t) - q_j - \xi_j(t))$$

or, schematically,

$$\frac{d\zeta(t)}{dt} = A_x(\zeta(t))$$

where $A_x(\zeta(t))_i = (p_i + \eta_i, \sum_{j \neq i} F(q_i + \xi_i - q_j - \xi_j)$.

Let F be a bounded function vanishing outside the interval $(-R, R)$ and let $\zeta \in \mathcal{Y}_x$. One can then verify that $A_x(\zeta) \in \mathcal{Y}_x$, (at this point the assumption $\nu = 1$ is essential), and moreover, there exist constants C, D (depending on $\|x\|$) such that

$$\|A_x(\zeta)\|_x \leq C + D\|\zeta\|_x \log_+ (\|\zeta\|_x)$$

for all $\zeta \in \mathcal{Y}_x$. This defines $A_x(\zeta)$ as a non-linear operator on the Banach space $\mathcal{Y}_x$.

The factor $\log_+(\|\zeta\|_x)$ comes directly from the choice of the logarithm as the function governing velocity and local density fluctuations. If we had taken a more rapidly increasing function we would have an estimate for $\|A_x(\zeta)\|_x$ which increases more rapidly with $\|\zeta\|_x$. We can now see the significance of the choice of the logarithm.

If $\frac{d\zeta(t)}{dt} = A_x(\zeta(t))$ and $\zeta(0) = 0$, it is at least plausible that the inequality

$$\|\zeta(t)\|_x \leq \int_0^t d\tau \, \|A_x(\zeta(\tau))\|_x$$

holds for $t \geq 0$. Using the estimate

$$\|A_x(\zeta)\|_x \leq C + D \|\zeta\|_x \log(\|\zeta\|_x) ,$$

we get

$$\|\zeta(t)\|_x \leq \int_0^t d\tau \left[C + D\|\zeta(\tau)\|_x \log_+ (\|\zeta(\tau)\|_x) \right].$$

Hence if $h(t)$ is the solution of the integral
equation
$$h(t) = \int_0^t d\tau \left[C + Dh(\tau) \log_+(h(\tau)) \right]$$

it is again plausible that

$$\| \zeta(t) \|_x \leq h(t)$$

for all $t \geq 0$ for which $h(t)$ is defined. Thus,
in order to show that $\| \zeta(t) \|_x$ cannot go to in-
finity in finite time, it suffices to prove that
$h(t)$ is defined for all t, i.e. does not go to
infinity in finite time. But

$$h'(t) = C + D \; h(t) \log_+(h(t))$$

and h is given implicitly by

$$t = \int_0^h \frac{ds}{C + Ds \log_+(s)} \; .$$

Since $1/s \log s$ is just barely non-integrable
at infinity, $h(t)$ does not go to infinity unless t does also. If we had chosen a function
increasing slightly faster than the logarithm,
the above estimates would not prevent the solution
for running off to infinity in finite time.

The main problem now is to get a local existence theorem for the differential equation

$$\frac{d\zeta}{dt} = A_x(\zeta) \; .$$

If the operator A_x satisfied a local Lipshitz
condition, this would simply be a matter of
applying standard theorems about differential
equations on Banach spaces. Although A_x does
not, in general, satisfy a Lipshitz condition,

we can use the essential features of the standard existence proof. That is, we define a
sequence of approximate solutions

$$\zeta_0(t) = 0, \quad \zeta_{n\,1}(t) = \int_0^t d\tau \, A_x(\zeta_n(t)), \quad n = 0,1,..$$

the integrals being computed component by component. In the usual proof, one uses the Lipshitz
condition to show that, for sufficiently small
$|t|$, the sequence $\zeta_n(t)$ converges uniformly in
norm to the solution of the differential equation.
In the present case, we do not have a Lipshitz
condition, but it turns out that A_x satisfies
a weaker sort of inequality, namely

$$\| A_x(\zeta) - A_x(\zeta') \|_{x,m} \leqslant K \log_+(m) \| \zeta - \zeta' \|_{x,\alpha m} \quad \text{for } \alpha > 1$$

where $\| \zeta \|_{x,m}$ is defined as

$$\| \zeta \|_{x,m} = \sup_i \left\{ \frac{|\xi_i| \vee |\eta_i|}{\log_+(|q_i|)} \; : \; q_i \in [-m,m] \right\}.$$

From this inequality one can guarantee,
again by using the rigorous estimate $\| \zeta_n(t) \| \leqslant h(t)$,
that each component ($\xi_{i,n}(t)$, $\eta_{i,n}(t)$) converges
uniformly as $n \rightarrow \infty$, and that the limiting functions ($\xi_i(t)$, $\eta_i(t)$) give, in fact, a solution
of the equations of motion. A similar modification of the standard arguments prove the uniqueness assertions of the theorem.

The technique of proof yields another property of the solution which is of interest from the
physical point of view. Suppose we choose some

finite set of particles, say those with labels
1,2,...,n. For each M we make a finite configuration x_M by deleting from x all particles
with $|q_i| > M$, i.e. $x_M = x \cap (-M,M)$. For the re-
maining configuration we can solve the equations
of motion, and in particular find the trajectories $q_{i,M}(t)$, $1 \leq i \leq n$ of the first n particles. Now as $M \rightarrow \infty$, we can show that the
$q_{i,M}(t)$ converge to the trajectories of the first
n particles in the solution for the initial configuration with infinitely many particles. In
other words, this result can be interpreted as a
kind of stability result asserting that for very
large systems the behaviour of the particles near
the origin becomes insensitive to the presence or
absence of particles far away from the origin,
provided that these particles do not have too
large velocities or local densities. This stability property plays a fundamental role in the
proof of the results of the next section, since
they give us a way to relate the statements about
the infinite systems, in which we are interested,
to the corresponding statements for finite systems.

6.   <u>Kinetic theory of one-dimensional systems</u>

The theorem stated in the preceding section
enables us to define a one-parameter group S^t of
evolution operators mapping $[\hat{x}]$ on itself. The
set $[\hat{x}]$ is the set of configurations for which one
can solve the equations of motion, and therefore

the mapping S^t is obtained by the following
natural definition: given an unlabelled configuration x in $[\hat{x}]$ we choose a labelled configuration x representing it and we put $S^t[x] = [x(t)]$,
where $x(t)$ is the solution of the equations of
motion with the initial condition $x(0) = x$. If
the mappings are measurable, they define a time
evolution of measures on $[\hat{x}]$ and, in particular,
of states of classical statistical mechanics which
are concentrated on $[\hat{x}]$ (i.e. for which $[x] \backslash [\hat{x}]$
has measure zero). This time evolution will now
be described.

   We first state the initial theorem on which
this study is based.

   The set $[\hat{x}]$ is a Borel or measurable subset
of $[x]$ and the time-evolution $(t, [x]) \rightarrow S^t [x]$
is a Borel mapping from $\mathbb{R} \times [\hat{x}]$ to $[\hat{x}]$.

   We are next interested in giving sufficient
conditions for a state ϱ of classical mechanics
to be concentrated on $[\hat{x}]$. What we have discussed, at the end of section 4, regarding fluctuations in the grand-canonical ensemble of a
system of free particles, makes the following
statement very reasonable (β and μ were arbitrary.)

   For a state ϱ in classical statistical
mechanics to be concentrated in $[\hat{x}]$ it is sufficient that

1.   it has a Maxwellian velocity distribution,
i.e. has correlation functions of all orders of
the form $\varrho(x_1,\ldots,x_n) = \varrho(q_1,\ldots,q_n)\exp\left\{-\beta\sum_{1\leq i\leq n}\frac{p_i^2}{2}\right\}$

and

2.   the correlation functions $\varphi(q_1,\ldots,q_n)$
admit a majorisation of the form

$$\varphi(q_1,\ldots,q_n) \leqslant \lambda^n$$

with λ independent of n, $q_1,\ldots,q_n$.

These two conditions are satisfied for many
interesting states and in particular for the
equilibrium states for a large class of potentials
(the so-called superstable potentials) and for any
temperature and chemical potential. This result,
proved by Ruelle, holds for any dimension ν of
the space. We remember that a potential Φ is
said to be stable if there exists a constant B,
such that for all n and all $q_1,\ldots,q_n$

$$\sum_{1\leqslant i<j\leqslant n} \Phi(q_i - q_j) > - \; nB \; .$$

One needs this condition of stability, together
with a condition of Φ decreasing at infinity (in
particular this should be verified if Φ is a
finite range potential), in order that thermodynamic functions at equilibrium exist. A potential Φ is said to be superstable if it is of
the form

$$\Phi = \Phi_1 + \Phi_2$$

where Φ_1 is a stable potential and Φ_2 is a
non-negative continuous function of bounded support, strictly positive at the origin.

Since the finite volume correlation functions

$$\varrho_\Lambda(x_1,\ldots,x_n) = \frac{\sum_{m\geq 0} e^{\beta\mu(n+m)}\int_{(\Lambda\times\mathbb{R})^m} e^{-\beta H(x_1,\ldots,x_n,y_1,\ldots,y_m)}\frac{dy_1\ldots dy_m}{m!}}{\sum_{m\geq 0} e^{\beta\mu m}\int_{(\Lambda\times\mathbb{R})^m} e^{-\beta H(y_1,\ldots,y_m)}\frac{dy_1\ldots dy_m}{m!}}$$

also satisfy in this case conditions 1) and 2),
one can construct equilibrium measures on $[\mathcal{X}]$
corresponding to the infinite volume correlation
functions $\varrho(x_1,\ldots,x_n)$ obtained by taking limits
of the $\varrho_\Lambda(x_1,\ldots,x_n)$ along subsequences of boxes
tending to infinity; the equilibrium measures
need not be unique (they can depend on the particular subsequence chosen, unless the activity
$e^{\beta\mu}$ is small in which case this uniqueness has
been proved).

Let ϱ be a state of classical statistical
mechanics which is concentrated on $[\hat{\mathcal{X}}]$. We denote by $\varrho_t = \varrho \circ S^t$ the state obtained from it
at time t by the time evolution. One naturally
expects that the equilibrium measures are stationary, i.e. $\varrho_t = \varrho$. This statement has been
proved for the equilibrium states we have just
described.

States obtained by taking infinite volume
limits of grand canonical ensembles with a given
superstable potential with compact support are
invariant under the time evolution defined by
that potential.

Finally we have also the property that

the entropy per unit volume is conserved by the
time evolution. The entropy per unit volume of
a translationally invariant state φ is defined
by the limit,

$S(\varphi)$ =

$$\lim_{\substack{a \to -\infty \\ b \to \infty}} \frac{1}{b-a} \sum_{n \geqslant 0} \int D_{(a,b),n}(x_1,\ldots,x_n) \log D_{(a,b),n}(x_1,\ldots x_n)$$
$$\times \frac{dx_1 \ldots dx_n}{n!}$$

which can be shown to exist when the distribution
functions $D_{(a,b),n}(x_1,\ldots,x_n)$ exist (i.e. if the
measures restricted to $[\mathsf{x}] \cap \Lambda$ are absolutely continuous with respect to Lebesgue measure). Other-
wise we put $S(\varphi)$ = $-\infty$ Here the condition of
superstability of the potential again plays an
essential role. We assume it in defining the
time evolution and consider translationally in-
variant state φ, which is concentrated on $[\hat{\mathsf{x}}]$
and which has finite mean kinetic energy and
finite mean square number of particles per unit
volume (this means that the corresponding expectation values are finite). Then we have

$$S(\varphi_t) \quad = \quad S(\varphi) \quad .$$

Let us now return to the questions raised in
section 1. A first question we can ask is whether
a state which at time t = 0 is described by a
set of correlation functions, can still be described by a set of correlation functions when
t $\neq$ 0. One also asks whether these correlation

functions satisfy the BBGKY hierarchy of equations.
One can hope to answer this question affirmative-
ly, but we have first to discuss the initial states
that we want to consider.

Let us suppose that the system is in equilibrium at temperature β^{-1} and chemical potential
μ and in an external potential h which is
localized in a finite region Λ_h. At time t=0
we switch off this external potential and the
system begins to evolve. We note that initial
states of the kind just described suffice, in
principle, for the study of transport properties,
since the physical situation we are considering
is the following: We have a system, the locally
perturbed part of our infinite system, which is
enclosed in a (finite) container in thermal equilibrium with a big reservoir; the non-perturbed
part of our infinite system plays the role of the
container and the big reservoir which fixes the
external conditions (temperature and chemical
potential) for the equilibrium. The system is
put in a non-equilibrium situation by some external means; then, at a certain instant, one sup-
presses the external cause of non-equilibrium
and looks at how the system tends to equilibrium
with the big reservoir. The fact that the system
is infinite and that the external perturbation
is localized in a finite region guarantees that
our system is in equilibrium at temperature
β^{-1} and chemical potential μ far enough from
Λ_h. We intuitively expect that the system will

evolve towards the equilibrium state with temperature β^{-1} and chemical potential μ.

The correlation functions $\varphi(x_1,\ldots,x_n;0)$
describing the state considered as initial in
the above discussion are obtained by the limit,
when $\Lambda \to \infty$, of the finite volume correlation
functions $\varphi_\Lambda(x_1,\ldots,x_n;0)$, which we have given
above, and in which we suppose the Hamiltonian
$H(x_1,\ldots,x_n)$, is replaced by the perturbed
Hamiltonian

$$H'(x_1,\ldots,x_n) = H(x_1,\ldots,x_n) + \sum_{1 \leq i \leq n} h(q_i).$$

We assume also, for the reasons that have been
discussed, that the potential interaction in H
is superstable. The time evolving correlations
$\varphi(x_1,\ldots,x_n;t)$, correspond to those of the
evolved state at time t, into which the initial
state $\varphi(x_1,\ldots,x_n;0)$ goes under the evolution
S^t coming from its internal dynamics (i.e.
without the external potential). We have still
the same expression as before for the finite
volume correlation functions $\varphi_\Lambda(x_1,\ldots,x_n;t)$;
we have only to replace $H'(x_1,\ldots,x_n)$ by

$$H'(S^{-t}(x_1,\ldots,x_n)) = H(x_1,\ldots,x_n) + \sum_{1 \leq i \leq n} h(S^{-t}q_i)$$

where we have suppressed the evolution operator
from $H(\ldots)$ because of the energy conservation.

Assuming h to be a non-negative function,
it has been proved that <u>the correlation functions
$\varphi(x_1,\ldots,x_n;t)$ exist for all t and satisfy
the BBGKY hierarchy of equations.</u> They provide
therefore an example of an existence theorem for
a solution of this system of equations, with
initial condition of the type described above.
With h not assumed to be non-negative it has
been shown that the correlation functions
$\varphi(x_1,\ldots,x_n;t)$ satisfy the BBGKY hierarchy in
the sense of distributions.

However, one has not been able to prove that
the correlation functions $\varphi(x_1,\ldots,x_n;t)$
approach, as t goes to infinity, the correlation
functions of the equilibrium state at temperature
β^{-1} and chemical potential μ , as it is intuitively expected. One has only shown that the
time averaged correlation functions

$$\bar{\varphi}(x_1,\ldots,x_n;T) = \frac{1}{T} \int_0^T dt\, \varphi(x_1,\ldots,x_n;t)$$

tend, as $T \to \infty$ to stationary correlation functions (without knowing if they are the equilibrium
ones), and that they satisfy the stationary BBKGY
hierarchy of equations.

7. <u>Ergodic properties of a one-dimensional hard
rods system</u>

We have seen in the preceding section that,
if Φ is a superstable, finite range potential,

and if the corresponding force $F(q) = -\Phi'(q)$
satisfies the hypothesis of the time evolution
theorem, then the equilibrium measures associated
with Φ are concentrated on $[\hat{\mathcal{X}}]$ and are invariant
under the corresponding time evolution. We have
thus exactly the mathematical situation which is
studied in ergodic theory, namely

1. $[\mathcal{X}]$ is a probabilistic measure space
2. φ is a probability measure on $[\mathcal{X}]$
3. S^t is a one-parameter group of isomorphisms
(mod 0) of $[\mathcal{X}]$ into itself leaving φ invariant.

   The triplet $([\mathcal{X}], \varphi, S^t)$ with the preceding
properties is what is called in ergodic theory a
"dynamical system". (We recall that (mod 0) means
"up to a set of measure zero".) It would be of
great interest to know the ergodic properties of
these objects, e.g., whether the system is ergodic,
mixing, a K-system, etc. Before discussing this
question, let us make some comments on the ergodicity property.
   The dynamical system $([\mathcal{X}], \varphi, S^t)$ is said to
be ergodic if for every $f([x]) \in L^1([\mathcal{X}], \varphi)$ we
have

$$\lim_{T \to \infty} \frac{1}{T} \int_0^T f(S^t[x])dt = \int f([x]) \, \varphi(d[x])$$

for almost every $[x]$ with respect to φ.

   In other words the ergodic hypothesis asserts
that the asymptotic time average of every summable
function is identical to its integral with respect

to the probability measure giving the statistics of the system. It is easy to see that this condition is equivalent to all invariants of the evolution operator being constant almost everywhere.

For finite classical systems the number of particles and the total energy are invariants under the evolution, so the canonical and grand-canonical ensembles cannot be ergodic states. However, the microcanonical ensemble is expected to be ergodic. In fact, the ergodicity of the hard spheres model has been proved by Y. Sinai. The reason why one may conjecture that equilibrium states of infinite classical systems are ergodic, is the following: the particle number and the total energy are not defined, so that one might think that constants are the only invariants. In the cases in which one can establish this ergodic conjecture one gets an answer to the question of approach to equilibrium in the sense that states φ_0 which are absolutely continuous with respect to the equilibrium measure φ approach (in the sense of weak convergence of the time average) that equilibrium state φ as they evolve in time, i.e.

$$\text{weak } \lim_{T \to \infty} \frac{1}{T} \int_0^T dt \, \varphi_t = \varphi \quad .$$

In order to verify this, take an arbitrary function $f \in L^1([X], \varphi)$. Because of the ergodicity property and the absolute continuity of φ_0 we have

$$\lim_{T \to \infty} \frac{1}{T} \int_0^T dt \int \varphi_t(d[x]) f([x]) =$$

$$= \lim_{T \to \infty} \frac{1}{T} \int_0^T dt \int \varphi_0(d[x]) f(S^t[x])$$

$$= \int \varphi_0(d[x]) \int \varphi(d[x]) f([x]) = \int \varphi(d[x]) f([x]).$$

We see that the condition of being an absolutely
continuous measure fixes the temperature and
chemical potential of the final state towards
which φ_0 evolves.

It is not known what are the ergodic properties of the system $([X], \varphi, S^t)$ we have been
discussing. Nevertheless, there do exist some
encouraging indications. The perfect gas and a
one-dimensional system of hard rods, have been
proved to be K-systems and consequently ergodic.

There are several reasons why the case of
particles with hard cores is particularly interesting. First, there is an aspect of interparticle
forces that has been ignored in the development
described in the preceding sections: the fact
that two particles tend to repel each other
strongly when they are close together. This
should, in principle, prevent many particles from
getting into a small volume and should help to
prevent the sort of catastrophes that lead to the
breakdown of the equations. This effect has probably to be taken into account to attack the
problem of time evolution in more than one dimension. If we consider a system of particles with
hard cores plus a finite range ordinary interaction, then we see, directly from the definition

of the system, that each particle feels the effect,
of only finitely many other particles. However,
this system presents other difficulties to which
the techniques we have described do not apply;
nevertheless, it has been reported by O. Lanford
that a proof of the existence of time-evolution
has been found, by the Sinai group, for a one-
dimensional system of particles with hard cores.
On the other hand, the study of the equilibrium
states of the one-dimensional system with hard
cores is much more complete (see G. Gallavotti
and S. Miracle) than in the case of point particles; in particular it has been shown that
the equilibrium state corresponding to a given
temperature and chemical potential is unique.

After these preliminaries let us now describe the model in which we are interested. We
consider a semi-infinite system on the positive
semi-axes, in which the particles are hard rods
(or have a hard core) of length a. The collisions between the rods and with a perfectly
reflecting wall at $q = -a/2$, will be the only
interactions considered. The dynamics of this
system consists entirely of an interchange of
the velocities of pairs of neighbouring particles
upon collision, inversion of the velocity at
origin and free motion between collisions.

The space $\mathbf{X}_a$ of labelled configurations
of this system will be the space of sequences
$x_a = (x_1, x_2, \ldots)$ with the conditions

$$q_i \geqslant 0, \quad |q_i - q_j| \leqslant a \quad \text{if } i \neq j.$$

Such sequences form a subset of the space $\mathcal{X}$ considered before, and one can define analogously
the space $[x_a]$ of unlabelled configurations of
hard rods and provide it with a topological and
measure structure. Let $P(\mu, \beta)$ and $\mathcal{G}^a_{\mu, \beta}$
be the thermodynamic pressure and equilibrium state
of the system at temperature β^{-1} and chemical
potential μ. Now, let $[\mathcal{X}]$ denote the space of
locally finite configurations in $\mathbb{R}^+$ of particles
without hard core. We call $\mathcal{G}_{\mu,\beta}$ the equilibrium
state of this semi-infinite perfect gas. One can
show that there exists an isomorphism π between
the two measured spaces $([\mathcal{X}], \mathcal{G}^a_{\mu,\beta})$ and
$([\mathcal{X}], \mathcal{G}_{\mu,\beta})$ with $\mu' = -\mu a \beta^2 \frac{\partial}{\partial a} P(\mu, \beta)$. This is
done in the following way. We first construct
a mapping from $[\mathcal{X}_a]$ into $[\mathcal{X}]$. Let $x_a = (x_1, x_2 \dots)$
be an element of $[\mathcal{X}_a]$ after ordering the particles
according to the increasing abscissae. We set

$$\pi(x_a) = (x_1', x_2', \dots) \text{ with } q_i' = q_i - (i-1)a, \ p_i' = p_i$$

Of course, $\pi(x_a)$ does not belong to $[\mathcal{X}]$ for every
$x_a \in [\mathcal{X}_a]$; for instance, if we consider $q_i = i a$,
then $q_i' = a$ for every i, and consequently
$\pi(x_a)$ is not a locally finite configuration.
Moreover π is not injective. However, there exist
Borel subsets $[\hat{\mathcal{X}}_a]$ of $[\mathcal{X}_a]$ and $[\tilde{\mathcal{X}}]$ of $[\mathcal{X}]$ such
that π is a Borel isomorphism from $[\hat{\mathcal{X}}_a]$ to $[\tilde{\mathcal{X}}]$.
Furthermore, $\mathcal{G}^a_{\mu,\beta}([\mathcal{X}_a] \setminus [\hat{\mathcal{X}}_a]) = 0$ and
$\mathcal{G}_{\mu',\beta}([\mathcal{X}] \setminus [\tilde{\mathcal{X}}]) = 0$. This allows us to conclude
that π is an isomorphism (mod 0) between the
measure spaces $([\mathcal{X}_a], \mathcal{G}^a_{\mu,\beta})$ and $([\mathcal{X}], \mathcal{G}_{\mu',\beta})$.

The method of defining an isomorphism by restricting (in a way suggested for physical reasons) the set of configurations is already familiar to us from the discussion of the evolution operator. Let us make explicit that in this case one takes $[\hat{\mathbf{X}}_a] = ([\mathbf{X}_a] \mid [\mathbf{X}_a^c]) \cap [\mathbf{X}_a^1]$ where

$$[\mathbf{X}_a^c] = \left\{ x_a : \text{either for some } |q_i - q_j| = a \text{ and } (q_i - q_j)(p_i - p_j) < 0 \text{ or some } q_i = 0 \text{ and } p_i < 0 \right\}$$

$$[\mathbf{X}_a^1] = \left\{ x_a : \sup_i \frac{|p_i|}{\log_+(q_i)} \vee \sup_\alpha \frac{\alpha}{a \vee (\alpha - N_{[0,\alpha)}(x))} < \right.$$

$$< +\infty .$$

We remark that $[\mathbf{X}_a^c]$ is the set of configurations for which either two particles collide or one particle collides with the reflecting wall. We subtract $[\mathbf{X}_a^c]$ from $[\mathbf{X}_a]$ to make Π injective. On the other hand $[\mathbf{X}_a^1]$ is the natural domain of definition of the evolution operator S^t. The condition $\sup_i \frac{|p_i|}{\log_+(q_i)}$ has already been used, and the second bounding condition means that the residual volume between the rods in $[0,\alpha)$ is of the same order as α for large α, i.e. the configuration x does not have the "close-packing" density. Similarly we set $[\tilde{\mathbf{X}}] = ([\mathbf{X}] \setminus [\mathbf{X}^c]) \cap [\mathbf{X}^1]$ where $[\mathbf{X}^c]$ and $[\mathbf{X}^1]$ are the corresponding subsets of $[\mathbf{X}]$.

Now in $[\tilde{\mathbf{X}}]$ the evolution operator S^t is easily defined for the perfect gas; because of

the simple dynamics of the hard core model, the
evolution operator in $[\hat{\mathcal{X}}_a]$ will be $S_a^t = \Pi \circ^{-1} S^t \circ \Pi$.
We get then the following proposition.

Π is an isomorphism (mod O) between the
dynamical systems $([\mathcal{X}_a], \varphi_{\mu,\beta}^a, S_a^t)$ and $([\mathcal{X}], \varphi_{\mu,\beta}, S^t)$

From this it follows that the ergodic properties of the two dynamical systems are the same,
and then the following can be proved.

de Pazzis' theorem.   The dynamical system
$([\mathcal{X}_a], \varphi_{\mu,\beta}^a, S_a^t)$ is a K-system.

As we shall not prove this theorem here, we
will not give a precise definition of a K-system.
Let us only recall the following chain of implications among ergodic properties:
K-system $\Rightarrow$ countably multiple Lebesgue spectrum $\Rightarrow$
$\Rightarrow$ mixing $\Rightarrow$ ergodic
and refer for an exposition of their meanings to
Dr. Farquhar's lectures appearing in this same
series, where an interesting discussion of the
physical implications of such results is given.
In particular one sees that the hope expressed
after the Sinai result for a finite system of
hard spheres (from which it follows that the
transport coefficients given by the Kubo formulae
are zero in a finite system), that these formulae
can possibly yield transport coefficients in the
thermodynamic limit, is not realized in this
simple model (for instance, it is intuitively
clear that the diffusion coefficient must be zero,
because, if the particle x_1 could diffuse far from

the wall, the vacuum could occupy all the volume).

This absence of hydrodynamical description limits somewhat the interest of these models of infinite systems that one has been able to treat. Nevertheless, it would be interesting to know if the infinite hard rods system is ergodic. It is reported by O. de Pazzis that such a proof has been obtained by Sinai.

8. Mathematical formalism of quantum statistical mechanics

In the quantum mechanical case an n-particle state of the system is given by a wave function $\Psi_n(q_1,\dots q_n)$, i.e. a square integrable function of the positions of the n particles, an element of $L^2(\mathbf{R}^{n\nu})$ (where ν is the dimension of the space). According to whether the particles obey Bose or Fermi statistics, their wave function should be symmetric or antisymmetric for the interchange of particles, i.e. should belong to $L^2_{\pm}(\mathbf{R}^n)$. When one wishes to describe an arbitrary (unspecified) number of Bose or Fermi particles, the appropriate Hilbert space is the Fock space

$$\mathcal{H}_{\pm} = \bigoplus_{n \geqslant 0} L^2_{\pm}(\mathbf{R}^{n\nu})$$

where $L^2_{\pm}(\mathbf{R}^0)$ is identified with the set $\mathbf{C}$ of complex numbers. Explicitly $\psi \in \mathcal{H}_{+}$ is a sequence $\{\Psi_0, \Psi_1(q_1), \Psi_2(q_1, q_2),\dots\}$ with $\Psi_0 \in \mathbf{C}$ (corresponding to the probability amplitude of finding zero particles), $\Psi_n \in L^2_{\pm}(\mathbf{R}^{n\nu})$, and $\|\Psi\| = 1$ which we take to be defined by

$$\|\Psi\|^2 = |\Psi|_0^2 = \sum_{n \geqslant 1} \int |\Psi(q_1, \ldots, q_n)|^2 \frac{dq_1 \ldots dq_n}{n!}$$

For each wave function $f(q) \in L^2(\mathbb{R}^\nu)$ one defines
the anihilation operator $a(f)$ on $\mathcal{H}_\pm$ by

$$(a(f)\Psi)_n(q_1, \ldots, q_n) = \int dq \, f(q) \Psi_{n-1}(q, q_1, \ldots, q_n)$$

and the creation operator $a^*(f)$ as the adjoint of
$a(f)$. They satisfy the following commutation or
anticommutation relations

$$\left[a(f), a(g)\right]_\pm = \left[a^*(f), a^*(g)\right]_\pm = 0 \quad \left[a(f), a^*(g)\right]_\pm = (f, g)$$

The mapping $f \to a(f)$ we can write

$$a(f) = \int dq \, f(q) \, a(q),$$

introducing the "field" notation $a(q)$ (and similarly $a^*(q)$).

   The movement of the particles is defined such
that the vector state $\Psi_n(q_1, \ldots, q_n; o)$ at time $t=o$,
develops into the state $\Psi_n(q_1, \ldots, q_n; t)$ at time
t according to the Schrödinger equation

$$\frac{\partial \Psi_n(q_1, \ldots, q_n; t)}{\partial t} = H(x_1, \ldots, x_n) \Psi_n(q_1, \ldots, q_n; t)$$

(where p_i is interpreted as the operator $-i\frac{\partial}{\partial q_i}$. We
will denote by U^t the unitary operator of evolu-
tion in the Fock space which is induced by the
transformations $\Psi_n(q_1, \ldots q_n; o) \to \Psi_n(q_1, \ldots, q_n; t)$.

   In the above formalism each state describes
with probability 1 a finite number of particles.
Therefore, it is not adapted to statistical mechanics in which typically one wishes to study "large"
systems where the density per unit volume is finite.
It is for this reason that one introduces in

classical statistical mechanics the correlation
functions to describe the state of the system. In
quantum statistical mechanics the analogous concept
is the <u>reduced density matrix</u> $\rho(q_1,\ldots,q_n,q_1',\ldots q_m')$
which is defined as the expectation value of a pro-
duct of creation and anihilation operators

$$\rho(q_1,\ldots,q_n,q_1',\ldots,q_m') = \left\langle a^*(q_1)\ldots a^*(q_n)a(q_1')\ldots a(q_m')\right\rangle$$

where the state we have represented by $\langle\ldots\rangle$ can no
longer be interpreted as a vector state or density
matrix in a Fock space but rather as a limit when
$\Lambda \to \infty$ of the expectation values of the states of
the system enclosed in a box Λ (where we have a finite number of particles and the state is actually
a density matrix in the Fock space $\mathcal{H}_{\pm}(\Lambda)$ of wave
functions vanishing outside Λ).

In the classical case we have described such
states as probability measures or normalized posi-
tive linear functionals on the C^*-algebra $\mathcal{C}([\mathcal{X}])$
of continuous functions on the space of configura-
tions. Here we can also construct a C^*-algebra α
formed by the linear combinations and norm limits
of products $a^*(f_1)\ldots a^*(f_n)a(g_1)\ldots a(g_m)$ of creation
and annihilation operators. This applies directly
to the Fermi case in which the operators $a^*(f)$ and
$a(f)$ are bounded operators; in the Bose case the
same notions can be adopted but require more pre-
cautions. For simplicity, we shall restrict our
discussion in the following to systems of Fermi
particles.

<u>We define then a state of quantum statistical</u>

mechanics as a normalized positive functional (or
a state in the mathematical sense) over the C^*-al-
gebra $\mathcal{O}$. The fundamental difference with the classical case is that the algebra $\mathcal{O}$ is no longer com-
mutative.

For every bounded region Λ , ρ defines a state
on the subalgebra $\mathcal{O}(\Lambda)$ of $\mathcal{O}$ generated by the
creation and annihilation operators on $\mathcal{H}(\Lambda)$. This
state is normal on $\mathcal{O}(\Lambda)$ when it can be represented
by a density matrix D_Λ on $\mathcal{H}(\Lambda)$, i.e.

$$\rho(A) = \text{Tr}(D_\Lambda A) \quad \text{for every } A \in \mathcal{O}(A)$$

When ρ is in every Λ represented as a density matrix
state we say that ρ is a locally normal state.
Then a locally normal state can be equivalently
described by a family $\{D_\Lambda\}$ of density matrices
on $\mathcal{H}(\Lambda)$; this notion corresponds to the density
distributions we have found in classical systems.
It has been proved that if the state ρ has finite
mean density of particles per unit volume, then
it is locally normal.

Since the products of creation and annihilation
operators generate the algebra $\mathcal{O}$, one also characterizes the state by the following expectation
values:

$$\omega_{nm}(f_1,\ldots,f_n;g_1,\ldots,g_m)=\rho(a^*(f_1)\ldots a^*(f_n)a(g_1)\ldots a(g_m))$$

which are related to the reduced density matrices,
when they exist, by

$$\omega_{nm}(f_1,\ldots,f_n;g_1,\ldots,g_m) =$$
$$=\int \rho(q_1,\ldots,q_n;q_1',\ldots,q_m')\overline{f_1(q_1)}\ldots\overline{f_n(q_n)}g_1(q_1)\ldots g_m(q_m)$$
$$dq_1\ldots dq_n dq_1'\ldots dq_m'$$

   Having described the states, we next introduce
some definitions and notations that will be used
in the next section.

   The state ρ is defined to be <u>even</u> if
$$\omega_{nm}(f_1,\ldots,f_n; g_1,\ldots,g_m)=0$$
whenever n+m is odd. The state ρ is said to be
translationally <u>invariant if</u>

$$\rho(\alpha_a A) = \rho(A) \qquad A \in \alpha, a \in \mathbb{R}^\nu$$

where $\alpha_a, a \in \mathbb{R}^\nu$ is a group of automorphisms of the
algebra α generated by the translations, i.e. it
is defined by

$$\alpha_a a(f)=a(V_a f), \quad \alpha_a a^*(f)= a^*(V_a f)$$

where V_a is the unitary operator in $L^2(\mathbb{R}^\nu)$:
$:(V_a f)(q)= f(q-a)$. One knows that <u>each translatio-</u>
<u>nally invariant state is automatically even.</u>

   An even state over α is also completely determined by the <u>truncated functions</u>
$\omega^T_{nm}(f_1,\ldots,f_n,g_1,\ldots,g_m)$ which are defined recur-
sively by the following formulae

$$\omega_{nm}(f_1,\ldots,f_n;g_1,\ldots,g_m) =$$

$$\sum_\pi (-1)^{\sigma(\pi)} \omega^T_{r_1 s_1}(f_{i_1},\ldots,f_{i_{r_1}};g_{j_1},\ldots,g_{j_{s_1}})\ldots$$

$$\ldots\omega^T_{r_p s_p}(f_{k_1},\ldots,f_{k_{r_p}};g_{e_1},\ldots,g_{e_{s_p}})$$

where the sume is taken over all partitions π of
$\{f_1,\ldots,f_n; g_1,\ldots,g_m\}$ into disjoint subsets. The
f's and g's appear within each subset of a partition
π in the same order as they appear in the original

set, and σ (π) is the permutation of the set required to rearrange the f's and g's into the order
in which they appear in the partitioned subsets.

A translationally invariant state ρ is completely determined by the set of functions

$$\omega^T_{\{f_n\}\{g_m\}}(q_2-q_1,\ldots,q_{n+m}-q_n) =$$

$$\omega^T_{nm}(V_{q_1}f_1,\ldots,V_{q_n}f_n,V_{q_{n+1}}g_1,\ldots,V_{q_{n+m}}g_m)$$

and is defined to be <u>square integrable</u> if

$$\int d\xi_1\ldots d\xi_{n+m-1}\left| \omega^T_{\{f_n\}\{g_m\}}(\xi_1,\ldots,\xi_{n+m-1})\right|^2 < +\infty$$

for $n+m > 2$ and for all f_i, g_j such that $\tilde{f}_i, \tilde{g}_j \in \mathcal{D}$,
where $\tilde{f}$ means the Fourier transform of f and $\mathcal{D}$ is
the space of test functions of Schwartz. If ρ is
associated with a set of reduced density matrices
this condition means that the truncated reduced density matrices are square integrable functions. We
can also recall that the cluster property of a state
ρ is related to the decreasing at infinity of the
truncated functions $\omega^T_{\{f_n\}\{g_m\}}$, and that a square
integrable state then has this cluster property.

Finally, an even state on $\mathcal{O}$ is called <u>quasi-
free</u> if

$$\omega^T_{nm} = 0 \qquad\qquad \text{for } n-m > 2$$

and a <u>gauge invariant quasi-free state</u> if

$$\omega^T_{nm} = 0 \qquad\qquad \text{for } (n,m) \neq (1,1)$$

Further, if ρ is an arbitrary state over $\mathcal{O}$ we can
define the <u>associated gauge invariant quasi-free</u>
by putting
$$\hat{\rho}(a^*(f)\,a(g)) = \rho(a^*(f)a(g))$$

and $\widehat{\omega}\,^T_{nm}=0$ for $(n,m) \neq (1,1)$.

Let us end this section with a simple comment about the problem of time-evolution. It is clear that a hamiltonian operator for the infinite system does not exist, as the Hilbert spaces at our disposal describe the system in finite boxes. Let us call H_Λ the hamiltonian acting on $\mathcal{H}(\Lambda)$. Given an initial state ρ_0 we can examine it in a large box Λ, and let $D_\Lambda(o)$ be the corresponding density matrix. If we neglect the influence of what is outside Λ we have that the system in the box evolves according to

$$D_\Lambda(t) = U^t_\Lambda D_\Lambda(o)\, U^{-t}_\Lambda$$

where U^t_Λ is the unitary operator of evolution coming from H_Λ. As one expects for reasonable interactions some stability property of the evolution like the one described in section 5, one can hope to obtain the evolved state ρ_t at time t by taking in a suitable way the limit $\Lambda \to \infty$.

The simplest case occurs when the evolution can be given as a one-parameter group of automorphisms σ^t of the algebra $\mathcal{O}$. This happens in the following situation: Suppose that for each A localized in some subalgebra $\mathcal{O}(\Lambda_0)$ and that for each $\Lambda \supset \Lambda_0$, one has $U^{-t}_\Lambda A U^t_\Lambda \in \mathcal{O}(\Lambda)$ and, further, that when $\Lambda \to \infty$, $U^{-t}_\Lambda A U^t_\Lambda$ is norm convergent in $\mathcal{O}$, then one can define

$$\sigma^t A = \lim_{\Lambda \to \infty} U^{-t}_\Lambda A U^t_\Lambda$$

and it is simple to verify that

$$\rho_t(A) = \rho(\sigma^t A)$$

One cannot expect that this is the general

situation, as one could not define in the classical
case a similar time evolution of every infinite
configuration of the system.

However, there are physically interesting cases in which the time evolution automorphisms
exist. As we shall see, this is the case in free
systems. Also, for interacting quantum lattice
systems, the existence of the automorphisms σ^t
can be proved, and we have a well defined time
evolution; moreover, the equilibrium states are
time invariant and satisfy the Kubo-Martin-Schwinger
condition for this evolution. We shall not discuss
these results here but refer for their study to
Ruelle's book.

9. Approach to equilibrium of free quantum systems

We have seen that in non-interacting classical
systems, it is possible to analize the properties
of the states of the system as they change in time;
in particular we mentioned that the equilibrium
states of a free gas and of a gas of hard rods,
have strong ergodic properties with respect to time
evolution and, in fact, provide examples of K-systems. In this section we will analize the corresponding problem in free quantum systems. We shall consider only the case of Fermi particles, although
very similar results have been obtained in the case
of Bose particles.

From the Schrödinger equation without interactions it follows that if $\psi_n(q_1,\ldots,q_n)$ is an
n-particle wave function at time t=o it develops

into

$$(U^t\psi)_n(q_1,\ldots,q_n) =$$

$$\int dp_1\ldots dp_n\tilde{\psi}_n(q_1,\ldots,q_n)e^{i(p_1^2+\ldots+p_n^2)t} \times$$

$$\times e^{-i(p_1q_1+\ldots+p_nq_n)}$$

at time t, where $\tilde{\psi}_n$ is the Fourier transform of ψ_n. Then an easy calculation shows that σ^t has the following action on the generating elements of the algebra $\mathcal{O}$

$$\sigma^t a(f)=a(U^t f) \quad , \quad \sigma^t a^*(f)=a^*(U^t f)$$

and this definition extends to give a one-parameter group of automorphisms of $\mathcal{O}$. This allows us to define the free time development of an arbitrary state ρ over $\mathcal{O}$ by

$$\rho_t(A) = \rho(\sigma^t A).$$

With this we have exactly defined the time development of the system and we are now interested in whether the states ρ_t approach some limiting states, as $t\rightarrow\infty$. Not all states can have such a property and hence the problem is to find subsets of them which do. It appears that two general restrictions on the states are necessary to ensure the desired property. Firstly, one must impose some form of homogeneity, and, secondly, one must restrict the interparticle correlations. We have given in the preceding section some precise definitions of these sorts of requirements, which allow us to formulate the following theorem:

Theorem (Lanford-Robinson). If ρ is a square integrable (hence translationally invariant) state over $\mathcal{O}$ and $\hat{\rho}$ is the associated gauge invariant quasi-free state, then

$$\lim_{t \to \infty} \rho \, (\sigma^t A) = \overline{\rho} \, (A)$$

for every $A \in \alpha$.

We remark that this is the first case we en-
counter in which we can give a precise statement
saying that some final "equilibrium state" is
actually reached by the system as $t \to \infty$. However,
the proof of the theorem does not require complicated mathematical techniques.

One has to show that all the truncated functions
$\omega^T_{m,n}$ tend to zero as $t \to \infty$, except ω^T_{11} which is
conserved. Let us look first at the second order
ones, i.e. $\omega^T_{11}, \omega^T_{20}, \omega^T_{02}$ which according to their
definition coincide with $\omega_{11}, \omega_{20}, \omega_{02}$. We have
to show that

$$\rho(\sigma^t(a^*(f)a(g)) = \rho \, (a^*(f)a(g))$$

$$\lim_{t \to \infty} \rho \, (\sigma^t(a(f)a(g)) \; = \; 0$$

$$\lim_{t \to \infty} \rho \, (\sigma^t(a^*(f)a(g)) \; = \; 0$$

We note that, as $\| a(f) \| = \| a^*(f) \| = \| f \|$ one has

$$| \rho \, (a^*(f) \, a \, (g)) | \leq \| f \| \, \| g \|$$

so there exists an operator A on $L^2(\mathbb{R}^\nu)$ with
$\| A \| \leq 1$ such that

$$\rho \, (a^*(f)a(g)) = (f, \, Ag)$$

Now, as ρ is translationally invariant, A
commutes with the group of translation operators
$f \to V_a f$, and hence it is a multiplication operator
in the momentum space. Thus, there is a bounded

function $\tilde{\varphi}(p)$ such that

$$\varphi(a^*(f)a(g)) = \int dp \, \tilde{\varphi}(p) \, \overline{\tilde{f}}(p)g(p) =$$
$$= \varphi(a^*(U^t f)a(U^t g))$$

and the proposition for ω_{11} is proved.

Similarly there exists an operator B with $\|B\| \leq 1$, such that

$$\varphi(a(f)a(g)) = (\bar{f}, Bg)$$

Again B commutes with the translations and hence there is a bounded function $\tilde{\sigma}$, such that

$$\varphi(a(f)a(g)) = \int dp \, \tilde{\sigma}(p) \, \tilde{f}(-p) \, \tilde{g}(p)$$

But then we have

$$\varphi(a(U^t f)a(U^t g)) = \int dp \, \tilde{\sigma}(p)\tilde{f}(-p)\tilde{g}(p)e^{2ip^2 t}$$

which goes to zero as $t \to \infty$ by the Riemann-Lebesgue lemma. If φ is square integrable one may re-apply the Riemann-Lebesgue lemma to show that ω^T_{mn} tends to zero for $m+n > 2$.

The existence of square integrable states is assured by the work of J.Ginibre who shows that the equilibrium states for a large class of interacting systems have this property at low density.

A slightly stronger result can also be proved, viz.

$$\lim_{t \to \infty} \varphi(B\sigma^t(A)C) = \varphi(BC) \, \hat{\varphi}(A)$$

and has the interpretation that the locally perturbed state $\varphi'(A) = \varphi(BAC)/\varphi(BC)$ also evolves towards $\hat{\varphi}(A)$.

A priori one may be tempted to argue that the limit or equilibrium state reached by the system as $t \to \infty$, should be identifiable with the Gibbs equilibrium state compatible with the given energy

and particle density; this is clearly not the case.
The reason for this discrepancy is however immediate: The free evolution is pathological in the
sense that it allows too many constants of motion
(in particular, the gauge invariant two point functions remain constant in time). Nevertheless, one
can use the information gathered from the freely
evolving systems to check a number of general
principles. In particular, although the states we
are considering do not necessarily approach the
Gibbs state, one can show that the states which they
do approach have the maximum mean entropy compatible
with the constants of motion.

We have (maximum entropy principle)

$$S(\hat{\rho}) = \sup \left\{ S(\rho') : \hat{\rho}' = \hat{\rho} \right\}$$

and therefore

$$S (\lim_{t \to \infty} \rho_t) \geqslant S(\rho_t) = S(\rho)$$

For a translationally invariant locally normal
state ρ , the mean entropy is defined by

$$S(\rho) = \lim_{\Lambda \to \infty} - \frac{1}{V(\Lambda)} \, \text{Tr} \, (D_\Lambda \log D_\Lambda)$$

where $V(\Lambda)$ is the volume of Λ and the limit is
known to exist. The theorem is proved by using the
inequality

$$-\text{Tr}(D_\Lambda^{(1)} \log D_\Lambda^{(1)} - D_\Lambda^{(1)} \log D_\Lambda^{(2)}) \leq 0$$

where $D_\Lambda^{(1)}$ and $D_\Lambda^{(2)}$ are any density matrices. It is
also intuitively clear, since for a description of
the state ρ' one needs to know many intercorrela-
tions ω_{mn} which are different from zero, while $\hat{\rho}$
has only the $\omega_{11} \neq 0$ and they coincide with those

of ρ' ｜ Since ρ_t has density matrices given by $U_\Lambda^t \, D_\Lambda \, \bar{U}^t$, and the trace is invariant under unitary transformations, it follows that $S(\rho_t)=S(\rho)$, i.e. the entropy is a constant of motion. This, however, does not rule out the increase of the entropy in the limit $t \to \infty$ as S is usually only an upper semicontinuous function.

References

The author's references made in the text can be found in the following list; their titles are self-explanatory. Let us, however, indicate that classical one-dimensional systems are discussed in references (1-2-3-4), their ergodicity properties in (7-8), quantum free systems in (12-13-14) and quantum lattice systems in (10-11), while some equilibrium results, relevant in the first three developments just mentioned, can be found in (5-6),(9) and (15) respectively. (16-17-18-19) are general references and (20-21-22) are concerned with the axiomatic approach to time-evolution.

1) O.E. LANFORD: Comm.Math.Phys.9,176 (1968)
2) O.E. LANFORD: Comm.Math.Phys.11, 257 (1969)
3) G. GALLAVOTTI, O.E. LANFORD and J.L. LEBOWITZ: J. Math. Phys. 11, 2898 (1970)
4) O.E. LANFORD: S.I.A.M. Symposium on Mathematical Aspects of Statistical Mechanics, April 1971, New York
5) D. RUELLE: Comm.Math.Phys. 18, 127 (1970)
6) G. GALLAVOTTI: Nuovo Cim. 52B, 208 (1968)

7) O. de PAZZIS: Comm.Math.Phys. 22, 121 (1971)

8) Ya SINAI and K. VOLKOVISKIJ: Ergodic Properties
    of the perfect gas with infinite many degrees
    of freedom (unpublished)

9) G. GALLAVOTTI and S. MIRACLE: J. Math.Phys. 11
    147 (1969)

10) D.W. ROBINSON: Comm. Math. Phys. 7, 337 (1968)

11) M.B. RUSKAI: Comm. Math. Phys. 20, 193 (1971)

12) R. HAAG: Asymptotic Behaviour in Time of the
    States of a Free Fermi Gas.Publication of
    C.N.R.S. 1970

13) O.E. LANFORD and D.W. ROBINSON: Comm. Math.
    Phys. 24, 193 (1972)

14) P.W. ROBINSON: Amer. Math.Soc. Symposium on
    Statist. Mech. New York, April 1971

15) J. GINIBRE: J. Math. Phys. 9, 1120 (1968)

16) D. RUELLE: Statistical Mechanics, Benjamin,1969

17) V.I. ARNOLD and A. AVEZ: Problemes Ergodiques
    de la Mecanique Classique, Gauthier-Villars,
    1967

18) J.L. LEBOWITZ: I.U.A.P. Conference on Statist.
    Mech. at Chicago University, March 1971

19) Ya SINAI: Sov.Math.Dokl. 4, 1818 (1963)

20) D. RUELLE: Lecture Notes, Les Houches Summer
    School, 1970

21) R. HAAG, N.M. HUGENHOLTZ and M. WINNINK:
    Comm. Math. Phys. 5, 215 (1967)

22) M. TAKESAKI: Tomita's Theory of Modular Hilbert
    Algebras and its Applications. Springer, 1970

NON-EQUILIBRIUM STATISTICAL MECHANICS

R. Balescu

Faculte des Sciences,
Université Libre de Bruxelles
Association Euratom, Etat Belge

1. The distribution vector
2. The dynamics of correlations
3. Dynamics of ideal systems
4. Vacuum and correlations
5. Formal solution of the complete Liouville equation
6. Programme for a kinetic theory of irreversibility
7. Corstruction of the operator Π
8. Subdynamics and kinetic theory
9. Translation of the abstract formalism and final conclusions

1. The distribution vector

As we all know, statistical mechanics deals with the behaviour of large assemblies of particles (or other kinds of entities, such as oscillators ...). In the present lectures we shall only consider systems governed by the laws of classical mechanics: however all the methods discussed here can easily be extended to the quantum mechanical case as well.

In this framework the system is fully described by the distribution function in phase-space: $F(x_1 \ldots x_N)$, where $x_j = (\underline{q}_j, \underline{p}_j)$ is a set of canonical variables describing the particle labelled j, $j = 1,2,\ldots, N$. The evolution in time of the system is determined by the Liouville equation

$$\partial_t F \;=\; L\,F \qquad\qquad (1.1)$$

where L is the Poisson-bracket operator with the Hamiltonian

$$L \;=\; \big[H,\ldots\big]_p \qquad\qquad (1.2)$$

The Liouville equation of a truly non-trivial system is extremely complicated. On the other hand it is important to note that a knowledge of the full distribution function is, in a sense, a luxury, for this function contains a lot of information which is absolutely inaccessible to the macroscopic observer. We may therefore say that the purpose of statistical mechanics is a detailed analysis of the relevant parts of the general

mechanical problem, and a progressive elimination
of the superfluous elements of the theory. The
final aim should be a reduction of the description
of dynamical systems in which, as far as possible,
only concepts of direct relevance to the assigned
macroscopic purpose should be conserved in a consistent formalism.

This "economical" approach is typical of the
well-known kinetic equations, of which the Boltzmann and Landau equations are typical examples.
The whole dynamics is here described in terms of
a reduced one-particle distribution function and
this could be achieved in the original derivations
only by means of semi-phenomenological probabilistic arguments.

Our purpose in these lectures will be to outline a general analysis of the time-evolution
process, and to show how and under what conditions
such a reduction can be achieved in the framework
of an exact dynamical theory.

In order to fix our ideas, we shall always
assume here that the Hamiltonian is of the form

$$H = \sum_{j=1}^{N} H^{o}(x_j) + \sum_{j<n=1}^{N} V(x_j, x_n) \quad (1.3)$$

The important feature here is that the first
term consists of a sum of one-particle functions,
i.e. the kinetic energies:

$$H^{o}(x_j) = p_j^2/2m \quad (1.4)$$

whereas the second term involves interactions of
pairs of particles which we may take as

$$V(x_j, x_n) = V(|\underline{q}_j - \underline{q}_n|) \qquad (1.5)$$

Corresponding to this separation, the Liouvillian
is also represented as

$$L = \sum_j L^o_{\ j} + \sum_{j<n} \sum L'_{\ jn} \qquad (1.6)$$

with

$$L^o_j = -\underline{v}_j \cdot \underline{\nabla}_j$$

$$L'_{jn} = (\underline{\nabla}_j V_{jn}) \cdot \underline{\partial}_{jn} \qquad (1.7)$$

where the following abbreviations are used:

$$\underline{v}_j = \underline{p}_j/m$$

$$\underline{\nabla}_j = \partial/\partial\underline{q}_j$$

$$\underline{\partial}_{jn} = \underline{\partial}_j - \underline{\partial}_n = \partial/\partial\underline{p}_j - \partial/\partial\underline{p}_n \qquad (1.8)$$

The Liouville equation is, of course, a very
complicated equation, involving a function of very
many variables. In statistical mechanics we are
usually interested in a mathematical limiting procedure, which consists of letting the number of
particles N and the volume of the system, v,
go to infinity, as follows:

$$\text{T-limit} \begin{cases} N \to \infty \\ v \to \infty \\ N/v = n = \text{finite constant} \end{cases} \qquad (1.9)$$

By this "thermodynamic limit" we are able to
extract from the mathematical expressions the

physical quantities which have a local meaning
(intensive variables).

The distribution function $F(x_1 \ldots x_N)$ is not
well suited for a direct application of the thermodynamic limit. Instead, we may describe the
system by a set of <u>reduced distribution functions</u>

$$f_s(x_1 \ldots x_s) = \frac{N!}{(N-s)!} \int dx_{s+1} \ldots dx_N$$

$$\tag{1.10}$$

$$F(x_1 \ldots x_s \ x_{s+1} \ldots x_N), \quad s=0,1,\ldots, \ N.$$

In a "normal system", by definition, these functions tend towards finite and well-behaved functions as $N \rightarrow \infty$ in the thermodynamic way.

As the function $F(x_1 \ldots x_N)$ is normalized
to one, we see that the reduced distributions are
normalized as follows, in the thermodynamic limit:

$$\text{T-lim } N^{-s} \int dx_1 \ldots dx_s \ f_s(x_1 \ldots x_s) \ = \ 1 \quad (1.11)$$

The reduced distribution functions obey the
well-known BBGKY-hierarchy:

$$\partial_t f_s(x_1 \ldots x_s) = \sum_{j=1}^{s} L_j^o \ f_s(x_1 \ldots x_s)$$

$$+ \sum_{j \langle n=1}^{s} L_{jn}' \ f_s(x_1 \ldots x_s) \qquad (1.12)$$

$$+ \sum_{j=1}^{s} \int dx_{s+1} \ L_{j,s+1}' \ f_{s+1}(x_1, \ldots, x_{s+1})$$

Many authors have tried to study approximations to the hierarchy by introducing more or less justified truncation assumptions. We will rather study these equations globally, and show at the end of our analysis how a considerable reduction sets in quite naturally.

In order to arrive at such a global view, we regard the various functions f_s as components of a formal "distribution vector"

$$\mathcal{f} = \left\{ f_s(x_1 \ldots x_s); \quad s = 0,1,2,\ldots \right\} \quad (1.13)$$

(the set is infinite in the thermodynamic limit). As a function of time this vector satisfies a generalized Liouville equation:

$$\partial_t \mathcal{f}(t) = \mathcal{L}\, \mathcal{f}(t) \equiv (\mathcal{L}^\circ + \mathcal{L}')\, \mathcal{f}(t) \qquad (1.14)$$

where the operator $\mathcal{L}$ now appears as a matrix. Indeed, we may write the components of eq. (1.14) as

$$\partial_t \, f_s(x_1 \ldots x_s) = \sum_{s'=0}^{\infty} \langle s | \mathcal{L}^\circ + \mathcal{L}' | s' \rangle$$

$$f_{s'}(x_1 \ldots x_{s'}) \quad (1.15)$$

These equations must, of course, be identical with the BBGKY equations so we may identify the matrix elements as

$$\langle s | \mathcal{L}^\circ | s' \rangle = \delta_{s',s} \left\{ \sum_{j=1}^{s} L_j^\circ \right\}$$

$$\langle s| \mathcal{L}' |s' \rangle = \delta_{s',s} \left\{ \sum_{j\,<\,n=1}^{s} \sum L'_{jn} \right\}$$

$$+ \ \delta_{s',s+1} \left\{ \sum_{j=1}^{s} \int dx_{s+1} \ L'_{j,s+1} \right\}$$

$$(1.16)$$

An important feature of these equations is the diagonal character of the unperturbed operator $\mathcal{L}^{\circ}$; the interaction operator $\mathcal{L}'$ on the other hand has diagonal as well as off-diagonal contributions. A convenient representation of the various terms is by means of diagrams. We draw s superposed, labelled lines to represent the function f_s. An interaction operator L'_{jn} will be represented as a vertex connecting the lines j and n. If the operator is non-diagonal, i.e. $\int dx_n \ L'_{jn}$, we do not draw the line n to the left of the vertex. These rules are illustrated for the case s = 2:

$$(\partial_t - L_1^{\circ} - L_2^{\circ})f_2(x_1,x_2) = L'_{12}f_2(x_1,x_2)$$

$$+ \int dx_3 (L'_{13}+L'_{23})f_3(x_1,x_2,x_3)$$

$$(1.17)$$

which is represented diagrammatically as in fig.1.

$$(\partial_t - L_1^0 - L_2^0)f_2(x_1, x_2)$$

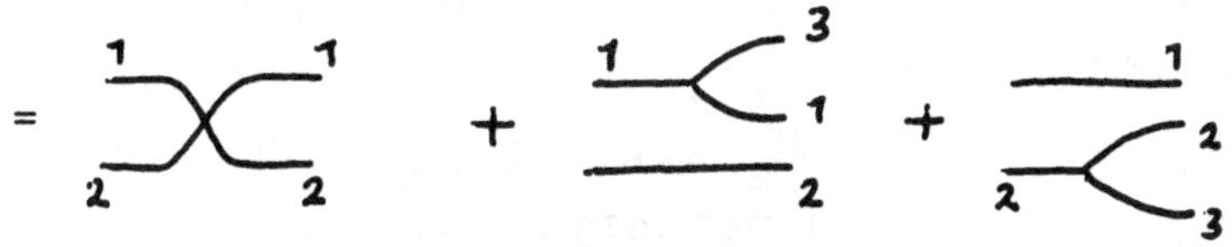

Fig. 1.

2.   The dynamics of correlations

For many purposes in kinetic theory the description in terms of reduced distribution functions
is not yet sufficiently refined. It is often
essential to distinguish between states where the
particles are statistically independent and states
where they are mutually correlated (in various
ways). This distinction comes out quite clearly
in the well-known cluster representation of the
distribution functions.

$$f_1(1) \quad\quad = \quad f_1(1)$$

$$f_2(1,2) \quad = \quad f_1(1)\, f_1(2) + g_2(12)$$

$$f_3(1,2,3) \quad = \quad f_1(1)\, f_1(2)\, f_1(3) + f_1(1)\, g_2(23)$$

$$+ f_1(2)\, g_2(13) + f_1(3)\, g_2(12)$$

$$+ g_3(123), \quad \text{etc.} \quad\quad\quad (2.1)$$

The functions g_s are called (irreducible) correlation functions. Each term in this representation
describes a certain type of correlation between
the particles: it corresponds to a given par-

tition of the set of s particles into non-overlapping subsets.

   As a matter of notation, we may attribute a "correlation index" Γ_s to every partition of the set s, and call $\pi_s(1,\ldots s; \Gamma_s)$ the corresponding correlation pattern; by convention we associate $\Gamma_s = 0_s$ with the completely factorized pattern. Then the cluster representation is written compactly as

$$f_s(1\ldots s) \;=\; \sum_{\Gamma_s} \pi_s(1\ldots s; \Gamma_s) \qquad (2.2)$$

Alternatively, we may use a more explicit notation, where the particular partition is shown by semicolons, e.g.

$$\pi_3(123; [0_3]) \;=\; \pi_3(1; 2; 3) \;=\; f_1(1)\, f_1(2)\, f_1(3).$$

   Our aim is to study the dynamics of the separate correlation patterns. This is easily done by considering successively s = 1,2,... in (1.12) and combining the equations with the expansion (2.1). Thus we have

$$(\partial_t - L_1^0)\pi_1(1) \;=\; \int dx_2 \Big\{ L_{12}'\, \pi_2(1;2) + L_{12}'\, \pi_2(12) \Big\} \tag{2.3}$$

Next, noting that $\pi_2(1;2) = \pi_1(1)\,\pi_1(2)$, etc.

$$(\partial_t - L_1^o - L_2^o) \, \pi_2(1;2)$$

$$= \int dx_3 \left\{ L_{13}' \, \pi_3(1;2;3) + L_{13}' \, \pi_3(2;13) \right.$$

$$\left. + L_{23}' \, \pi_3(1;2;3) + L_{23}' \, \pi_3(1;23) \right. \qquad (2.4)$$

We now combine eq. (1.17) with (2.1), and subtract term by term eq. (2.4), obtaining

$$(\partial_t - L_1^o - L_2^o) \, \pi_2(12) \;=\; L_{12}' \Big[\pi_2(1;2) + \pi_2(12) \Big]$$

$$+ \int dx_3 \left\{ L_{13}' \, \pi_3(1;23) + L_{23}' \, \pi_3(2;13) \right.$$

$$\left. + (L_{13}' + L_{23}') \, \pi_3(3;12) + (L_{13}' + L_{23}') \, \pi_3(123) \right\} \quad (2.5)$$

Clearly this process can be continued systematical-ly to obtain a <u>linear hierarchy of equations for the correlation patterns.</u> In order to derive these equations, we made specific use of the factoriza-tion properties of the patterns π_s. However, at this stage we may switch the stress of our argu-ment, thus obtaining a more general definition of the correlation patterns. We now <u>define</u> the set of <u>dynamical correlation patterns</u> $p_s(x_1 \ldots x_s; \Gamma_s)$ as <u>any</u> solution of the set of differential equations starting with (2.3-5), and which we write as

$$\partial_t \, p_s(x_1 \ldots x_s; \Gamma_s)$$

$$= \sum_{s'=0}^{\infty} \sum_{\Gamma_{s'}'} \langle s, \Gamma_s | \mathcal{L}^o + \mathcal{L} | s', \Gamma_s' \rangle p_{s'}(x_1 \ldots x_{s'}; \Gamma_{s'}')$$
$$(2.6)$$

Clearly, by construction, this set of equations
admits a solution factorized in the cluster
fashion. Thus, if initially $p_s(\Gamma_s) = \pi_s(\Gamma_s)$,
this property is maintained in time. But this
is only one among many possible initial conditions.
The matrix elements appearing in eq. (2.6) are
considered to be known quantities, because they
can be constructed systematically although the
algebraic procedure soon becomes quite tedious.
We can, however, devise a very simple graphical
method for this purpose.

To this end, we agree to represent the correlation patterns on the right hand side, i.e.
the states on which the operators act, by a set
of labelled lines (as before), plus a set of
connections corresponding to the initial correlations (if any). Thus eq. (2.3) corresponds
to fig.2.

$$(\partial_t - L_1^o)p_1(1) = $$

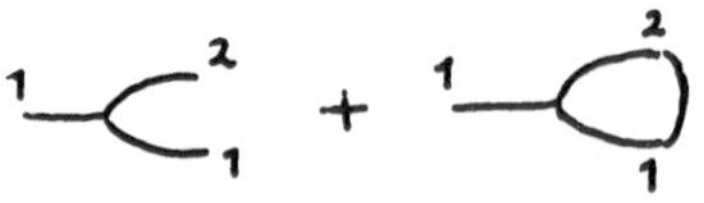

Fig. 2.

Similarly, eqs. (2.4) and (2.5) are represented
in figs. 3 and 4.

$$(\partial_t - L_1^o - L_2^o)p_2(1;2) = $$

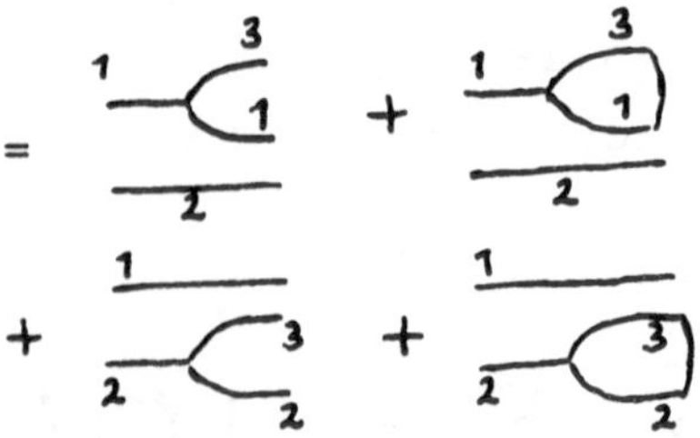

Fig. 3.

$$(\partial_t - L_1^0 - L_2^0)p_2(12) =$$

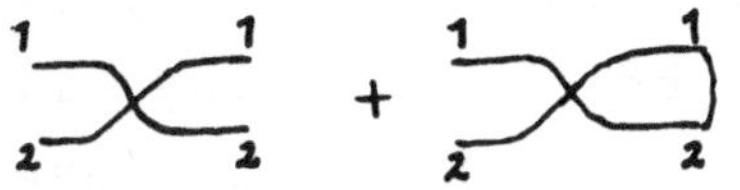

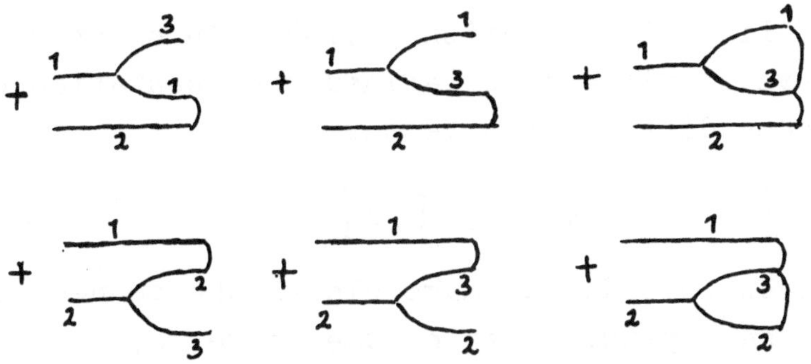

Fig. 4.

There is a fundamental topological difference between the diagrams of figs. 3 and 4. For the former, the diagrams fall into two disconnected pieces, whereas in the latter they are fully connected. Thus the diagrams contributing to the uncorrelated pattern $p_2(1;2)$ are made up of two distinct components ending on the left, respectively, with the lines labelled 1 and 2. This property gives the clue to the general rule for obtaining the equation for a pattern $\pi_s(\Gamma_s)$:

1) draw all the diagrams contributing to $\partial_t f_s$

2) replace each diagram by a sum of diagrams corresponding to all possible patterns describing the "initial state" (on the right).

3) among all the diagrams pick out those which consist of connected pieces corresponding precisely to the partition Γ_s.

We now come back to our global picture. The

set of e quations (2.6) is of course equivalent
to the BBGKY hierarchy (1.13). We can consider
it as just another representation of the generalized Liouville equation (1.14). In other words,
we have simply a more refined representation of
the same distribution vector $\mathcal{f}$:

$$\mathcal{f} = \left\{ p_s(x_1 \ldots x_s; \Gamma_s) \right\} \tag{2.7}$$

Moreover in order to make closer the connection
with the physical idea of the patterns, we require the following normalization

$$\text{T-lim} \quad N^{-s} \int dx_1 \ldots dx_s \; p_s(x_1 \ldots x_s; [0_s]) = 1$$

$$\text{T-lim} \quad N^{-s} \int dx_1 \ldots dx_s \; p_s(x_1 \ldots x_s; [\Gamma_s]) = 0,$$

$$\Gamma_s \neq 0_s \tag{2.8}$$

This is clearly realized when $p_s \equiv \pi_s$, as can
be seen from eqs. (2.1) and (1.11).

We note that the matrix $\mathcal{L}^0$ is now diagonal
not only in the number of particles s, but <u>also</u>
in the correlation index Γ_s. In other words, <u>the</u>
<u>free-motion Liouvillian $\mathcal{L}^0$ cannot change the type</u>
<u>of correlation within a set of particles.</u>
On the other hand, the interaction Liouvillian can
cause a variety of transitions between various
types of correlation patterns. A few examples are
shown in fig. 5: A) an uncorrelated pattern
$p_3(1;2;3)$ goes over into an uncorrelated pattern
$p_2(1;2)$; B) An uncorrelated pattern $p_s(1;2;3)$
goes over into a correlated pattern $p_3(3;12)$

(creation of correlations);   C) a correlated
pattern $p_3(2;13)$ goes over into the uncorrelated
pattern $p_2(1;2)$ (destruction of correlations).

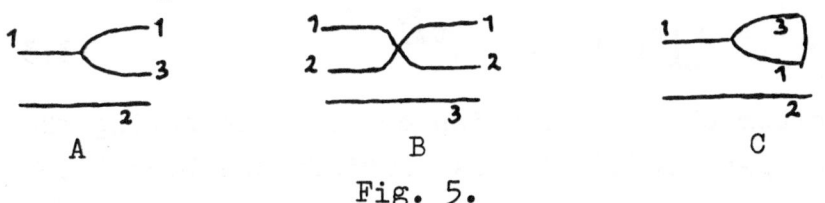

A   B   C

Fig. 5.

We now have a quite fascinating picture of the
exact process of evolution, as a set of transitions among correlation patterns of various
types. Hence the name of <u>dynamics of correlations</u>
given to this picture of mechanics.

3.   Dynamics of ideal systems

The simplicity of a system of non-interacting
particles is such that the problem of its evolution in time can be solved exactly. Such a system
is therefore a natural starting point for the
general theory, although not very interesting in
itself. For an ideal system the generalized
Liouville equation is simply written as

$$\partial_t f(t) \; = \; \mathcal{L}^0 f(t) \qquad\qquad (3.1)$$

The formal solution of the initial value problem
for eq. (3.1) can be written in the form

$$f(t) \; = \; \mathcal{U}^0(t)\, f(0) \qquad\qquad (3.2)$$

This equation <u>defines</u> the operator $\mathcal{U}^\circ(t)$, called the unperturbed <u>propagator</u>. It could be represented formally as

$$\mathcal{U}^\circ(t) \quad = \quad e^{t\mathcal{L}^\circ} \tag{3.3}$$

Another characterization consists in noting that the operator $\mathcal{U}^\circ(t)$ must satisfy the Liouville equation

$$\partial_t \mathcal{U}^\circ(t) \quad = \quad \mathcal{L}^\circ \mathcal{U}^\circ(t) \tag{3.4}$$

with the initial condition:

$$\mathcal{U}^\circ(0) \quad = \quad I \tag{3.5}$$

where I is the identity operator.

The simplicity of the unperturbed case lies mainly in the fact that the Liouvillian $\mathcal{L}^\circ$ is <u>diagonal</u>. Hence the components of eq. (3.1) are:

$$\partial_t p_s(x_1..x_s;\Gamma_s;t) = (\sum_{j=1}^{s} L_j^\circ)p_s(x_1...x_s;\Gamma_s;t) \tag{3.6}$$

We stress again the characteristic property of the free-Liouvillian of being a <u>sum</u> of one-particle operators. Combining now eqs. (3.2) and (3.3), we obtain

$$p_s(x_1...x_s;\Gamma_s;t) = \exp\left\{ t \sum_{j=1}^{s} L_j^\circ \right\} p_s(x_1...x_s;\Gamma_s;0)$$

$$\equiv U_{1..s}^\circ (t)\, p_s(x_1...x_s;\Gamma_s;\, 0) \tag{3.7}$$

This is the (s, Γ_s) component of the abstract

equation (3.2), and we identify

$$\langle s \; \Gamma_s | \mathcal{U}^0(t) | s' \; \Gamma'_{s'} \rangle = \delta_{s,s'} \delta_{\Gamma_s, \; \Gamma'_{s'}} \; U^0_{1..s}(t).$$

$$(3.8)$$

Moreover, using eq. (1.7), we get the explicit solution

$$p_s(\underline{q}_1\underline{p}_1 \cdots \underline{q}_s\underline{p}_s; \Gamma_s; t)$$

$$= \exp\left\{-t \sum_j \underline{v}_j \cdot \underline{\nabla}_j\right\} p_s(\underline{q}_1\underline{p}_1 \cdots \underline{q}_s\underline{p}_s; \Gamma_s; t)$$

$$= p_s(\underline{q}_1 - \underline{v}_1 t, \; \underline{p}_1, \cdots, \underline{q}_s - \underline{v}_s t, \; \underline{p}_s; \Gamma_s; 0) \quad (3.9)$$

An alternative and often very useful way of expressing the solution of eq. (3.1) is in terms of the <u>resolvent operator</u>, which will presently be defined. This concept is introduced through the <u>Laplace transformation</u> of the time dependent functions. The Laplace transform $\tilde{f}(z)$ of a (sufficiently regular) function of time $f(t)$ is defined by

$$\tilde{f}(z) = \int_0^\infty dt \; e^{izt} f(t) \quad (3.10)$$

where z is a <u>complex</u> variable. The well-known inversion formula is:

$$f(t) = \frac{1}{2\pi} \int_C dz \; e^{-izt} \; \tilde{f}(z) \quad (3.11)$$

where C is a contour in the complex z-plane, parallel to the real axis, and lying above all the singularities of $\tilde{f}(z)$.

In particular, we define the unperturbed resolvent operator $\mathcal{R}^0(z)$ as the Laplace transform of the unperturbed propagator $\mathcal{U}^0(t)$:

$$\mathcal{U}^0(t) = (2\pi)^{-1} \int_C dz\, e^{-izt} \mathcal{R}^0(z) \qquad (3.12)$$

Hence the solution to the initial-value problem for (3.1) is

$$f(t) = (2\pi)^{-1} \int_C dz\, e^{-izt} \mathcal{R}^0(z)\, f(0) \quad (3.13)$$

Formally, we may represent the resolvent as follows

$$\mathcal{R}^0(z) = (-\mathcal{L}^0 - iz)^{-1} \qquad (3.14)$$

with matrix elements

$$\langle (s)\, \Gamma_s \mid \mathcal{R}^0(z) \mid (s')\, \Gamma'_{s'} \rangle$$

$$= \delta_{ss'}\, \delta_{\Gamma_s \Gamma'_{s'}} \left[-\sum_{j=1}^{s} L_j^0 - iz \right]^{-1} \quad (3.15)$$

The resolvent $\mathcal{R}^0(z)$ and the propagator $\mathcal{U}^0(t)$ thus provide two equivalent methods for solving the Liouville equation.

4. <u>Vacuum and correlations</u>

We note that, among all possible correlation patterns of s particles, one pattern is privileged. This is the pattern

$$p_s(x_1 \ldots x_s; [0_s]) \equiv p_s(x_1; x_2; \ldots ; x_s) .$$

This pattern corresponds to the maximum partition of the set s or, physically, to completely un-correlated particles. Its importance can be understood from its normalization property: it is normalized to <u>one</u>, whereas all other patterns are normalized to zero (see eq. 2.8). As $p_s([0_s])$ carries the whole normalization of f_s <u>it can never vanish</u>. On the contrary, correlations $p_s([\Gamma_s])$ with $\Gamma_s \neq 0_s$ may or may not be present in a given statistical state of the system.

We now collect all the correlation patterns $p_s([0_s])$, for $s = 0,1,2, \ldots,$ and arrange them into an ordered set, of the same form as the dis-tribution vector f . This set is clearly a sub-set of f . It will be called the <u>vacuum component</u> of the distribution vector (or simply the <u>vacuum</u>)[*] and will be denoted by Vf . The remaining corre-lation patterns form a complementary subset, called the <u>correlation component</u> of the distribution vector (or simply the <u>correlations</u>) and denoted by Cf . We thus write:

$$f = Vf + Cf \qquad\qquad (4.1)$$

---

[*] This terminology is inspired from quantum field theory. In the latter it denotes a state with-out particles; here it denotes a state without correlations.

where

$$V \mathcal{f} \;=\; \{ p_s(x_1 \ldots x_s; [0_s]) \} \qquad (4.2)$$

$$C \mathcal{f} \;=\; \{ p_s(x_1 \ldots x_s; [\Gamma_s]) : \Gamma_s \neq 0_s \} \qquad (4.3)$$

Note that the components of $C\mathcal{f}$ for $s = 0$ and $s = 1$ are identically null: there are no cor- relations of less than two particles.

We now note that the separation (4.1) can be performed by acting on the distribution vector with two <u>operators</u> V and C. The effect of these operators is to select a subset out of the set $\mathcal{f}$ according to the definitions (4.2-3). Let us study some properties of these operators.

We first note that these operators are <u>linear</u>: given two distribution vectors $\mathcal{f}^{(1)}$ and $\mathcal{f}^{(2)}$ we have:

$$V(a\mathcal{f}^{(1)} + b\,\mathcal{f}^{(2)}) \;=\; a\,V\mathcal{f}^{(1)} + b\,V\mathcal{f}^{(2)} \qquad (4.4)$$

where $a\,\mathcal{f}^{(1)} + b\,\mathcal{f}^{(2)}$ is the set whose components are*

$$a\mathcal{f}^{(1)} + b\mathcal{f}^{(2)} = \{ a\,p_s^{(1)}(x_1 \ldots x_s; [\Gamma_s])$$
$$+ b\,p_s^{(2)}(x_1 \ldots x_s; [\Gamma_s]) \}.$$

The linearity follows from the fact that the dynamical correlation patterns $p_s(x_1 \ldots x_s; [0_s])$

---

* We may note that such a linear combination has a physical meaning only if $a+b=1$. Otherwise the normalization properties of $a\mathcal{f}^{(1)} + b\,\mathcal{f}^{(2)}$ would be violated.

are defined as the solutions of a set of linear differential equations. Any linear combination of particular solutions of these equations is still a solution.

Eq. (4.1) implies that

$$V + C = I \qquad (4.5)$$

This relation simply means that the two subsets are complementary. We next note the properties

$$V^2 = V$$
$$C^2 = C$$
$$VC = 0$$
$$CV = 0 \qquad (4.6)$$

These equations simply express the fact that the two subsets do not overlap. If the components $\Gamma_s = 0_s$ are selected by a first application of V, a subsequent application of V will not change the result; a subsequent application of C however yields zero, because there are no correlated components in the subset $V\mathcal{f}$.

The properties (4.4-6) entitle us to call V and C (somewhat loosely) <u>projection operators</u> (or <u>projectors</u>) on the vacuum and on the correlations, respectively. We can write an explicit representation of the matrix elements of these operators:

$$\langle (s); [\Gamma_s] \mid V \mid (s'); [\Gamma'_{s'}] \rangle = \delta_{s's} \, \delta_{\Gamma'_{s'}, 0_s} \, \delta_{\Gamma_s, 0_s}$$
$$(4.7)$$

$$\langle (s); [\Gamma_s] \mid C \mid (s'); [\Gamma'_{s'}] \rangle = \delta_{s's} \delta_{\Gamma_s \Gamma'_{s'}} (1 - \delta_{\Gamma_s, 0_s})$$

$$(4.8)$$

It is easily verified that these definitions are precisely equivalent to the operational definitions (4.2-3).

One of the most important properties of the correlation patterns is their <u>invariance under the unperturbed motion</u>. Let us clearly explain what is meant by this sentence. An individual correlation pattern generally changes its shape under the unperturbed motion, as can be seen from eq. (3.9). But it does not change its character (specified by the partition index Γ_s). Starting from a given $p_s([\Gamma_s])$, the unperturbed motion cannot generate a $p_s([\Gamma'_s])$ with $\Gamma'_s \neq \Gamma_s$. Conversely, the evolution of $p_s([\Gamma_s])$ is unaffected by the other correlation patterns. All these properties are consequences of the diagonal character of $\mathcal{L}^\circ$.

The invariance property of individual correlation patterns applies a fortiori to the sets of correlation patterns $V\mathcal{f}$ and $C\mathcal{f}$. Indeed it is easily seen that the equations of evolution of the vectors $V\mathcal{f}$ and $C\mathcal{f}$ are:

$$\partial_t V\mathcal{f} = \mathcal{L}^\circ V\mathcal{f}$$
$$\partial_t C\mathcal{f} = \mathcal{L}^\circ C\mathcal{f}$$

$$(4.9)$$

These equations bring out in a clear way the

complete mutual independence of the components $V\!\!f$ and $C\!\!f$. Hence the separation (4.1), obtained by means of the projectors V and C, defines a <u>decomposition of the space of the dis-</u> <u>tribution vectors f into two subspaces which are</u> <u>invariant under the unperturbed motion.</u> Any element of the subspace $\{V\!\!f\}$ <u>remains</u> in $\{V\!\!f\}$ under the effect of the motion.

　　This property can also be expressed in a different, quite suggestive way. Consider an arbitrary matrix operator $\mathcal{P}$ acting on the distribution vectors f . By making use of (4.5) this operator can be uniquely decomposed in the following fashion.

$$\mathcal{P} = I\mathcal{P}I = (V+C)\mathcal{P}(V+C)$$

hence

$$\mathcal{P} = V\mathcal{P}V + V\mathcal{P}C + C\mathcal{P}V + C\mathcal{P}C \qquad (4.10)$$

We express this by saying that every operator has a vacuum-to-vacuum part, a correlation-to-vacuum part, a vacuum-to-correlation part and a correlation-to-correlation part.

　　In particular, the unperturbed Liouville equation (3.1) can be written as

$$\partial_t f(t) = (V\mathcal{L}^\circ V + V\mathcal{L}^\circ C + C\mathcal{L}^\circ V + C\mathcal{L}^\circ C)\, f(t).$$

Applying now the operators V or C to both sides and using eqs. (4.6) we obtain, respectively,

$$\partial_t \, v \int(t) \; = \; v \, \mathcal{L}^\circ v \int(t) + v \, \mathcal{L}^\circ c \int(t)$$

$$\partial_t \, c \int(t) \; = \; c \, \mathcal{L}^\circ v \int(t) + c \, \mathcal{L}^\circ c \int(t) \tag{4.11}$$

These equations must be identical to eqs. (4.9). It therefore follows that

$$v \, \mathcal{L}^\circ c \quad = \quad 0$$
$$c \, \mathcal{L}^\circ v \quad = \quad 0$$
$$v \, \mathcal{L}^\circ v \quad = \quad \mathcal{L}^\circ v$$
$$c \, \mathcal{L}^\circ c \quad = \quad \mathcal{L}^\circ c \tag{4.12}$$

These equations are equivalent to the following set:

$$\mathcal{L}^\circ v \quad = \quad v \, \mathcal{L}^\circ$$
$$\mathcal{L}^\circ c \quad = \quad c \, \mathcal{L}^\circ \tag{4.13}$$

as can easily be seen from eqs. (4.6). Hence, the operators $\mathcal{L}^\circ$ and V (and therefore also $\mathcal{L}^\circ$ and C) commute with each other. Eq. (4.13) provides the most general expression of the decoupling of the subspaces $\{v \int\}$ and $\{c \int\}$. Finally, let us mention the following trivial consequences of eq. (4.13)

$$\mathcal{U}^\circ(t) \, V \quad = \quad v \, \mathcal{U}^\circ(t)$$
$$\mathcal{U}^\circ(t) \, C \quad = \quad c \, \mathcal{U}^\circ(t) \tag{4.14}$$

and similarly

$$\mathcal{R}^\circ(z) \, V \quad = \quad v \mathcal{R}^\circ(z)$$
$$\mathcal{R}^\circ(z) \, C \quad = \quad c \mathcal{R}^\circ(z) \tag{4.15}$$

These equations hold for arbitrary values of the
parameters t and z.

The main significance of these results is
the following. The problem of the complete time
evolution of f(t) has been split into two independent problems, i.e. the study of V f(t) and
the study of C f(t). If it so happens that in a
given problem we are interested only in the value
of V f(t), we can completely forget about
C f(t) in its evaluation. The component V f(t)
obeys its own "subdynamics" (as does C f(t)).

The reduction of the dynamical problem is
trivial in the case of the unperturbed motion,
because the problem can in any case be solved
exactly. The very unexpected and highly non-
trivial result is that a similar reduction leading
to independent subdynamics exists even in a system
of interacting particles.

5.   Formal solution of the complete Liouville
equation

The solution of the complete Liouville equation (1.14), including interactions, can again be
expressed in terms of a propagator U(t), defined
by the relation

$$f(t) \quad = \quad U(t)\, f(0) \qquad\qquad (5.1)$$

Alternatively this propagator can be defined as
the solution of the equation

$$\partial_t \mathcal{U}(t) \quad = \quad (\mathcal{L}^0 + \mathcal{L}')\, \mathcal{U}(t) \qquad (5.2)$$

with the initial condition

$$\mathcal{U}(0) \quad = \quad I \qquad\qquad (5.3)$$

We might be tempted to adopt the formal solution

$$\mathcal{U}(t) \quad = \quad e^{t(\mathcal{L}^0 + \mathcal{L}')} \qquad\qquad (5.4)$$

However, this way of writing is not useful, as we do not know how to define the exponential of the operator $\mathcal{L}$. In order to obtain a useful form, we note that eq. (5.2), together with the initial condition (5.3), is equivalent to the following integral equation:

$$\mathcal{U}(t) \quad = \quad \mathcal{U}^0(t) + \int_0^t dt\, \mathcal{U}^0(t-\tau)\, \mathcal{L}'\, \mathcal{U}(t) \qquad\qquad (5.5)$$

This equation can be solved by successive iterations:

$$\mathcal{U}(t) \quad = \quad \mathcal{U}^0(t) + \int_0^t dt\, \mathcal{U}^0(t-\tau)\, \mathcal{L}'\, \mathcal{U}^0(t)$$

$$+ \int_0^t d\tau_1 \int_0^{\tau_1} d\tau_2\, \mathcal{U}^0(t-\tau_1)\, \mathcal{L}'\, \mathcal{U}^0(\tau_1 - \tau_2)$$

$$\mathcal{L}'\, \mathcal{U}^0(\tau_2) \quad \cdots \qquad (5.6)$$

A similar expansion can be written for the resolvent operator $\mathcal{R}(z)$. As the convolutions go over into ordinary products under Laplace transformation, we find

$$\mathcal{R}(z) \quad = \quad \sum_{n=0}^{\infty} \mathcal{R}^0(z) \left[\mathcal{L}'\, \mathcal{R}^0(z) \right]^n \qquad (5.7)$$

This expansion coincides with the perturbation expansion in terms of the interaction strength. Of course, in order to have a precise meaning, the series should be convergent. This property cannot be proven in general; we shall leave it open, treating (5.7) as a kind of "raw material" to be further transformed.

It will soon become clear that the vacuum and correlation projectors V and C must play an important role in the theory, but in the representations (5.5) and (5.7) the projectors do not appear at all. Hence, in order to start our program we first need a representation of the propagator $U(t)$ and of the resolvent $R(z)$ where the projectors V and C are explicitly displayed.

It is not difficult to introduce these operators into the theory. Indeed, combining eq. (5.7) with (4.5) we have[*]

$$R(z) = R^o(z) + R^o(z)\, L'\, (V+C)\, R(z) \qquad (5.8)$$

Let us eliminate the correlation term in the r.h.s. of this equation. To do so, we project both sides on the correlation subspace:

$$CR = CR^o + CR^o L'VR + CR^o L'CR \qquad (5.9)$$

---

[*] One should never forget, in the coming calculations, that $L'V \neq VL'$, $L'C \neq CL'$, but $R^oV = VR^o$, $R^oC = CR^o$.

This is a linear inhomogeneous equation for the component $c\mathcal{R}$ which can be formally solved by successive iterations:

$$c\mathcal{R} = \sum_{m=0}^{\infty} (c\mathcal{R}^{\circ} \mathcal{L}')^m \left[c\mathcal{R}^{\circ} + c\mathcal{R}^{\circ} \mathcal{L}' \, v\mathcal{R} \right]$$

Substituting this result into the r.h.s. of eq. (5.8) we obtain

$$\mathcal{R} = \mathcal{R}^{\circ} + \mathcal{R}^{\circ} \sum_{m=0}^{\infty} \mathcal{L}' (c\mathcal{R}^{\circ} \mathcal{L}')^m \, c\mathcal{R}^{\circ}$$

$$\text{(5.10)}$$

$$+ \mathcal{R}^{\circ} \sum_{m=0}^{\infty} \mathcal{L}' (c\mathcal{R}^{\circ} \mathcal{L}')^m \, v\mathcal{R}$$

We now introduce the following operator, which appears naturally in this equation:

$$\tilde{\mathcal{E}}(z) = \sum_{m=0}^{\infty} \mathcal{L}' \, (c\mathcal{R}^{\circ}(z) \mathcal{L}')^m \qquad \text{(5.11)}$$

This operator, which plays a fundamental role in the theory, will be called the <u>irreducible evolution operator</u>. It has the same general structure as the resolvent (5.7) with the important difference that the projector C prevents transitions through the vacuum in intermediate states. The operator $\tilde{\mathcal{E}}(z)$ can also be defined by the following equation:

$$\tilde{\mathcal{E}}(z) = \mathcal{L}' + \mathcal{L}' \, c\mathcal{R}^{\circ}(z) \, \tilde{\mathcal{E}}(z) \qquad \text{(5.12)}$$

The resolvent equation (5.10) can now be rewritten as

$$\mathcal{R}(z) = \mathcal{R}^o(z) + \mathcal{R}^o(z)\,\tilde{\mathcal{E}}(z)\;c\mathcal{R}^o(z)$$
$$+ \mathcal{R}^o(z)\,\tilde{\mathcal{E}}(z)\;v\mathcal{R}(z) \tag{5.13}$$

We transform this equation further, to bring it into a particularly useful form. Multiplying both sides by $(-iz)V$ we obtain

$$(-iz)v\mathcal{R} = (-iz)\mathcal{R}^o v - iz\mathcal{R}^o v\tilde{\mathcal{E}}c\mathcal{R}^o - iz\mathcal{R}^o v\tilde{\mathcal{E}}v\mathcal{R} \tag{5.14}$$

As a consequence of the definition (3.16) of the unperturbed resolvent, we have the following very useful identity:

$$-iz\mathcal{R}^o(z) = 1 + \mathcal{L}^o\mathcal{R}^o(z) \tag{5.15}$$

After some simple algebra and making use of this identity we find:

$$v\mathcal{R}(z) = (-iz)^{-1}\left[v + v\tilde{\mathcal{E}}(z)c\mathcal{R}^o(z)\right] \tag{5.16}$$
$$+ (-iz)^{-1}\left[v\tilde{\mathcal{E}}(z)v + \mathcal{L}^o v\right]v\mathcal{R}(z)$$

The iterative solution of this equation is:

$$v\mathcal{R} = \sum_{n=0}^{\infty}(-iz)^{-n-1}\left[v\tilde{\mathcal{E}}v + \mathcal{L}^o v\right]^n(v + v\tilde{\mathcal{E}}\mathcal{R}^o c) \tag{5.17}$$

If we now combine eq. (5.17) with (5.13), we can write the complete resolvent in the form:

$$\mathcal{R}(z) = \sum_{n=0}^{\infty}(-iz)^{-n-1}(v + c\mathcal{R}^o\tilde{\mathcal{E}}v)(v\tilde{\mathcal{E}}v + \mathcal{L}^o v)^n$$
$$\cdot(v + v\tilde{\mathcal{E}}\mathcal{R}^o c) + c\mathcal{R}^o + c\mathcal{R}^o\tilde{\mathcal{E}}\mathcal{R}^o c \;. \tag{5.18}$$

The main result to be retained from these

somewhat tedious calculations is the following.
Suppose we decompose the resolvent operator in
the form of eq. (4.10)

$$\mathcal{R} = V\mathcal{R}V + V\mathcal{R}C + C\mathcal{R}V + C\mathcal{R}C . \qquad (5.19)$$

It follows from eq. (5.18) that <u>the four components
are not independent: they can all be expressed in
terms of the single component</u> $V\mathcal{R}(z)V$ as

$$V\mathcal{R}(z)C \; = \; V\mathcal{R}(z)V \cdot V\tilde{\mathcal{E}}(z)\mathcal{R}^{\circ}(z) \; C$$

$$C\mathcal{R}(z)V \; = \; C\mathcal{R}^{\circ}(z)\tilde{\mathcal{E}}(z)V \cdot V\mathcal{R}(z)V$$

$$C\mathcal{R}(z)C \; = \; C\mathcal{R}^{\circ}(z)\tilde{\tilde{\mathcal{E}}}(z)V \cdot V\mathcal{R}(z)V$$

$$\cdot V\tilde{\mathcal{E}}(z)\mathcal{R}^{\circ}(z)C + C\mathcal{R}^{\circ}(z)$$

$$+ C\mathcal{R}^{\circ}(z)\tilde{\mathcal{E}}(z)\mathcal{R}^{\circ}(z)C \quad (5.20)$$

As for the V - V component, it is represented as:

$$V\mathcal{R}(z)V \; = \; \sum_{n=0}^{\infty} (-iz)^{-n-1} \left[V\mathcal{L}^{\circ} + V\tilde{\mathcal{E}}(z)V \right]^{n} \qquad (5.21)$$

This very peculiar structure of the resolvent
will play a considerable role in the theory.

6.   <u>Program for a kinetic theory of
          irreversibility</u>

If we look back at the equations of evolution
considered in kinetic theory (Boltzmann, Fokker-
Planck, Landau), we discover a certain number of
common features.

(a)   They are <u>closed</u> equations for the one-particle distribution function.

(b)   They lead irreversibly to a well-defined thermal equilibrium state.

The property (a') can be reformulated in a more general way in terms of the correlation patterns:

(a)   They are closed equations for the vacuum component of the distribution vector.

Indeed, if the factorized form is chosen for the vacuum patterns, a closed equation for the vacuum component reduces to a (non-linear) closed equation for the one-particle distribution function.

Now, we know from the equations of the dynamics of correlations that property (a) cannot be true for the distribution vector describing a system of interacting particles. We know from section 2 that the Liouville operator $\mathcal{L}'$ has matrix elements connecting the vacuum components $p_s([0_s])$ to the correlated components $p_{s'}([\Gamma_s])$. This can be seen in a compact form when the Liouville equation is projected on the vacuum:

$$\partial_t \, V \mathcal{f} \;=\; V(\mathcal{L}^\circ + \mathcal{L}')\mathcal{f}$$

or

$$\partial_t \, V \mathcal{f} \;=\; \mathcal{L}^\circ V\mathcal{f} + V\mathcal{L}'V\mathcal{f} + V\mathcal{L}'C\mathcal{f} \quad (6.1)$$

Hence, the equation for the vacuum component $V\mathcal{f}$ involves the correlations $C\mathcal{f}$, and vice versa. We have localized here the crux of the difficulty: as long as this apparent contradiction is not

understood the problem of irreversibility remains
open. It is clear that this question is directly
related to Boltzmann's "Stosszahlansatz".

As the Stosszahlansatz cannot be true in
exact dynamics, the next simplest guess we can
make is the following: we may assume that the
distribution vector can be decomposed into two
terms:

$$f(t) \quad = \quad \bar{f}(t) \; + \; \hat{f}(t) \qquad\qquad (6.2)$$

the splitting being such that the term $\bar{f}(t)$ has
the characteristics of a kinetic theory. In particular, its vacuum part should obey a closed
kinetic equation, describing irreversible approach
to equilibrium, while the remainder $\hat{f}(t)$ should
be unimportant, at least for the problems studied
in kinetic theory.

It must be realized however that, without
further specification, the separation (6.2) into
a kinetic and non-kinetic part is trivial and does
not prove anything. It is always possible to
write a number as the sum of a prescribed number
plus the difference between the original and the
prescribed numbers. In order to make eq. (6.2)
the basis of a true theory, we must require it to
reflect an <u>intrinsic and self-consistent structure</u>
which is forced upon us (rather than the desired
result being forced by us into the theory).

As a first property we may ask that the
separation (6.2) have a <u>geometrical meaning</u>.
Considering the space of all possible distribution
vectors $\{f\}$, it may be possible to find two time-

independent operators, say Π and $\hat{\Pi}$, which
separate the space into two complementary subspaces, one containing all kinetic parts, the
other containing all non-kinetic parts. These
operators, applied to an arbitrary element of the
space, $f(t)$, would automatically perform the
separation (6.2):

$$\Pi f(t) \quad = \quad \bar{f}(t)$$
$$\hat{\Pi} f(t) \quad = \quad \hat{f}(t) \qquad\qquad (6.3)$$

The two terms would then appear as true components
of the vector $f(t)$. For self-consistency, the
operators must have the properties of "projection
operators":

$$\Pi^2 = \Pi$$
$$\hat{\Pi}^2 = \hat{\Pi}$$
$$\Pi\hat{\Pi} = \hat{\Pi}\Pi = 0 \qquad\qquad (6.4)$$

together with the completeness relation:

$$\Pi + \hat{\Pi} = \quad I \qquad\qquad (6.5)$$

These properties ensure that the two subspaces
$\{\bar{f}\}$ and $\{\hat{f}\}$ are complementary and not overlapping.

   But the most important property, which would
really convince us of the deep nature of the theory,
should be an <u>invariance property</u>. Here is what we
mean (see fig. 6). Consider a system described
at time zero by a distribution vector $f(0)$. We
split it into two components according to (6.2)

and (6.3):

$$f(0) = \bar{f}(0) + \hat{f}(0)$$
$$\bar{f}(0) = \Pi f(0)$$
$$\hat{f}(0) = \hat{\Pi} f(0)$$

and let the system evolve. At time t its distribution vector is obtained by the action of the propagator $\mathcal{U}(t)$ on $f(0)$:

$$f(t) = \mathcal{U}(t) f(0)$$

The vector $f(t)$ is then decomposed according to (6.2) with the result:

$$f(t) = \bar{f}(t) + \hat{f}(t)$$
$$\bar{f}(t) = \Pi f(t) = \Pi \mathcal{U}(t) f(0)$$
$$\hat{f}(t) = \hat{\Pi} f(t) = \hat{\Pi} \mathcal{U}(t) f(0) \qquad (6.6)$$

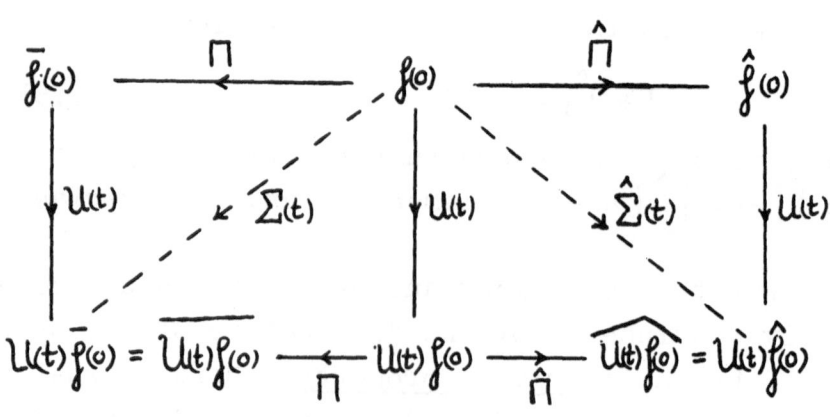

Fig. 6.

On the other hand, we may wonder what happens
to the separate components $\bar{f}(0)$ and $\hat{f}(0)$ when
they are taken as an initial condition and when
their evolution is followed in time. Clearly:

$$\bar{f}(0) \longrightarrow U(t)\,\bar{f}(0) \;=\; U(t)\,\Pi f(0)$$
$$\hat{f}(0) \longrightarrow U(t)\,\hat{f}(0) \;=\; U(t)\,\hat{\Pi} f(0) \qquad (6.7)$$

At this point it should be clear that a theory
can only be self-consistent if, at time t, the
Π-component of $f(t)$ coincides with the result
of the evolution of the initial Π-component; in
other words, if the right hand sides of eqs. (6.6)
and (6.7) are identical. If this were not the case,
the separation would depend on the time at which
it is performed; there would be a privileged instant of time. But such an instant could not be
singled out by any special physical property: hence
such a theory would be physically untenable. Com-
paring eqs. (6.6) and (6.7) we see that the condition of <u>invariance of the separation (6.2) under
time translation</u> is expressed in the following
simple form

$$\Pi\,U(t) \;=\; U(t)\,\Pi$$
$$\hat{\Pi}\,U(t) \;=\; U(t)\,\hat{\Pi} \qquad (6.8)$$

<u>The operators Π and $\hat{\Pi}$ must commute with the
propagator $U(t)$ for all values of t</u>:

As a direct consequence of this commutation
relation, the components $\bar{f}(t)$ and $\hat{f}(t)$ obey
<u>separate</u> equations of evolution: there is no
mixing between the $\{\bar{f}\}$ and the $\{\hat{f}\}$ subspaces.

Indeed, using eqs. (5.2) and (6.8):

$$\partial_t \bar{f}(t) = \partial_t \Pi \mathcal{U}(t) f(0) = \partial_t \mathcal{U}(t) \Pi f(0)$$
$$= \mathcal{L} \mathcal{U}(t) \Pi f(0)$$

or

$$\partial_t \bar{f}(t) = \mathcal{L} \bar{f}(t) \quad . \tag{6.9}$$

Similarly

$$\partial_t \hat{f}(t) = \mathcal{L} \hat{f}(t) \quad . \tag{6.10}$$

Hence the components $\bar{f}$ and $\hat{f}$ evolve independently in time. We will say that each component obeys its own <u>subdynamics</u>.

We now note that, in the particular case of a <u>non-interacting system</u>, we have already found a pair of operators having precisely all the properties listed above: they are the operators V and C. Indeed it was shown in section 4 that the vacuum and correlation parts of f are disconnected, that V and C are projection operators, and that they commute with the unperturbed propagator. Hence, it is reasonable to expect that, if an operator Π having all the required properties can be constructed, it should reduce to the operator V when the interactions are switched off:

$$\Pi \xrightarrow[\text{interactions}]{\text{no}} V$$
$$\hat{\Pi} \xrightarrow{\hspace{2cm}} C \tag{6.11}$$

Summarizing this discussion, we may start
on the following programme. Given a system of
interacting degrees of freedom, to construct an
operator Π having the following properties:
(It is clearly sufficient to construct Π : the
operator $\hat{\Pi}$ is then simply $I - \Pi$).

(A) Π commutes with $U(t)$.
(B) Π reduces to V in the limit of no
         interactions.
(C) Π is idempotent (i.e. Π is a projector).
(D)   The vacuum part of $\Pi \rho(t)$ obeys a closed
         equation of evolution.
(E)   The stationary solution of this evolution
         equation coincides with the equilibrium
         distribution vector.
(F)   The complementary component $\hat{\Pi} \rho (t)$ is
         irrelevant in certain well-defined
         problems.

This programme will be developed below but
we may already announce the great surprise coming
out of this investigation. We will see that
condition (A), combined with (B), is so strong
that it completely determines a unique operator
Π [*]). Hence, we are given no freedom: either
this unique operator does possess the remaining

---

[*] More precisely: The given operator V is
uniquely continued into an operator Π . This
remark is relevant, because there are many
possible realizations of the operator V, obeying its defining properties. This is an important element of flexibility of the theory.

properties (C) to (F), or it does not. This unexpected feature completely changes our strategy:
we can no longer make a choice among a number of
items "on the market" and take the one that
pleases us most. On the contrary, the requirement of invariance offers us a single possibility:
it is now up to us to <u>prove</u> that this possibility
has anything to do with kinetic theory.

We cannot insist enough on the fact that the
requirements (A) and (B) have nothing to do with
irreversibility. Hence, if we do succeed (and we
shall!) in proving that the unique operator Π
defined by conditions (A) and (B) possesses all
the properties (C) - (F), we will be in the
presence of a truly objective theory of irreversibility. The result can be reformulated as
follows: <u>The only invariant decomposition of the</u>
<u>distribution vector with respect to time transla-</u>
<u>tion, is one in which the component</u> $\tilde{\rho}(t)$ <u>has</u>
<u>a time evolution of kinetic type.</u>

Let us conclude this section with the following remark. It would be completely unreasonable
if we could construct an operator Π for <u>all</u>
possible systems of interacting particles. A
system of three interacting particles does not
approach equilibrium; a system of particles
interacting through unscreened long-range forces
(e.g. gravitational forces) most probably does not
reach equilibrium either. There must be some restriction on the validity of our scheme. Sure
enough, we will find that a non-trivial operator

Π exists only if certain additional conditions
are met by the system, conditions which involve
the thermodynamic limit on one hand and the
nature of the interactions on the other.

7.   Construction of the operator Π

We want to construct our operator Π acting
on the set of distribution vectors f, and satisfying conditions (A) and (B) of our programme,
i.e. eqs. (6.8) and (6.11). The former implies:

$$\Pi R(z) \quad = \quad R(z)\Pi \tag{7.1}$$

for all values of z within the domain of regularity of $R(z)$. In order to get a first hint
towards the solution, we use the representation
(5.18) of the resolvent.

The disadvantage of eq. (7.1) is that it
provides a relation between $R(z)$ and the operator
Π which, by definition, is independent of z.
It would be helpful for the solution to derive
a relation involving only operators independent
of z and the particular structure of the resolvent is a useful guide for this operation. It
is clear that the value z = 0 plays a privileged
role: all terms in our equation (except one) have
an explicitly exhibited singularity (i.e. a multiple pole) in z = 0. Of course, we do not know,
a priori, the nature of the other z-dependent
operators appearing in the equation, and in particular the location of their singularities. We

are therefore forced to make additional assumptions at this point. These assumptions may seem
arbitrary at first sight; let us calm the reader's
apprehensions. What appears here as a rigid
postulate is actually the crystallization of long
experience of many people with these operators.
We will show later that there exist non-trivial
systems which do satisfy them.

Coming back to our problem, the simplest
case is realized when, in each term of eq. (5.18),
the only singularity in $z = 0$ is the multiple
pole $(-iz)^{-n-1}$. This is so if the following conditions are satisfied.

$$\left.\begin{array}{l} v\,\tilde{\mathcal{E}}(z)v \\[4pt] v\,\tilde{\mathcal{E}}(z)\,\mathcal{R}^{o}(z)c \\[4pt] c\,\mathcal{R}^{o}(z)\,\tilde{\mathcal{E}}(z)v \\[4pt] c\,\mathcal{R}^{o}(z) \\[4pt] c\,\mathcal{R}^{o}(z)\,\tilde{\mathcal{E}}(z)\,\mathcal{R}^{o}(z)c \end{array}\right\}$$
are regular functions
of z in the neigh-
bourhood of $z = 0$
and do not vanish in
$z = 0$.

$$(7.2)$$

These <u>auxiliary conditions define</u> a class of
dynamical systems and these are the only ones to
be considered here.

We may now use these conditions as follows.
We integrate all the terms of eq. (7.2) (combined with (5.18)) over z along a small circle
centred at the origin.

We then use the Cauchy residue theorem in
the form:

$$\frac{1}{2\pi} \oint dz \; \frac{1}{(-iz)^{n+1}} f(z) = \lim_{z \to 0} \left\{ \frac{1}{n!} \partial^n f(z) \right\} \quad (7.3)$$

where we use the abbreviation:

$$\partial \equiv i \frac{\partial}{\partial z} \qquad\qquad (7.4)$$

We write $\lim_{z \to 0} \left\{ \partial^n f(z) \right\}$ rather than $\left\{ \partial^n f(z) \right\}_{z=0}$
for the following technical reason.. The function
$f(z)$ may have singularities other than poles,
particularly branch-cuts in the upper half-plane
or on the real axis. If this happens, the value
in $z = 0$ may be ambiguous and must be specified
more precisely. The ambiguity disappears by
realizing that the resolvent, being a Laplace
transform, is regular in a half-plane lying above
the contour C of eq. (3.12). As we now let z
approach the real axis, we must be careful, if we
cross any cut, to choose the particular branch
which is the analytic continuation of the function $f(z)$ originally defined in its domain of
regularity. To sum up, $\lim_{z \to 0}$ must be understood
as: "limit for $z \to 0$ from above." We do not want
to insist too much on these technical points, although they may become important in some problems.
     With these operations, eq. (7.2) becomes

$$\Pi \cdot \left\{ \lim_{z \to 0} \sum_{n=0}^{\infty} (n!)^{-1} \, \partial^n \left[v + c \mathcal{R}^\circ \tilde{\mathcal{E}} v \right] \times \right.$$

$$\left. \times \left[v \mathcal{L}^\circ + v \tilde{\mathcal{E}} v \right]^n \left[v + v \tilde{\mathcal{E}} \mathcal{R}^\circ c \right] \right\}$$

$$= \left\{ \lim_{z \to 0} \sum_{n=0}^{\infty} (n!)^{-1} \, \partial^n \left[v + c \mathcal{R}^\circ \tilde{\mathcal{E}} v \right] \times \right.$$

$$\left. \times \left[v \mathcal{L}^\circ + v \tilde{\mathcal{E}} v \right]^n \left[v + v \tilde{\mathcal{E}} \mathcal{R}^\circ c \right] \right\} \cdot \Pi$$

$$(7.5)$$

We first discard the two trivial solutions:

$$\Pi = 0, \qquad \Pi = I$$

since neither of these satisfies the boundary condition (6.11). We areleft with the obvious solution:

$$\Pi = \lim_{z \to 0} \sum_{n=0}^{\infty} (n!)^{-1} \, \partial^n \left[v + c \mathcal{R}^\circ(z) \tilde{\mathcal{E}}(z) \; v \right]$$

$$\left[v \mathcal{L}^\circ + v \tilde{\mathcal{E}}(z) v \right]^n \left[v + v \tilde{\mathcal{E}}(z) \mathcal{R}^\circ(z) c \right]$$

$$(7.6)$$

It remains to be shown that this operator satisfies the boundary condition (6.11). In order to take the limit formally, we replace the interaction Liouvillian $\mathcal{L}'$ by $\lambda \mathcal{L}'$, where λ is a dimensionless scaling parameter, and then let λ go to zero. From eq. (5.11) we see first that

$$\lim_{\lambda \to 0} \tilde{\mathcal{E}}(z) = 0 \qquad (7.7)$$

Hence,

$$\lim_{\lambda \to 0} \Pi = \lim_{z \to 0} \sum_{n=0}^{\infty} (n!)^{-1} \partial^n V(\mathcal{L}^{\circ})^n V$$

$$= \lim_{z \to 0} (0!)^{-1} \partial^{\circ} V(\mathcal{L}^{\circ})^{\circ} V = V.$$

The second step follows from the fact that $\mathcal{L}^{\circ}$
is independent of z, so that only the term
$n = 0$ in the sum can give a non-vanishing contribution.

The construction described above does not
constitute a proof that Π satisfies our initial
requirement (6.8). It simply suggests that the
operator defined by eq. (7.6) is a possible candidate for Π. It must now be shown separately
that this operator does commute with $\mathcal{U}(t)$ for
all values of t. We omit here the details of
the proof (see ref. 3). The idea is to show
first that the product $\Pi \mathcal{U}(t)$ can be written
as an operator $\Sigma(t)$:

$$\Pi \, \mathcal{U}(t) \quad = \quad \Sigma(t) \tag{7.8}$$

with

$$\Sigma(t) = \lim_{z \to 0} \sum_{n=0}^{\infty} (n!)^{-1}(t+\partial)^n \left[V + c \mathcal{R}^{\circ}(z) \, \tilde{\mathcal{E}}(z) V \right]$$

$$\left[V \mathcal{L}^{\circ} + V \tilde{\mathcal{E}}(z) V \right]^n \left[V + V \tilde{\mathcal{E}}(z) \mathcal{R}^{\circ}(z) c \right] \tag{7.9}$$

In a second step we show that we also have

$$\mathcal{U}(t) \, \Pi \quad = \quad \Sigma(t) \qquad\qquad (7.10)$$

Eqs. (7.8, 10) constitute a proof of eq. (6.8).

Finally, the following uniqueness theorem can be proved. Besides the operator defined by eq. (7.6), there is no other operator, constructed by means of $\mathcal{L}^o$, $\lambda\mathcal{L}'$, V and analytic in λ, commuting with $\mathcal{U}(t)$ and reducing to V as $\lambda \rightarrow 0$. The condition for the validity of this theorem is that $\mathcal{L}^o$ does not commute with $\mathcal{L}'$. Hence, as we stated in section 6, every realization of the projector V is continued into an operator Π and when the latter exists, it is unique.

Before closing this section, we note that the operator $\Sigma(t)$ defined by eqs. (7.8-10) provides an alternative definition of the component $\bar{f}(t)$:

$$\bar{f}(t) \;=\; \Pi f(t) \;=\; \Pi \mathcal{U}(t) f(0) = \Sigma(t) \, f(0).$$

If we also introduce the operator

$$\hat{\Sigma}(t) \;=\; \mathcal{U}(t) - \Sigma(t)$$

we obtain the following important formulae:

$$
\begin{aligned}
\bar{f}(t) &= \Sigma(t) \, f(0) \\
\hat{f}(t) &= \hat{\Sigma}(t) \, f(0)
\end{aligned}
\qquad (7.11)
$$

Whereas the definition (6.3) is of a geometrical type (because it provides an operator which splits every element $f(t)$ into two

components), the definition (7.11) is of a
dynamical type. Indeed, it provides two operators
which, applied to the initial condition $f(0)$
give directly the two separate components at time
t (see fig. 6). They play the same role as the
propagator $U(t)$; actually they provide an invariant decomposition of this operator:

$$U(t) = \sum(t) + \hat{\sum}(t) \qquad (7.12)$$

We finally note the obvious relations

$$\sum(0) = \Pi , \qquad \hat{\sum}(0) = \hat{\Pi} . \qquad (7.13)$$

To complete the point C of our programme,
we have now to prove that Π is idempotent, i.e.
the first eq. (6.4). The proof is far from
trivial and will be found in ref. 4. Suffice it
to say that every detail of the expression (7.6)
plays an important role in this proof.

We have now completed the part of the programme which is sufficient for demonstrating the
existence of separate subdynamics for the two
components $\bar{f}(t)$ and $\hat{f}(t)$ of the distribution
vector. The very existence of such a clean separation in a large system of interacting particles
is a quite unexpected feature of the time-
evolution process. It now remains to be seen how
these properties are connected with kinetic
theory.

8. Subdynamics and kinetic theory

Until now the connection between ordinary kinetic theory and the kinetic component $\mathcal{f}(t)$ of the distribution vector is not yet clear. We shall now study this point.

First we need a few definitions. The inverse Laplace transform of the irreducible evolution operator $\tilde{\mathcal{E}}(z)$ will be called $\mathcal{E}(t)$:

$$\mathcal{E}(t) = (2\pi)^{-1} \int_C dz\, e^{-izt}\, \tilde{\mathcal{E}}(z) \qquad (8.1)$$

It should be noted that the zeroth order term in $\tilde{\mathcal{E}}(z)$ is independent of z (see eq. (5.11); its Laplace transform is therefore $\mathcal{L}'\, \delta(t)$. It is sometimes convenient to separate this special term from $\mathcal{E}(t)$ and write

$$\mathcal{E}(t) = \mathcal{L}'\, \delta(t) + \mathcal{G}(t) \qquad (8.2)$$

where the operator $\mathcal{G}(t)$ is (usually) non-singular as a function of time.

We will also denote by a special symbol the following frequently occurring operator.

$$c\,\mathcal{C}(t)V = (2\pi)^{-1} \int_C dz\, e^{-izt}\, c\,\mathcal{R}^o(z)\,\tilde{\mathcal{E}}(z)V \qquad (8.3)$$

$c\,\mathcal{C}(t)V$ will be called the creation operator because its action may be understood as "creating correlations" out of the vacuum.

The following important relation is easily derived from eqs. (5.11) and (8.2):

$$V \mathcal{G}(t)V = V \mathcal{L}' c \mathcal{C}(t)V \qquad (8.4)$$

We now come back to the study of the kinematic component $\overline{\mathcal{g}}(t)$. This component, being an element of the set of distribution functions, can always be decomposed into a vacuum and a correlation part according to eq. (4.1):

$$
\begin{aligned}
\overline{\mathcal{g}}(t) &= V \overline{\mathcal{g}}(t) + c \overline{\mathcal{g}}(t) \\
&= V \Pi \mathcal{g}(t) + c \Pi \mathcal{g}(t)
\end{aligned}
\qquad (8.5)
$$

We must now clearly understand the following crucial point. If an <u>arbitrary</u> distribution vector is decomposed into its vacuum and correlation components, the latter are, of course, completely independent of each other. But the vector $\overline{\mathcal{g}}(t)$ in eq. (8.5) is <u>not</u> arbitrary: rather, it is <u>the component in a well-defined subspace</u> of an arbitrary vector. One may therefore expect its vacuum and correlation components to be interrelated in a well-defined way.

Let us illustrate this statement by means of a simple geometrical analogy in ordinary two-dimensional cartesian space (see fig. 7). Consider a pair of orthogonal reference axes, labelled OV and OC and another pair of orthogonal axes, with the same origin, labelled OΠ and O$\hat{\Pi}$; the angle between the axes OΠ and OV we call θ.

Let a be a vector originating at O and

making an angle α with OV. It has components $\bar{a}$ and $\hat{a}$ along the axes O$\sqcap$ and O$\hat{\sqcap}$, respectively

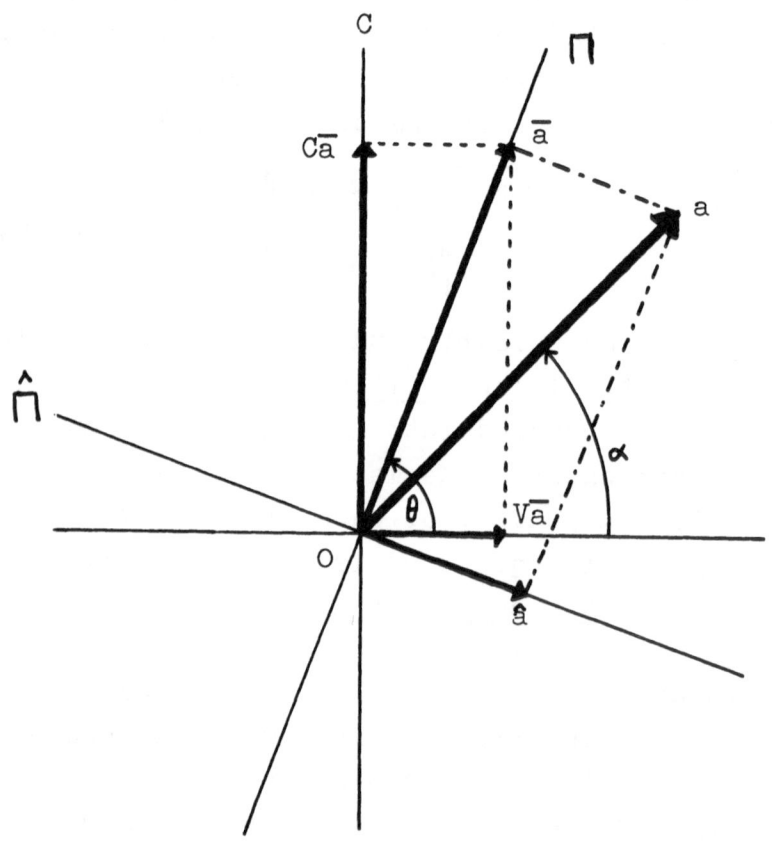

Fig. 7.

(this corresponds to the decomposition (8.5)). Its component $\bar{a}$ may be further resolved into components $\overline{Va}$ and $\overline{Ca}$ along OV and OC.

Clearly:

$$C\overline{a} \quad = \quad \tan \Theta \; (V\overline{a}) \quad .$$

This relation shows that the ratio $C\overline{a} \,/\, V\overline{a}$ is independent of the orientation α of the original vector a ; it only depends on the fixed relative orientation of the reference axes $O\Pi$ and OV. Thus, as we said above, the correlation component of the Π-component of a bears a fixed relationship to the vacuum component of the Π-component of a, whatever the vector a. On the contrary, the C- and V-components of the original vector a are interrelated by

$$Ca \quad = \quad \tan \alpha \, .(Va) \quad .$$

Their ratio is different for every different orientation α : this is what we mean by stating that they are mutually independent.

Coming back now to our original problem, it can be shown that the correlation part of the kinetic distribution vector $\overline{f}(t)$ is related to its vacuum part by the following functional relationship:

$$C\overline{f}(t) \quad = \quad \int_{0}^{\infty} d\tau \; C\, G(\tau) \; V\overline{f}(t-\tau) \tag{8.6}$$

We shall not give the detailed proof of this property here (see ref. 3). Suffice it to say that it is a consequence of the characteristic symmetry of the resolvent pointed out in (5.20).

The determination of the vector $\bar{f}(t)$ is now reduced to the computation of the single component $V\bar{f}(t)$. The latter is, in principle, completely defined by eqs. (7.11) and (7.9) but this defini- tion, in terms of an infinite power series in t is not convenient for practical purposes, because the convergence of this series is very slow, especially for large values of t. We now prove that the relevant component can be characterized in a different way, as the solution of an integro-differential equation.

From eq. (6.9) we know already that the kinetic component $\bar{f}(t)$ obeys the Liouville equation. Its vacuum component therefore obeys:

$$\partial_t V\,\bar{f}(t) \;=\; \mathcal{L}^0 V\,\bar{f}(t) + V\,\mathcal{L}'V\,\bar{f}(t) + V\,\mathcal{L}'C\,\bar{f}(t)$$
$$(8.7)$$

There is however a serious difference between eq. (8.7) and the apparently similar Liouville equation (6.1), where the components $V\,f(t)$ and $C\,f(t)$ are mutually independent: eq. (6.1) must be completed with an equation for $\partial_t C\,f(t)$ which is coupled to $V\,f(t)$. On the contrary, the correlation component of $\bar{f}(t)$ is an explicit, fixed functional of the vacuum component $V\,\bar{f}(t)$. Eq. (8.6) must be added to eq. (8.7) as a constraint to be satisfied at all times.

The correlation component can now be eliminated from eq. (8.7) by means of the con- straint:

$$\partial_t v \bar{f}(t) = \mathcal{L}^o v \bar{f}(t) + v \mathcal{L}' v \bar{f}(t)$$
$$+ \int_0^\infty d\tau \ v \mathcal{L}' c \mathcal{G}(\tau) v \bar{f}(t-\tau)$$

which, by eq. (8.4) can be transformed into

$$\partial_t v \bar{f}(t) = v(\mathcal{L}^o + \mathcal{L}')v \bar{f}(t)$$
$$+ \int_0^\infty d\tau \ v \mathcal{G}(\tau)v \bar{f}(t-\tau) \qquad (8.8)$$

Hence the constraint (8.6) transforms the coupled
Liouville equations into <u>a single closed equation</u>
<u>for the vacuum component of the kinetic part of</u>
<u>the distribution vector.</u>   This achieves the point
D of our programme of section 6.   Eq. (8.8) will
be called the <u>General Kinetic Equation</u>.   This name
is justified by the fact that the known kinetic
equations (Boltzmann, Landau,....) turn out to
be particular cases of eq. (8.8).

   We now see that the kinetic equation entered
the general theory quite smoothly, without any
"cheating" or contradiction with mechanics.   At
no point did we use any probabilistic argument:
the kinetic equation appears as an <u>exact</u> equation
of evolution for the vacuum component of an invariant part of the distribution vector.   The only
restriction on its validity is eq. (7.2), a purely
mechanical condition.   Unfortunately, nothing
general can be said at present about the validity
of that equation: each particular case must be

examined per se.

It can further be shown that eq. (8.8) possesses a quite remarkable exponential solution:

$$V \bar{f}(t) = \left\{ \exp(t\, V\, \Gamma\, V) \right\} V \bar{f}(0) \qquad (8.9)$$

where $V\, \Gamma\, V$ is a time-independent operator, having only a V – V component.

It then follows that $V \bar{f}(t)$ obeys a simple equation:

$$\partial_t\, V \bar{f}(t) = V\, \Gamma\, V\, \bar{f}(t) \qquad (8.10)$$

This equation must, of course, be equivalent to eq. (8.8). However, it is no longer an integro-differential equation, but rather a purely differential equation (in time). All the memory effects expressed explicitly in eq. (8.8) are now included in the new operator $V\, \Gamma\, V$. The price to be paid is of course in the added complexity of this operator as compared to $V\, \mathcal{G}(\tau)V$. We shall omit here the derivation of this operator (see ref. 3) and give only its final form:

$$V\, \Gamma\, V = V(\mathcal{L}^0 + \mathcal{L}')V + \sum_{n=1}^{\infty} V\, \Gamma_n\, V \qquad (8.11)$$

$$V\, \Gamma_n V = \int_0^\infty d\tau_2 \int_0^\infty d\bar{\tau}_4 \cdots$$

$$\cdots \int_0^\infty d\tau_{2n} \int_0^{-\tau_2} d\tau_3 \int_0^{\tau_3 - \tau_4} d\tau_5 \cdots \int_0^{\tau_{2n-3} - \tau_{2n-2}} d\tau_{2n-1}$$

$$V \, \mathcal{G}(\tau_2) V \, \tilde{\mathcal{U}}(-\tau_2 - \tau_3) V \, \mathcal{G}(\tau_4) \quad V \, \tilde{\mathcal{U}}(\tau_3 - \tau_4 - \tau_5) V \, \mathcal{G}(\tau_6) V \ldots$$

$$\ldots \, V \, \tilde{\mathcal{U}}(\tau_{2n-3} - \tau_{2n-2} - \tau_{2n-1}) V \, \mathcal{G}(\tau_{2n})$$

$$V \, \tilde{\mathcal{U}}(\tau_{2n-1} - \tau_{2n}) V \qquad (8.12)$$

where

$$V \, \tilde{\mathcal{U}}(t) V \quad = \quad \exp\left[t \, V(\mathcal{L}^0 + \mathcal{L}') V \right] \qquad (8.13)$$

We may note that the operator $V \, \Gamma \, V$ simplifies in the case of spatially homogeneous systems. It can be shown that these systems are characterized by the absence of $V - V$ elements of the Liouvillian (see section 9):

$$V \, \mathcal{L}^0 \; = \; 0 \, , \qquad V \, \mathcal{L}' \, V \; = \; 0 \; .$$

Hence the propagator $V \, \tilde{\mathcal{U}}(t) V$ reduces to the constant V, and eq. (8.12) becomes

$$V \, \Gamma_n V \; = \; \int_0^\infty d\tau_2 \int_0^\infty d\tau_4 \ldots \int_0^\infty d\tau_{2n} \int_0^{-\tau_2} d\tau_3 \int_0^{\tau_3 - \tau_4} d\tau_5 \ldots$$

$$\ldots \int_0^{\tau_{2n-3} - \tau_{2n-2}} d\tau_{2n-1}$$

$$V \, \mathcal{G}(\tau_2) \, V \, \mathcal{G}(\tau_4) V \ldots V \, \mathcal{G}(\tau_{2n}) V \qquad . \qquad (8.14)$$

The integrations over the odd variables can be done explicitly, yielding a polynomial in the even τ's. The final result can be expressed, if so desired, in terms of the Laplace transform

$\tilde{\psi}(z)$ of the operator $V \mathcal{G}(\tau)V$. The result is

$$V \, \Gamma \, V \quad = \quad \Omega \, \tilde{\psi}(0) \tag{8.15}$$

where the operator Ω is a complicated function of $\tilde{\psi}(0)$ and its derivatives with respect to $(-iz)$:

$$\Omega = 1 + \tilde{\psi}'(0) + \frac{1}{2}\tilde{\psi}''(0)\,\tilde{\psi}(0) + \tilde{\psi}'(0)\,\tilde{\psi}'(0) + \ldots \tag{8.16}$$

The operators Ω and $\tilde{\psi}(0)$ have a simple physical interpretation. Consider the kinetic equation (8.8) (for homogeneous systems) and suppose we neglected the retardation in the distribution vector under the integral. We would then obtain

$$\partial_t \, V \bar{f}(t) \approx \int_0^{\infty} d\tau \, V \mathcal{G}(\tau)V \, \bar{f}(t)$$

i.e.

$$\partial_t \, V \bar{f}(t) \approx \tilde{\psi}(0) \, V \bar{f}(t) \tag{8.17}$$

The neglect of the retardation means that the collision described by $\tilde{\psi}(0)$ is considered as an instantaneous process. We know however that (8.17) is an incorrect equation. Hence we see that the operator Ω in equation (8.15) describes the correction to eq. (8.17) accounting for the finite duration of the collision process.

Finally, we note that the relation between correlation and vacuum components of the kinetic distribution vector can also be transformed into a time-independent functional relation.

We substitute (8.9) in (8.6) to obtain

$$C \bar{f}(t) = \int_0^\infty d\tau \, C \, \mathcal{G}(\tau) V \{ \exp(t-\tau) V \, \Gamma \, V \} V \, \bar{f}(0)$$

and hence

$$C \bar{f}(t) = C \, \mathcal{C} \, V \, \bar{f}(t) \qquad\qquad (8.18)$$

with

$$C \, \mathcal{C} \, V = \int_0^\infty d\tau \, C \, \mathcal{G}(\tau) V \, \exp(-\tau V \Gamma V) \, . \quad (8.19)$$

Eq. (8.18) relates the correlation components $C \sum (t)$ to the vacuum components $V \sum (t)$ evaluated at the $\underline{\text{same time}}$, through the action of the $\underline{\text{time-independent}}$ operator C V.

If so desired, this operator can be expanded in the same fashion as $V \Gamma V$:

$$C \, \mathcal{C} \, V = \sum_{n=1}^{\infty} C \, \mathcal{C}_n \, V$$

the coefficient $C \, \mathcal{C}_n \, V$ being obtained by replacing the first factor $V \, \mathcal{y}(\tau_2) V$ in eq. (8.12) by $C \, \mathcal{G}(\tau_2) V$.

To sum up, we may now characterize the kinetic component of the distribution vector by the set of two equations

$$C \bar{f}(t) = C \, \mathcal{C} \, V \, \bar{f}(t)$$

$$\partial_t V \, \bar{f}(t) = V \, \Gamma \, V \, \bar{f}(t) \qquad\qquad (8.20)$$

When these equations are translated into the concrete correlation patterns language, it is easily shown that the set of kinetic equations for the

vacuum patterns $\overline{p}_s(1;2;\ ..;s)$ are compatible in
the T-limit with the factorization:

$$\overline{p}_s(x_1;x_2;\ldots;x_s) = \prod_{j=1}^{s} \overline{f}_1(x_j) \qquad (8.21)$$

Hence dynamics in the Π - subspace is completely
determined by the knowledge of a function $\overline{f}_1(x;t)$
depending on the coordinates of a single particle
and obeying a (non-linear) kinetic equation. This
is a formidable reduction of the initial dynamical
problem, which involved the determination of a
function of N ($\sim 10^{23}$) variables. With remarkable intuition, Bogoliubov had postulated this
reduction in 1946 and his famous "synchronization
assumption" stated precisely that in the "kinetic
regime" the correlations become functionals of the
one-particle distribution which, in turn, obeys
a kinetic equation. We now see that the
Bogoliubov ansatz appears as an exact property
of the kinetic part of the distribution vector.

    Finally, we state without proof the follow-
ing properties:
A) The equilibrium distribution is a Π-function.
B) The linear response to an external field is
    determined by only the Π-component of the
    distribution vector.
C) The time integral (from 0 to ∞) of any auto-
    correlation function is determined by only
    the Π-component of $\mathcal{f}$.
    These properties (which can be proved quite
simply) show that the subdynamics of Π-space is

rigorously sufficient for the study of all equilibrium thermodynamics and of all transport processes, i.e. for the most important problems of statistical mechanics.

9. Translation of the abstract formalism and final conclusions

We have obtained our results in a general compact form by the systematic use of the concept of a distribution vector. In order to come back to explicit equations for the reduced distribution functions, all we have to do is to write the components of the general equations in the correlation patterns formalism, and select the relevant matrix elements of the operators $\mathcal{G}$, $\mathcal{C}$, ... to a given order of approximation. We shall illustrate this procedure in the simplest case.

A weakly coupled gas is described by the hamiltonian (1.3) in which we write $V(r)$ as $\lambda V(r)$, λ being a formal parameter, which at the end of the calculations is put equal to 1. We then also have $\mathcal{L}' \rightarrow \lambda \mathcal{L}'$ and $L'_{ij} \rightarrow \lambda L'_{ij}$. We now wish to write down the kinetic equation (8.10) for the kinetic one-particle distribution $\bar{p}_1(x_1;t)$, to the leading order in λ, i.e. through order λ^2.

We first note from the definitions (8.2) and (5.11) that

$$v \, \mathcal{G}(\tau) v = \lambda^2 v \, \mathcal{L}' c \, \mathcal{U}^0(\tau) \, \mathcal{L}' v + O(\lambda^3) \, .$$

Similarly, from (8.13)

$$V \tilde{u}(\tau)V \ = \ V u^{\circ}(\tau)V + O(\lambda) \quad .$$

From (8.12) we conclude that $V \Gamma_n V$ is of order λ^{2n}. Hence in our problem only the term of order λ^2 in $V \Gamma_n V$ contributes to the kinetic equation, which is therefore written as

$$\partial_t V \bar{f}(t) = V(\mathcal{L}^{\circ} + \lambda \mathcal{L}')V \bar{f}(t)$$

$$+ \lambda^2 \int_0^\infty d\tau \ V \mathcal{L}'C u^{\circ}(\tau) \mathcal{L}' u^{\circ}(-\tau)V \bar{f}(t) \tag{9.1}$$

We now write the equation for the component $\bar{p}_1(x_1;t)$. The term in $\mathcal{L}^{\circ}$ is diagonal and poses no problem:

$$\langle 1|V \mathcal{L}^{\circ}V | 1;2;\ldots;s \rangle = \delta_{s,1} L_1^{\circ} \quad .$$

Similarly, there is only one matrix element of $\mathcal{L}'$ connecting the 1-particle vacuum to a vacuum state on the right (see fig. 2):

$$\langle 1 | V \mathcal{L}'V | 1;2 \rangle = \int dx_2 \ L'_{12} \quad .$$

In the third term we have to go from the 1-particle vacuum to a correlated state, which is possible in only one way (fig. 2)

$$\langle 1 | V \mathcal{L}'C | 12 \rangle = \int dx_2 \ L'_{12}$$

In the second transition we have to go from the correlated two-particle state to a vacuum state, which is also possible in only one way (fig. 4).

$$\langle 12 \mid c \mathcal{L}'V \mid 1; 2 \rangle \quad = \quad L'_{12} \ .$$

Hence, the one-particle equation (9.1) is, explicitly:

$$\partial_t \ \bar{p}_1(x_1) \ = \ L_1^o \ \bar{p}_1(x_1) + \lambda \int dx_2 \ L'_{12} \ \bar{p}_2(x_1; \ x_2)$$

$$\lambda^2 \int_0^\infty d\tau \int dx_2 \ L'_{12} \ U_{12}^o(\tau) \ L'_{12} U_{12}^o(-\tau)$$
$$\bar{p}_2(x_1; \ x_2) \qquad\qquad (9.2)$$

Proceeding similarly for the other vacuum components one soon finds that a particular solution of the resulting hierarchy (in the T-limit) is eq. (8.20). Using now eqs. (1.7), (3.7) and (3.9) we find after some algebra the explicit kinetic equation

$$\partial_t \bar{f}_1(q_1 p_1; t)$$

$$= -\underline{v}_1 \cdot \underline{\nabla}_1 \ \bar{f}_1(q_1 p_1; t) + \left[\underline{\nabla}_1 V(q_1; t) \right] \cdot \underline{\partial}_1 \ \bar{f}_1(q_1 p_1; t)$$

$$+ \int_0^\infty d\tau \int dp_2 \ dq_2 \ \underline{\partial}_{12} \cdot \left\{ \underline{\nabla}_1 V(\underline{r}_{12}) \ \underline{\nabla}_1 V(\underline{r}_{12} - \underline{g}_{12}\tau) \right\}$$

$$\cdot \left[\underline{\partial}_{12} - \tau \underline{\nabla}_{12} \right] \bar{f}_1(q_1 p_1; t) \ \bar{f}_1(q_2 p_2; t)$$
$$(9.3)$$

with

$$\bar{V}(\underline{q}_1; t) \ = \ \int dq_2 \ dp_2 \ V(\underline{q}_1 - \underline{q}_2) \ \bar{f}(q_2 p_2; t) \quad (9.4)$$

where $\underline{r}_{12} = \underline{q}_1 - \underline{q}_2$ and $\underline{g}_{12} = \underline{v}_1 - \underline{v}_2$.

One recognizes here the general and well-known features of a kinetic equation. On the

right hand side we have: the flow term, the
Vlasov term (describing an average self-consistent
field) and the collision term. In a spatially
homogeneous system, the function $\bar{f}_1(\underline{q}_1\underline{p}_1;t)$
depends on the momentum alone

$$\bar{f}_1(\underline{q}_1\underline{p}_1;t) \quad = \quad n \,\bar{\phi}(\underline{p}_1;\, t)$$

and eq. (9.3) reduces to:

$$\partial_t \,\bar{\phi}(\underline{p}_1;t) \quad = \quad n \int d\underline{p}_2 \,\underline{\partial}_{12} \cdot \,\underline{\underline{G}} \cdot \,\underline{\partial}_{12} \,\bar{\phi}(\underline{p}_1;t)\bar{\phi}(\underline{p}_2;t)$$

$$(9.5)$$

with

$$\underline{\underline{G}} \; = \; \int_0^\infty d\tau \int d\underline{r} \; \underline{\nabla}_1 \; V(\underline{r}) \; \underline{\nabla}_1 V(\underline{r} - g\tau) \quad .$$

This is the familiar Landau equation, of common
use in plasma physics.

Kinetic equations valid in less trivial
situations can be derived quite systematically
from the general formalism. Of course, the
variety of intermediate correlation states becomes
quite complex as one goes to higher approximations.
However these various processes can be studied
systematically by use of a diagram technique.
One represents every matrix element of the rele-
vant operators ($\mathcal{G}$, $\mathcal{C}$, Γ, $\mathcal{C}$) by a succession
of the elementary vertices of figs. 2, 3,... .
Quite simple rules can then be **derived**, establishing a classification of these contributions
in terms of elementary topological properties of
these diagrams. We have no time to go into the

details of this technique here (see ref. 1).
Suffice it to say that the translation of any
abstract expression into a concrete kinetic
equation, to a given degree of approximation, is
a quite automatic process and becomes simply a
matter of patience.

   Is this to say that all non-equilibrium
problems are solved? The answer is clearly
"no"! The theory presented here provides a
formal framework. It tells us precisely the
following: "The kinetic equation and the kinetic
correlations, to any approximation, are given by
eqs. (8.19), provided that the formal mathematical objects exist." We are therefore led
back to the conditions (7.2) which express this
proviso. What do we know in this respect at
present?

   A few examples can be worked out in detail,
by which it can be shown that the theory is certainly not void, but nothing can be said in
general. Moreover, there have appeared some
"hot points" in the last few years. A typical
one arose while attempting the rather natural
idea of expanding the collision operator Γ in
powers of the density, n.   In terms of a transport coefficient, such as the viscosity, one
tried the rather obvious idea of finding a series
expansion:

$$\eta = \eta_0 + \eta_1\, n + \eta_2\, n^2 + \ldots\ .$$

The result of this attempt is disappointing: the
coefficient η_2 is infinite! Does this endanger
the general theory? The answer is no! Indeed,
it has been shown (ref. 6) that by rearranging
the terms and performing partial summations one
can find a finite result of the form

$$\eta = \eta_0 + \eta_1 \, n + \eta_2' \, n^2 \, \ln(n) + \ldots .$$

Hence, the particular type of approximation was
wrong, not the general theory! It is simply not
true that the transport coefficients are analytic
functions of n - and there is no compelling
a priori reason why they should be.

The more recent problem of the "long tails"
in the autocorrelation functions is another
example of difficulties of this kind. We shall
hear more about these problems in other lectures
at this school.

Let us conclude that, while there are still
many unsolved problems, it is certainly true that
a beautiful general structure emerges from the
analysis of the time-evolution of a many-body
problem. This structure provides a sound formal
basis from which further developments can proceed.

REFERENCES

The correlation patterns formalism was developed in

1)   R. Balescu, Physica 56, 1 (1971).

The idea of subdynamics stems from the following paper

2)   I. Prigogine, C. George and F. Henin, Physica 45, 418 (1969).

The presentation and notation for the subdynamics theory as presented in these lectures are very close to:

3)   R. Balescu and J. Wallenborn, Physica 54, 477 (1971).

Many proofs omitted in these lectures will be found in this paper.

A complete proof of the property $\Pi^2 = \Pi$ is found in

4)   C. George, Bull. Cl. Sciences, Acad. Roy. Belg., 56, 386 (1970).

The relation between transport coefficients and Π-subdynamics was studied in:

5)   R. Balescu, L. Brenig and J. Wallenborn, Physica 52, 29 (1971).

The logarithmic term in the density expansion of the transport coefficients was found by

6)   K. Kawasaki and J. Oppenheim, Phys. Rev. 139, A 1763 (1965).

HYDRODYNAMICAL CONCEPTS IN STATISTICAL PHYSICS

P. Résibois

Faculté des Sciences

Université Libre de Bruxelles

I. STATEMENT OF THE PROBLEM

la) Correlation functions:

The analysis of time-dependent correlation functions is one of the fundamental problems of non-equilibrium statistical mechanics. In general, these functions are defined by [*]

$$\mathcal{G}^{AB}(r,t) = \frac{1}{n}\left\langle \hat{A}(r,t)\ \hat{B}(r,o) \right\rangle \qquad (I.1)$$

where $\hat{A}(r,t)$ and $\hat{B}(r,t)$ denote the Heisenberg representation of the operators $\hat{A}(r)$ and $\hat{B}(r)$, while the bracket designates the equilibrium average; n is the average particle density of the system.

To be more precise, let us consider an N-particle system with Hamiltonian H_N; in the classical case, we have

$$\hat{A}(r,t) = e^{iL_N t}\,\hat{A}(r) \qquad (I.2)$$

where L_N is the Liouville operator:

$$L_N = i\left\{ H_N, \dots \right\} \qquad (I.3)$$

($\{\ ,\ \}$ denotes the Poisson bracket) while:

$$\left\langle \dots \right\rangle = \frac{\int dr^N dp^N \dots e^{-\beta H_N}}{\int dr^N dp^N e^{-\beta H_N}} \qquad (I.4)$$

with $\beta = 1/k_B T$ (k_B: Boltzmann constant; T absolute temperature).

Similar formulas hold in the quantum mechanical case, but we shall not need them in this course, which will be restricted to classical mechanics.

To illustrate these rather formal definitions,

---

[*]We write vector symbols only when necessary to avoid ambiguity

let us consider two specific cases where they
play an important role:

      (a) <u>Transport coefficients</u>: It is a well-
known fact that the linear transport coefficients
  can be expressed through the so-called Green-
Kubo formulas [1];

$$X = \lim_{t \to \infty} \lim_{\Omega} \frac{1}{\Omega k_B T} \int_0^t dt \langle \hat{J}^X(t) \, \hat{J}^X(o) \rangle \qquad (I.5)$$

Here Ω is the volume of the system, $\lim_{\Omega}$ denotes
the thermodynamic limit $N \to \infty$, $\Omega \to \infty$, $N/\Omega = n$ finite
and $\hat{J}^X$ is the flow operator associated with X.

    For example, in the case of electrical conduc-
tivity $X \equiv \sigma$, we have:

$$\hat{J}^\sigma = \sum_i e_i v_i \qquad (I.6)$$

where e_i and v_i respectively denote the charge and
the velocity of particle i. Similarly, for the
shear viscosity $X \equiv \eta$, we have

$$\hat{J}^\eta = \sum_i \hat{J}_i^\eta \qquad (I.7)$$

$$\hat{J}_i^\eta = V_{i,x} V_{i,y} - \frac{1}{2} \sum_j r_{ij,x} \frac{\partial V}{\partial r_{ij,y}} \qquad (\ast)(I.8)$$

    Let us stress that the integrand of (I.5) is
precisely of the form (I.1) with $\hat{J}^X = \hat{A}(r) = \hat{B}(o)$,
independent of r .

    (b) <u>Neutron and light scattering</u>: In
1954, Van Hove showed that, when a thermal neutron
beam strikes a macroscopic target, the inelastic
neutron cross section $\sigma_{q,\omega}$ for momentum transfer
<u>q and energy</u> transfer ω is also related to some

$(\ast)$We always take the mass of the particles
equal to 1.

correlation function of the type (I.1) (see Fig.1).

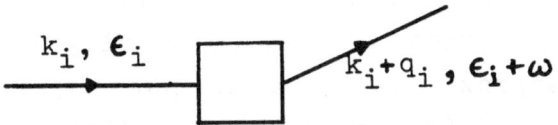

$$k_i, \epsilon_i \qquad\qquad k_i + q_i, \epsilon_i + \omega$$

Fig. 1: Neutron scattering

Indeed, when the target is a fluid (nuclear scattering), he found that the coherent part of the scattering is given by:

$$\sigma_{q,\omega}|_{coh.} \ \propto \ \text{Re} \ \mathcal{G}^{\delta n \, \delta n}_{q,\omega} \qquad\qquad (I.9)$$

where δn denotes the number density fluctuation:

$$\delta \hat{n}(r) = \sum_i \delta(r - r_i) - n \qquad\qquad (I.10)$$

In eq. (I.9), $\mathcal{G}^{AB}_{q,\omega}$ denotes the Fourier-Laplace transform of $\mathcal{G}^{AB}(r,t)$ with respect to space and time:

$$\mathcal{G}^{AB}_{q,\omega} = \int_0^\infty dt \ \exp(-i\,\omega t) \ \mathcal{G}^{AB}_q(t) \qquad\qquad (I.11)$$

and

$$\mathcal{G}^{AB}_q(t) = \int_\Omega dr \ \exp iqr \ \mathcal{G}^{AB}(r,t) \qquad\qquad (I.12)$$

This result is also valid for the inelastic scattering of light.

Although these examples clearly show the interest of these correlation functions, we should immediately stress their rather formal nature: indeed, as can be seen from (I.2) and (I.4), these formulas still involve a full many-body problem which cannot be treated exactly, except for very unrealistic models.

It is true that one can establish rigorously a few very general properties of these functions

(like the fluctuation-dissipation theorem, the
Kramers-Kronig relation (see Ref. 1)); yet, the
explicit evaluation of (I.1) usually involves
drastic approximations which can, at best, only
be controlled in limiting cases (like the dilute
gas limit, for example).

1b) <u>Long wave length, low frequency behaviour</u>
<u>of the correlation functions and linearised hydro-</u>
<u>dynamics</u>:

Due to these difficulties, it is worth-
while to see whether we can find some situations
in which the method and concepts of macroscopic
physics can help us in understanding the behaviour
of these correlation functions.

The simplest case where this is possible is the
long time, large distance behaviour of the so-called
Van Hove correlation function $\mathcal{G}^{\delta n \delta n}(r,t)$. Indeed,

$$\mathcal{G}^{\delta n \, \delta n}(r,t) = \frac{1}{n} \left\langle \delta \hat{n}(r,t) \, \delta \hat{n}(o,o) \right\rangle \qquad (I.13)$$

can be interpreted as the average particle density
fluctuation at r and t knowing that such a density
fluctuation was established at the origin at time
zero. Let us look at this function for long times
and large separations; equivalently, we consider
its Fourier transform [*]

$$\mathcal{G}^{\delta n \, \delta n}_{q,\omega} = \frac{1}{N} \left\langle \hat{n}_{q,\omega} \, \hat{n}_{-q}(o) \right\rangle \qquad (I.14)$$

in the limit of small q and ω . In such

---

[*] In writing (I.14), we have used the translation
invariance of the system and we have used a notation analogous to (I.11) (I.12) for the Fourier
transform of the density fluctuation.

circumstances, it is legitimate to assume that
all physical properties are slowly varying; we
are thus allowed to identify the <u>microscopic</u>
density operator $\delta\hat{n}$ which appears in (I.13,14),
with the smeared out local <u>macroscopic</u> density
fluctuation δn. This hypothesis introduces a
major simplification into the problem because this
macroscopic density is governed by the linearised
hydrodynamic equations $^{2a)3)}$; we write the Fourier-Laplace transform of these equations as:

$$-i\omega n_{\underline{q},\omega} +iq n v_{-\underline{q}\omega} = n_{\underline{q}} \quad (t=o) \tag{I.15}$$

$$-i\omega v_{\underline{q},\omega} +i\underline{q}\, \frac{1}{n}\, (\frac{\partial p}{\partial n})_T \; n_{\underline{q},\omega} \;+i\underline{q}\, \frac{1}{n}\, (\frac{\partial p}{\partial T})_n \; T_{\underline{q},\omega}+$$

$$\left\{ \frac{\eta}{n}\, q^2 v_{-\underline{q},\omega} + \frac{(\zeta + \frac{1}{3}\eta)}{n}\underline{q}[\underline{q}\cdot v_{-\underline{q},\omega}]\right\} = v_{-\underline{q}}(t=o) \tag{I.16a}$$

$$- i\omega T_{\underline{q},\omega} +i\underline{q}\, \frac{T}{nC_v}(\frac{\partial p}{\partial T})_n v_{\underline{q},\omega}+ \frac{\kappa}{n\,C_v}\, q^2 T_{\underline{q},\omega}=$$

$$= T_{\underline{q}}\; (t=o) \tag{I.16b}$$

Here $v_{q\omega}$ and $T_{q\omega}$ denote the Fourier-Laplace
transforms of the local velocity field and temperature; η, ζ and κ are the shear viscosity, bulk
viscosity and thermal conductivity; all thermodynamic symbols have their usual meanings and, in
order to avoid ambiguity, the vector symbols and
the scalar products (the dots) have been explicitly displayed.

Multiplying eqs.(I.15,16) by $\frac{1}{N}n_{-q}(o)$,and averaging over the equilibrium distribution, we get a
set of five coupled algebraic equations for

$\mathcal{G}_{q,\omega}^{\delta n,\ \delta n}$, $\mathcal{G}_{q,\omega}^{g,\ \delta n}$ and $\mathcal{G}_{q,\omega}^{T,\ \delta n}$ in terms of the initial conditions:

$$\mathcal{G}_{q}^{\delta n,\ \delta n}\ (t=o)= \frac{1}{N}\langle\hat{n}_{q}(o)\hat{n}_{-q}(o)\rangle \qquad (I.17a)$$

$$\mathcal{G}_{q}^{g,\ \delta n}\ (t=o)= \frac{1}{N}\langle\hat{v}_{q}(o)\hat{n}_{-q}(o)\rangle \qquad (I.17b)$$

$$\mathcal{G}_{q}^{T,\ \delta n}\ (t=o)=-\frac{1}{N}\langle\hat{T}_{q}(o)\hat{n}_{-q}(o)\rangle \qquad (I.17c)$$

In order to get an explicit formula for these initial correlations, we first note that, being interested in the behaviour of $\mathcal{G}_{q,\omega}^{\delta n\ \delta n}$ in the limit of small q, we can replace the right-hand sides of (I.17) by their $q\to o$ limit. For (I.17a), we can then use the well-known fluctuation theorem of statistical mechanics: [2b)

$$\lim_{q\to o}\ \frac{1}{N}\langle\hat{n}_{q}(o)\hat{n}_{-q}(o)\rangle =k_{B}T(\frac{\partial n}{\partial p})_{T}\ =$$

$$= nk_{B}T\,\chi_{T} \qquad (I.18)$$

where χ_{T} denotes the isothermal compressibility. For (I.17b), we can simply use the absence of correlations between momentum and space coordinates to get:

$$\lim_{q\to o}\ \frac{1}{N}\langle\hat{v}_{q}(o)\ \hat{n}_{-q}(o)\rangle = 0 \qquad (I.19)$$

We cannot immediately apply similar considerations to (I.17c) for the local temperature cannot be associated simply with a microscopic operator. However, we should remember that we are working here within the frame of linearised hydrodynamics (which rests, from the statistical point of view,

on the concept of local equilibrium: see Chapter
II); this presupposes that the following thermodynamic relation holds:

$$\hat{T}_q = \frac{1}{n\,C_v} \left[\hat{\epsilon}_q - (\frac{\partial e}{\partial n})_T \, \hat{n}_q \right] \qquad (I.20)$$

where $e(n,T)$ is the thermodynamic energy density
and $\hat{\epsilon}_q$ is the energy density operator.

The right hand side of (I.20), defined in
terms of the microscopic energy and particle
densities, provides us with a statistical definition of $\hat{T}_q$. Using (I.18) and the analogous relation:

$$\lim_{q\to o} \frac{1}{N} \langle \hat{\epsilon}_q(o)\, \hat{n}_{-q}(o) \rangle = k_B T (\frac{\partial e}{\partial p})_T =$$

$$= (\frac{\partial e}{\partial n})_T \, \chi_T \qquad (I.21)$$

we immediately have

$$\frac{1}{N} \langle \hat{T}_q(o)\, \hat{n}_{-q}(o) \rangle = 0 \qquad (I.22)$$

On combining these results, the solution of the
system of equations for $\zeta_{q,\omega}^{\delta n,\delta n}$, $\zeta_{q,\omega}^{g,\delta n}$ and $\zeta_{q,\omega}^{T,\delta n}$
is straightforward, though tedious. Great simpli-
fications occur when we take into account that q
is small; the result is:

$$\zeta_{q,\omega}^{\delta n,\delta n} \Big|_{\substack{q\to o \\ \omega\to o}} = nk_B T\, \chi_T \left[(1- \frac{C_v}{C_p}) \frac{1}{-i\omega +q^2 \kappa/nC_p} \right. +$$

$$\left. + \frac{1}{2} \frac{C_v}{C_p} \sum_{\pm} \frac{1}{-i\omega \pm icq + \Gamma q^2} \right] \qquad (I.23)$$

Here we have introduced the sound velocity:

$$c = \sqrt{\frac{C_p}{C_v}\, (\frac{\partial p}{\partial n})_T} \qquad (I.24)$$

and the sound absorption coefficient:

$$\Gamma = \frac{1}{2n}\left[4\eta/3 + \zeta + \left(\frac{1}{C_v} - \frac{1}{C_p}\right)\kappa\right] \qquad (I.24)$$

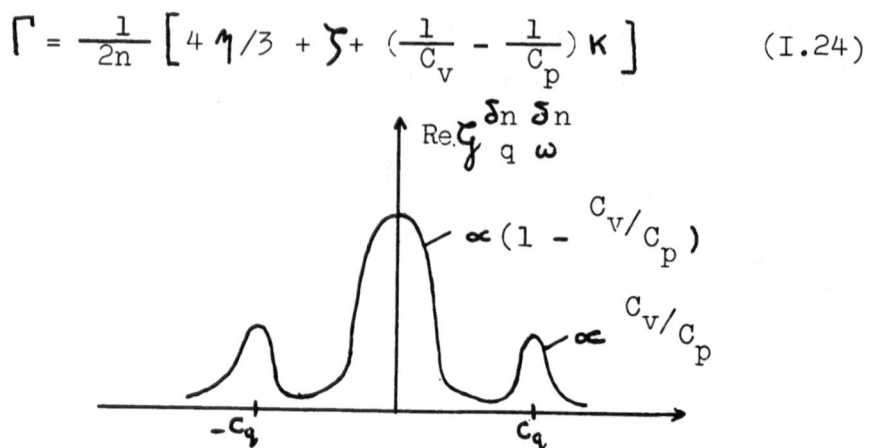

Fig.2: The real part of the Van Hove function in the small q limit.

Eq. (I.23) shows that, in the hydrodynamic limit, the density fluctuations propagate in two different ways (see also Fig. 2):

1) Through <u>adiabatic sound propagation</u>: this corresponds to the two peaks centered at $\pm$ cq (Brillouin scattering) with a width Γq^2.

2) Through <u>thermal diffusion</u>: the density fluctuation also generates an entropy fluctuation which diffuses in the system and leads to the central peak (Rayleigh diffusion) with a width $q^2\kappa/nC_p$.

Let us point out that the features predicted by the above theory are well verified by light scattering experiments.

1c) <u>Plan of the course</u>:

The simple example treated in the preceding section raises two types of questions, which we want to answer in the rest of this course:

    - Which kind of microscopic justification

can be given to the macroscopic arguments presented
in the derivation of (I.23). In particular, what
is the statistical foundation of the idea of hydrodynamical mode? This type of question will be answered in Chapter II, where we show how recent developments of the kinetic theory of fluids naturally
lead to a microscopic definition of the hydrodynamical modes.

     - Are there other situations, hopefully
more complex than in Section 1b), where the concept
of hydrodynamical mode can lead to information on
the behaviour of correlation functions? As a first
hint to this question, we should remember that
hydrodynamics is a theory valid for long times
and large distances; we should thus investigate
situations where such scales of time and space
play a dominant role.

    As a first example, we consider in Chapter III
the asymptotic time behaviour of the Green-Kubo
integrands (see (I.5)).

$$\langle \hat{J}^x (t)\, \hat{J}^x (o) \rangle \qquad\qquad (I.25)$$

Indeed, it has been recently discovered that these
functions, although spatially homogeneous, show a
slow power law decay for long times [4]; this results
from the propagation of coupled hydrodynamical
modes in the fluid.[5] In order to tackle this
problem, we shall set up a semi-macroscopic formalism [6], first used for critical dynamics.

    In Chapter IV, we illustrate the microscopic
equivalent of this semi-macroscopic treatment by
looking at another example where hydrodynamical

concepts are expected to play a crucial role; we
present there the kinetic theory of the transport
coefficients in a Van der Waals fluid [7]: here,
the large scales in space and time are due to the
long range character of the attractive part of
Van der Waals potential.

II.   HYDRODYNAMICAL MODES IN STATISTICAL PHYSICS

II.a)   Hydrodynamical modes in the dilute gas:

In order to illustrate the concept of
hydrodynamical modes in statistical physics, we
first discuss it for the dilute gas. In this case,
the Fourier transform of the one particle distribution function $f_1(r,v;t)$, namely:

$$f_q(v;t) = \int dr\, e^{iqr} f_1(r,v;t) \qquad (II.1)$$

satisfies the well-known Boltzmann equation [8],
which for small deviations from absolute equilibrium
can be written in linearised form. We have then
taking q along the x axis:

$$\partial_t f_q(v;t) + iqv_x f_q(v;t) = C^{\ell B} f \qquad (II.2)$$

where $C^{\ell B}$ is the following linear integral operator:

$$C^{\ell B} f_q(v_1;t) = n \int dv_2 \int d\bar{\Omega}\, \sigma(\bar{\Omega};|v_{12}|)|v_{12}|$$

$$\left[f_q(v_1^1;t)\, \varphi^{eq}(v_2^1) + \varphi^{eq}(v_1^1)\, f_q(v_2^1;t) - \right.$$

$$\left. - f_q(v_1;t)\, \varphi^{eq}(v_2) - \varphi^{eq}(v_1)\, f_q(v_2;t) \right] \qquad (II.3)$$

In this equation, $\sigma(\bar{\Omega};|v_{12}|)$ is the two body colli-
sion cross section and $\bar{\Omega}$ denotes the scattering
angles; v_1^1, v_2^1 and v_1, v_2 respectively denote the

velocities of particles 1 and 2 before and after
the collision process; finally, φ^{eq} is the Maxwellian velocity distribution.

Clearly, in order to solve the time dependent
problem (II.2), it is sufficient to solve the
eigenvalue problem:

$$(C^{\ell B}-iqv_x)\,\psi_n^{Bq}(v) = \lambda_n^{Bq}\,\psi_n^{Bq}(v) \qquad\qquad (II.4)$$

Here λ_n^{Bq} and ψ_n^{Bq} are the eigenvalue and the ei-
genfunction characterized by the index n. We have
put the superscript q to keep in mind the q-depen-
dence of the eigenvalues and eigenfunctions.

We are interested in the eigenvalue problem
(II.4) in the hydrodynamic limit, i.e. for q→o.
It is thus very natural to solve (II.4) by a perturbation calculus, considering $(-iqv_x)$ as a small
perturbation.

To this end, we first discuss the properties
of $C^{\ell B}$. Consider the eigenvalue problem:

$$C^{\ell B}\,\phi_n(v) = \lambda_n^{Bo}\,\phi_n(v) \qquad\qquad (II.5)$$

From the symmetry property:

$$\int dv\left[\varphi^{eq}(v)\right]^{-1}f(v)C^{\ell B}g(v)=$$
$$= \int dv\left[\varphi^{eq}(v)\right]^{-1}g(v)C^{\ell B}f(v) \quad (II.6)$$

it is easily verified that the eigenfunctions
$\phi_n(v)$ can be made orthonormal according to:

$$\int\left[\varphi^{eq}(v)\right]^{-1}\phi_n(v)\ \phi_{n'}(v)\ dv= \delta_{n,n'}^{Kr}, \qquad (II.7)$$

Indeed, this property is automatically fulfilled
if $\lambda_n^{Bo}\neq\lambda_{n'}^{Bo}$ and, if the eigenvalues are degenerate,
we can always apply the Schmidt orthogonalization
method.[9]

Although it is, in general, very difficult to find these eigenfunctions, it is readily verified from (II.3) that the five collision invariants $1, \underline{v}, v^2$ immediately provide us with five eigenfunctions ϕ_α ($\alpha \in 1,\ldots 5$) with zero eigenvalue.[*] Explicitly, we have:

$$\phi_1(v) = \varphi^{eq}(v) \qquad\qquad\qquad (a)$$

$$\phi_i(v) = \frac{v_i}{\sqrt{k_B T}}\, \varphi^{eq}(v) \quad (i=2,3,4) \quad (b) \quad (II.8)$$

$$\phi_5(v) = (\tfrac{2}{3})^{\frac{1}{2}}\left[\frac{v^2}{2k_B T} - \frac{3}{2}\right]\varphi^{eq}(v) \qquad (c)$$

Moreover, we make the assumption that the other eigenvalues λ_n^{Bo} ($n \notin \alpha$) have no accumulation point at the origin; physically, this implies a separation of time scales between the relaxation processes ($n \notin \alpha$) and the hydrodynamical phenomena.

In order to solve the eigenvalue problem (II.4) by a perturbation method for those five eigenvalues λ_α^{Bq} which tend to zero when $q \to o$, a little care is needed because the set (α) is degenerate for $q=o$ ($\lambda_\alpha^{Bo} = o$). We have thus first to remove this degeneracy by solving <u>exactly</u> the problem:

$$(C^{\ell B} - iqv_x)\,\psi_\mu^o(v) = \bar{\lambda}_\mu \psi_\mu^o(v) \quad (\mu \in 1\ldots 5) \qquad (II.9)$$

within the <u>subspace</u> spanned by the set (α).

Using an abstract linear vector space notation by considering $f(v)$ as the velocity space representation of the vector $|f\rangle$:

---

[*] In the following, greek subscripts will always describe an element of the set $(1,\ldots 5)$

$$\mathfrak{f}(v) = \langle v | \mathfrak{f} \rangle \qquad\qquad\qquad \text{(II.10)}$$

and defining, in this abstract space, the scalar product between $|\mathfrak{f}\rangle$ and $|g\rangle$ by:

$$\langle \mathfrak{f} | g \rangle = \int dv \left[\varphi^{eq}(v) \right]^{-1} \mathfrak{f}(v) g(v), \qquad\qquad \text{(II.11)}$$

we write instead of (II.9):

$$(c^{\ell B} - iqv_x) | \psi_\mu^\circ \rangle = \bar{\lambda}_\mu | \psi_\mu^\circ \rangle \qquad\qquad \text{(II.12)}$$

and expand $| \psi_\mu^\circ \rangle$ as a series in the $|\alpha\rangle$:

$$| \psi_\mu^\circ \rangle = \sum_{\mu'=1}^{5} c_{\mu\mu'} | \mu' \rangle \qquad\qquad\qquad \text{(II.13)}$$

where we have used the notation: $|\phi_\alpha\rangle \equiv |\alpha\rangle$ (II.13)
Inserting (II.13) into (II.12), using the ortho-gonality property (II.7) and $c^{\ell B} |\alpha\rangle = 0$, we immediately obtain the following set of five linear equations:

$$-iq \sum_\nu \varepsilon_{\mu\nu} \, c_{\mu\nu} = \bar{\lambda}_\mu c_{\mu\mu'} \quad (\mu, \nu, \mu' \in (\alpha)) \;\; \text{(II.14)}$$

where

$$\varepsilon_{\mu\nu} = \langle \mu | v_x | \nu \rangle \qquad\qquad\qquad \text{(II.15)}$$

These coefficients $\varepsilon_{\mu\nu}$ are readily evaluated; they all vanish except:

$$\varepsilon_{12} = \varepsilon_{21} = \sqrt{kT} \;\; ; \;\; \varepsilon_{25} = \varepsilon_{52} = \left(\tfrac{2}{3}kT\right)^{\frac{1}{2}} \qquad \text{(II.16)}$$

From this result, simple algebra leads to:

$$\bar{\lambda}_{1,2} = \pm \, ic^{(0)}q \qquad | \psi_{1,2}^\circ \rangle = \tfrac{1}{\sqrt{2}} \left[\sqrt{\tfrac{3}{5}} |1\rangle \pm |2\rangle + \sqrt{\tfrac{2}{5}} |5\rangle \right] \text{(a)}$$

$$\bar{\lambda}_{3,4} = 0 \qquad\qquad | \psi_{3,4}^\circ \rangle = |3,4\rangle \qquad \text{(b)} \qquad \text{(II.17)}$$

$$\bar{\lambda}_5 = 0 \qquad\qquad | \psi_5^\circ \rangle = \sqrt{\tfrac{2}{5}} \left[|1\rangle + \sqrt{\tfrac{3}{2}} |5\rangle \right] \qquad \text{(c)}$$

where $c^{(0)}$ is the sound velocity of the perfect gas:

$$c^{(o)} = \sqrt{5k_B T/3} \tag{II.18}$$

Direct calculation also shows that:

$$\langle \psi_\mu^o \mid \psi_{\mu'}^o \rangle = \delta_{\mu\mu'}^{Kr} \tag{II.19}$$

$$\langle \psi_\mu^o \mid (-iqv_x) \mid \psi_\nu^o \rangle = \bar{\lambda}_\nu \, \delta_{\mu,\nu}^{Kr} \tag{II.20}$$

Notice that the spectrum (II.17) remains degenerate; however (II.20) insures that there is no matrix element of the perturbation $(-iqv_x)$ between these states: this is all that is needed in order to develop perturbation calculus.

With these preliminaries, a perturbative solution of (II.4) is readily obtained. We expand the eigenvalues and eigenfunctions:

$$\lambda_\alpha^{Bq} = q \, \lambda_\alpha^{B(1)} + q^2 \, \lambda_\alpha^{B(2)} + \dots \quad (\lambda_\alpha^{Bo} = o \text{ for } \alpha \in (1\dots5)) \tag{II.21}$$

$$\mid \psi_\alpha^{Bq} \rangle = \mid \psi_\alpha^o \rangle + q \mid \psi_\alpha^B \rangle^{(1)} + \dots \tag{II.22}$$

and use the following unperturbed basis:

$$\begin{aligned} \mid \psi_\alpha \rangle &= \mid \psi_\alpha^o \rangle \quad \alpha = 1, 2 \dots 5 \\ \mid \psi_n \rangle &\equiv \mid n \rangle \quad\quad n \notin (\alpha) \end{aligned} \tag{II.23}$$

which obviously satisfies:

$$\langle \psi_n \mid \psi_{n'} \rangle = \delta_{nn'}^{Kr} \quad (\underline{\text{all}} \ n, n') \tag{II.24}$$

and is assumed to be complete.

The calculation proceeds exactly as in the quantum mechanical perturbation problem [10], and we merely give the results; we find:

$$q \, \lambda_\alpha^{B(1)} = \bar{\lambda}_\alpha \tag{II.25}$$

and

$$\lambda_\alpha^{B(2)} = \sum_{n \neq (\alpha)} \langle \psi_\alpha \mid v_x \mid \psi_n \rangle \frac{1}{\lambda_n^{Bo}} \langle \psi_n \mid v_x \mid \psi_\alpha \rangle \tag{II.26}$$

The restriction $n \neq (\alpha)$ can be eliminated if we subtract from the third factor of (II.26) the following quantity (see (II.20)):

$$\langle \psi_n | - g_\alpha | \psi_\alpha \rangle = - g_\alpha \, \delta^{Kr}_{n\alpha} \qquad (II.27)$$

where we have written $\bar{\lambda}_\alpha = iqg_\alpha$.

From the formal relation:

$$\lim_{\varepsilon \to 0} \sum_{all\ n} | \psi_n \rangle \frac{1}{\lambda^o_n - \varepsilon} | \psi_n \rangle = \frac{1}{c^{\ell B} - \varepsilon} \qquad (II.28)$$

we can then write (II.26) as:

$$\lambda^{B(2)}_\alpha = \lim_{\varepsilon \to 0} \langle \psi_\alpha | v_x \frac{1}{c^{\ell B} - \varepsilon} (v_x + g_\alpha) | \psi_\alpha \rangle \qquad (II.29)$$

From the explicit formulas (II.17) for the eigenfunctions $| \psi_\alpha \rangle$, we can write (II.29) in a more familiar form. After some simple algebra, involving the rotational invariance of the operator $c^{\ell B}$, we get:

$$\lambda^{Bq}_{1,2} = \pm\, i\, c^{(o)} q - \frac{1}{2} q^2 \left[\frac{4 \eta^B}{3n} \right. +$$

$$+ \left. \left(\frac{1}{c^{(o)}_v} - \frac{1}{c^{(o)}_p} \right) \frac{\kappa^B}{n} \right] + O(q^3) \quad (a)$$

$$\lambda^{Bq}_{3,4} = - \frac{\eta^B}{n} q^2 + O(q^3) \qquad (b) \qquad (II.30)$$

$$\lambda^{Bq}_5 = - \frac{\kappa^B}{n c^{(o)}_p} q^2 + O(q^3) \qquad (c)$$

where $C_v^{(o)}$ and $C_p^{(o)}$ respectively denote the perfect-gas specific heats at constant volume and pressure:

$$C_v^{(o)} = 3k_B/2 \qquad\qquad C_p^{(o)} = 5k_B/2 \qquad (II.31)$$

while the transport coefficients η^B and κ^B are formally defined by:

$$\eta^B = -\frac{n}{k_B T} \lim_{\epsilon \to 0} \int dv \; v_x v_y \; \frac{1}{c^{\ell B}_{-\epsilon}} \; v_x \; v_y \; \varphi^{eq}(v) \qquad (II.32)$$

$$\kappa^B = -\frac{n}{k_B T^2} \lim_{\epsilon \to 0} \int dv \; v_x \; \frac{v^2}{2} \; \frac{1}{c^{\ell B}_{-\epsilon}} (\frac{v^2}{2} - \frac{5kT}{2}) v_x \; \varphi^{eq}(v)$$
$$(II.33)$$

A comparison between these latter formulas and the results of the well-known Chapman-Enskog method would show that we can identify η^B and κ^B with the shear viscosity and thermal conductivity calculated in the Boltzmann approximation, but we shall not do this here.

Let us nevertheless point out that eq. (II.30) has precisely the same structure - for the dilute gas case - as the five eigenvalues of the macroscopic hydrodynamical equations (I.15)[*] (compare, in particular, (II.30a) and (II.30c) with the denominators in (I.23)): this allows us to interpret (II.30) as the microscopic equivalent - in the dilute gas limit - of the hydrodynamic modes. Similarly, eq. (II.17) gives us - to zeroth order in q - a microscopic expression for the corresponding eigenfunctions: eq. (II.17a) describes two propagating sound modes, eq. (II.17b) represents the two shear modes and eq. (II.17c) gives us the entropy mode already encountered in our phenomenological discussion of the Van Hove function $\varphi^{\delta n, \delta n}_{q, \omega}$.

---

[*] In the dilute gas the bulk viscosity $\varsigma = 0$.

II.b) <u>Hydrodynamical modes in a dense fluid:</u>

In the preceding section, we have discussed in a fairly detailed way the case of the dilute gas. We have done this because this example illustrates most of the features of the general problem without introducing the (mostly technical) difficulties related to many-body effects.

In this section, we briefly discuss the case of arbitrary density: we will however content ourselves with pointing out the main differences from the dilute gas and we refer the reader to the original papers [11)12)] for the detailed calculations.

In a dense system, the equation obeyed by the one particle distribution function $f_q(v;t)$ takes the following form:

$$\partial_t\, f_q(v;t) + iqv_x\, f_q(v;t) =$$

$$\int_0^t G_q(v;\tau)\, f_q\,(v;t-\tau)d\tau + \mathcal{D}_q(v;t) \tag{II.34}$$

This linear equation, valid for small deviations from absolute equilibrium, generalizes the Boltzmann equation (II.2) in two respects:

1) the collisions are now described by a non-Markoffian operator $G_q(v;\tau)$, which takes into account the finite duration of the collision processes.

2) we also find an inhomogeneous term, $\mathcal{D}_q(v;t)$, which describes the effect of the initial correlations on the evolution of the one-particle distribution function.

Eq. (II.34) can however be considerably simplified when we limit ourselves to the hydrodynamical
regime; indeed, this regime corresponds to the limit of long times, in which case we may neglect
the inhomogeneous term because one can show that

$$\mathcal{D}_q(v;t) = 0 \qquad t \gg \tau_c \qquad\qquad (II.35)$$

Moreover, in this hydrodynamic regime we expect
the distribution function $f_q(v;t)$ to evolve slowly
in time; we may then expand $f_q(v, t-\tau)$ according
to:

$$f_q(v;t-\tau) = \sum_{n=0}^{\infty} \frac{(-1)^n}{n!} \, \tau^n \, \partial_t^n \, f_q(v;t) \qquad (II.36)$$

Inserting (II.35,36) into (II.34) and taking into
account that $G_q(v;\tau)$ decays to zero for times
much smaller than t, it is a matter of straight
forward algebra to show that $f_q(v;t)$ satisfies the
following generalized Boltzmann equation:[13)17)33)]

$$\partial_t \, f_q(v;t) + iqv_x \, f_q(v;t) = c_q^\ell \, f_q(v;t) \qquad (II.37)$$

$$t \to \infty$$
$$q \to o$$

where

$$c_q^\ell = i \, \psi_q(v;o) + i(\Omega_q - 1)\left[-qv_x + \psi_q(v;o)\right] \qquad (II.38)$$

In this latter formula, $\psi_q(v;z)$ is the Laplace
transform of the kernel $G_q(v;\tau)$:

$$G_q(v;\tau) = \frac{+1}{2\pi i} \oint_c dz \, \exp -iz\tau \cdot \psi_q(v;z) \qquad (II.39)$$

and Ω_q is defined by the following implicit relation:

$$\Omega_q = 1 + \sum_{n=1}^{\infty} \frac{(-1)^n}{n!} \, \frac{\partial^n \psi_q}{\partial z}\bigg|_{z=0} \left\{\Omega_q\left[-v_x q + \psi_q(v;o)\right]\right\}^{n-1} \Omega_q \qquad (II.40)$$

Although the formal analogy between (II.37) and (II.2) is striking, there are two important differences in the operators of the right hand sides of these equations:

1) $\dot{C}_q^{\ell}$ is explicitly q-dependent while $C^{\ell B}$ is not; in our perturbative analysis, we will have to write:

$$C_q^{\ell} = C_o^{\ell} + qC^{\ell(1)} + q^2 C^{\ell(2)} + \ldots \qquad (II.41)$$

2) C_q^{ℓ} (and C_o^{ℓ}) is not a symmetric operator, in contrast to $C^{\ell B}$ (see (II.6)). This property forces us to consider the eigenvalue problem:

$$(iqv_x - C_q^{\ell}) | \psi_n^q \rangle = \lambda_n^q | \psi_n^q \rangle \qquad (II.42)$$

together with the adjoint problem:[9]

$$\langle \bar{\psi}_n^q | (iqv_x - C_q^{\ell}) = \lambda_n^q \langle \bar{\psi}_n^q | \qquad (II.43)$$

Nevertheless, we will follow the general procedure already used in section II.a), and first consider the q=o eigenvalue problems:

$$C_o^{\ell} | \varphi_n \rangle = \lambda_n^o | \varphi_n \rangle \qquad (II.44)$$

and:

$$\langle \bar{\varphi}_n | C_o^{\ell} = \bar{\lambda}_n \langle \bar{\varphi}_n | \qquad (II.45)$$

As is well-known, the sets $| \varphi_n \rangle$ and $\langle \bar{\varphi}_n |$ can always be made biorthonormal:

$$\langle \bar{\varphi}_{n'} | \varphi_n \rangle = \delta_{nn'}^{Kr} \qquad (II.46)$$

As in the dilute gas case, the eigenfunctions of C_o^{ℓ} with zero eigenvalue play an important role in the theory. Noticing that, for q=o, (II.10) leads to:

$$c_o^{\ell} = \Omega_o \, \psi_o \, (v;o) \tag{II.47}$$

it is possible to use remarkable properties of the operators Ω_o and ψ_o (this latter $\underline{is}$ symmetric and admits of the five functions (II.8) as eigenfunctions with zero eigenvalue) to construct explicitly the $|\varphi_\alpha\rangle$ and $\langle \bar{\varphi}_\alpha |$ ($\alpha = 1...5$). We shall, however, not display them here; let us merely point out that the only difference between them and the $|\alpha\rangle$ ($\langle\alpha|$) introduced in section II.a) is that $\langle\bar{\varphi}_5|$ corresponds to the one-particle $\underline{total}$ energy, while $|\varphi_5\rangle = |5\rangle$ still describes the one-particle kinetic energy.

In developing a perturbation scheme from (II.42), we are confronted with the same difficulty as in (II.4) when choosing a basis: the degeneracy of the λ_α^o forces us first to diagonalize exactly the problem:

$$(qc^{\ell(1)} - iqv_x) \, \psi_\mu^o \, (v) = \bar{\lambda}_\mu \, \psi_\mu^o(v) \tag{II.48}$$

(and its adjoint) inside the subspace spanned by the $|\varphi_\alpha\rangle$ (and $\langle\bar{\varphi}_\alpha|$). It is a remarkable feature of $c^{\ell(1)}$ that this can be done solely in terms of the equilibrium properties of the system. For example, we have shown that there exist two eigenfunctions of (II.48) with eigenvalues $\lambda_{1,2} = \pm icq$ (c is the sound velocity); these right eigenfunctions are:

$$|\psi_{1,2}^o\rangle = \frac{1}{\sqrt{2}} \left[\frac{\sqrt{k_B T}}{c} \, |1\rangle \pm |2\rangle + \left(\frac{3k_B T}{2} \right)^{\frac{1}{2}} \frac{(\partial p/\partial T)_n}{nc \, C_v} \, |5\rangle \right]; \tag{II.49}$$

in the dilute gas limit, they correctly reduce to (II.17a).

Once a suitable basis has been obtained from (II.48), the calculation follows exactly the steps described for the Boltzmann gas. The resulting eigenvalues generalize (II.30) for a dense fluid, namely:

$$\lambda_{1,2}^{q} = \pm icq - \frac{1}{2}\frac{q^2}{n}\left[\frac{4}{3}\eta + \zeta + (\frac{1}{C_v} - \frac{1}{C_p})\kappa\right] + O(q^3)$$

(damped propagating sound modes) (a)   (II.50)

$$\lambda_{3,4}^{q} = -\frac{\eta}{n} q^2 + O(q^3)$$

(damped shear modes)   (b)

$$\lambda_{5}^{q} = \frac{-\kappa}{nC_p} q^2 + O(q^3)$$

(damped entropy mode)   (c)

The corresponding right- and left-eigenfunctions are given, to zeroth order in q, by formulas of the type indicated in (II.49); we will not need them explicitly here.

In analogy with (II.32,33), we have also obtained explicit microscopic formulas for the transport coefficients κ, η, ζ ; these expressions have been shown to be equivalent to the Green–Kubo formulas (I.5), including the potential terms.

II.c) Complementary remarks:

It is a remarkable feature of the present theory that the transport coefficients, including their potential parts, can be obtained from the generalized Boltzmann equation for the one particle distribution function only. Indeed in the

traditional approach to the calculation of transport
coefficients in a dense system,[15] the first step is
to solve eq. (II.37) by the Chapman-Enskog procedure for the purpose of,(1) evaluating the kinetic
part of the transport coefficients,

(2) determining the two-
particle distribution function, which can be
expressed as a functional of f_1:

$$f_2(r_1,r_2;v_1,v_2;t) = f_2(r_1,r_2;v_1,v_2 | f_1) \qquad (II.51)$$

From f_2, the potential part of the transport coefficients is then computed. By pushing the perturbation analysis of the kinetic equation to second
order in q, we found no need for the functional
relation (II.45).

However, it should be emphasized that the present method was not devised for an explicit calculation of these transport coefficients. On the contrary, the usefulness of the formalism lies in the
fact that, whenever an expression is encountered
of the type

$$(i\omega - iqv_x + C_q^\ell)^{-1} \qquad (II.52)$$

involving the inverse linearized generalized Boltzmann operator, it does not need to be computed but
can be replaced, at least for q and ω small by

$$\lim_{\substack{q \to 0 \\ \omega \to 0}} \frac{1}{i\omega - iqv_x + C_q^\ell} = \sum_{\mu \epsilon(\alpha)} |\psi_\mu^\circ\rangle \frac{1}{i\omega + \lambda_\mu^q} \langle \bar{\psi}_\mu^\circ | \qquad (II.53)$$

where the transport coefficients appearing in λ_μ^q
are replaced by the approximation adequate to the
problem at hand.

This situation is, for example, the one found
in the microscopic analysis of the long wave-length,
low frequency limit of the Van Hove correlation
function $\mathcal{G}^{\delta n, \delta n}_{q, \omega}$ [14]; the use of (II.47) is the
crucial step in the statistical justification of
(I.23). Moreover, because the expansion (II.47)
is valid irrespective of the conserved or non-con-
served nature of the quantities appearing on its
right or on its left, it can be applied to situations where the macroscopic hydrodynamic equations
are not directly applicable; for example, we have
been able to discuss within this formalism the ki-
netic energy density correlation function $\mathcal{G}^{\epsilon_\kappa \epsilon_\kappa}_{q, \omega}$
although the kinetic energy density is
not a conserved quantity in a dense fluid.[15] We
shall however not present these applications here,
leaving for Chapter IV another, non trivial,
illustration of the method.

III.   MODE-MODE COUPLING (I) AND THE LONG TIME
       BEHAVIOUR OF THE GREEN-KUBO INTEGRANDS:
       III.a) N-particle approach to hydrodynamical
modes:

          (i) Notation: We shall present here
the formalism of Kadanoff and Swift [18] for constructing an N-particle representation of the hydrodynamical eigenfunctions and eigenvalues. This
theory was initially elaborated for studying the
singularities of the transport coefficients near
the critical point. Although it has not been
remarked explicitly in the literature, this

formalism also provides us with a very convenient
scheme for analysing other long wave-length phenomena, in particular the slow time decay of the
Green-Kubo integrand, already alluded to in Chapter I.

We start with the elaboration of an adequate
notation:

The N-particle distribution function
$\rho_N(r^N, p^N; t)$ is considered as the phase-space repre-
sentation of a vector $|t\rangle$ such that:

$$\langle r^N, p^N | t\rangle = \rho_N(r^N, p^N; t) \qquad \text{(III.1)}$$

The Liouville equation obeyed by ρ_N is then
written:

$$(i\partial_t - L_N) | t\rangle = 0 \qquad \text{(III.2)}$$

where the Liouville operator L_N is defined by its
matrix elements in the phase space representation:

$$\langle r'^N, p'^N | L_N | r^N, p^N\rangle = i \sum_{a=1}^{N} \left(\frac{\partial H_N}{\partial r_a} \frac{\partial}{\partial p_a} - \frac{\partial H_N}{\partial p_a} \frac{\partial}{\partial r_a} \right) \times$$

$$\times \langle r'^N, p'^N | r^N, p^N\rangle \qquad \text{(III.3)}$$

Here the scalar product $\langle r'^N, p'^N | r^N, p^N\rangle$ is defined
by:

$$\langle r'^N, p'^N | r^N, p^N\rangle = \prod_{a=1}^{N} \delta(r_a - r'_a) \, \delta(p_a - p'_a) \qquad \text{(III.4)}$$

The canonical equilibrium state is denoted by $|\rangle$:

$$\langle r^N, p^N | \rangle \equiv \exp{-\beta H_N(r^N, p^N)} / Z_N \qquad \text{(III.5)}$$

where Z_N is the N-particle partition function;
similarly the bra $\langle |$ will be written for:

$$\langle | = \int dr^N dp^N \langle r^N, p^N | \qquad \text{(III.5')}$$

a "summation state" which is <u>not</u> the conjugate of $|\rangle$ but nevertheless satisfies the following "conjugate" property:

$$\langle l | L_N = 0 \qquad \text{and} \qquad L_N | \rangle = 0 \qquad (III.5'')$$

This summation state is very convenient for expressing statistical averages; for an arbitrary N-body operator X^{op}, its average value is given by

$$\langle X^{op} \rangle_t = \int dr^N dp^N X^{op}(r^N, p^N) \mathcal{P}_N(r^N, p^N; t)$$

$$= \langle | x^{op} | t \rangle \qquad (III.6)$$

The five operators $\hat{n}(r)$ (<u>particle density fluctuation</u>), $\hat{g}(r)$ (<u>momentum density fluctuation</u>) and $\hat{\epsilon}(r)$ (<u>energy density fluctuation</u>) play a crucial role in a hydrodynamical theory. They are defined by: (*)

$$\langle r'^N, p'^N | \hat{n}(r) | r^N, p^N \rangle = \left[\sum_{a=1}^{N} \delta(r - r_a) - n \right]$$

$$\langle r'^N, p'^N | r^N, p^N \rangle \qquad (III.7)$$

or more compactly:

$$\hat{n}(r) = \left[\sum_{a=1}^{N} \delta(r - r_a) - n \right] \qquad (III.8a)$$

Similarly, one has:

$$\hat{g}(r) = \sum_{a=1}^{N} p_a \delta(r - r_a) \qquad (III.8b)$$

$$\hat{\epsilon}(r) = \sum_{a=1}^{N} \left[\frac{p_a^2}{2} + \frac{1}{2} \sum_{a \neq b} V(r_{ab}) \right] \delta(r - r_a) - \bar{\epsilon} \quad (III.8c)$$

(*) It is convenient to work with the deviation from the equilibrium value, in order to avoid volume dependent singularities of the Fourier transform of these operators at q=o.

Similarly, we will need the <u>currents</u> associated
with these conserved quantities, defined by:

$$- \nabla \underline{\hat{j}}\ (r) = i\left[L_N, \hat{n}\ (r)\right] \qquad (a)$$

$$- \nabla \underline{\hat{\tau}}(r) = i\left[L_N, \underline{\hat{g}}\ (r)\right] \qquad (b) \qquad\qquad (III.9)$$

$$- \nabla \underline{\hat{j}}^{\epsilon}\ (r)= i\left[L_N, \hat{\epsilon}(r)\right] \qquad (c)$$

We immediately get from (III.8a, 9a):

$$\left\langle r'^N, p'^N \mid \underline{\hat{j}}\ (r) \mid r^N, p^N \right\rangle = \sum_{a=1}^{N} \underline{p}_a \delta(r-r_a)$$

$$\left\langle r'^N, p'^N \mid r^N, p^N \right\rangle \qquad (III.10a)$$

Similarly, we have:

$$\underline{\hat{\tau}}\ (r)= \sum_{a=1}^{N}\left[\underline{p}_a : \underline{p}_a - \frac{1}{2}\sum_{b\neq a}\frac{\partial V}{\partial \underline{r}_{ab}} : \underline{r}_{ab}\right]\delta(r-r_a) \quad (III.10b)$$

$$\underline{\hat{j}}^{\,\epsilon}\ (r)= \sum_{a=1}^{N}\left[\frac{\underline{p}_a \cdot \underline{p}_a^{2}}{2} + \frac{\underline{p}_a}{2}\sum_{b\neq a}V(r_{ab})\right.$$

$$\left. - \frac{1}{2}\sum_{b\neq a}(\underline{p}_a \cdot \frac{\partial V}{\partial \underline{r}_{ab}})\ \underline{r}_{ab}\right]\delta(r-r_a) \qquad (III.10c)$$

Instead of the energy density $\hat{\epsilon}\ (r)$, we will
often use the entropy density operator; this is
defined by the following relation:

$$\hat{s}\ (r) = \frac{1}{T}\left[\hat{\epsilon}\ (r)- \frac{h}{n}\ \hat{n}(r)\right] \qquad (III.11)$$

where h is the enthalpy density and T the absolute
temperature. As we shall only deal with small
deviations from absolute equilibrium these two quantities — referring to this equilibrium state — will
always be well defined.

The corresponding entropy current is defined by:

$$\hat{j}^{\ s}(r) = \frac{1}{T}\left[\hat{j}^{\ \epsilon}(r) - \frac{h}{n}\hat{j}\ (r)\right] \qquad (III.12)$$

(ii) <u>Local equilibrium states</u>: Local
equilibrium states, characterized by <u>linear</u> deviations of the conserved quantities away from absolute
equilibrium, play an important role in any statistical theory of hydrodynamics. They are defined in
terms of the Fourier transform of the conserved
quantities, generically denoted by $\bar{a}_\alpha(r)(\alpha \in (1...5))$:

$$\bar{a}_{\alpha,q} = \int dr\ e^{iqr}\bar{a}_\alpha(r) \qquad (III.13)$$

More precisely, we write these local equilibrium
states as:

$$|\alpha,q\rangle = a_{\alpha,q}|\ \rangle$$
$$\langle\alpha,q| = \langle|a_{\alpha,-q} \qquad (III.14)$$

where the $a_{\alpha,q}$ are linear combinations of the
$\bar{a}_{\alpha,q}$:

$$a_{\alpha,q} = \sum_{p=1}^{5} C_{\alpha p}\ \bar{a}_{p,q} \qquad (III.15)$$

choosen in such a way that the states $|\alpha,q\rangle$ are
mutually orthogonal: (*)

$$\langle\alpha,q|\alpha',q'\rangle \equiv \delta^{Kr}_{\alpha\alpha'}\ \Omega\delta^{Kr}_{q,q'}\text{(discrete case)}$$
$$= \delta^{Kr}_{\alpha\alpha'}\ \delta(q-q')\ 8\pi^3 \qquad (III.16)$$
$$\text{(continuous case).}$$

We take thus:

(*) Although, physically, we are working in an
infinite volume, in which the continuous normalization should be taken, it is often simpler in the
calculation to consider a large but finite volume,
with periodic boundary conditions.

$$a_{1,q} = C_{15} \, \hat{s}_q$$

$$a_{2,q} = C_{25} \, \hat{s}_q + C_{21} \, \hat{n}_q \qquad \text{(III.17)}$$

$$a_{i,q} = \bar{c} \, \hat{g}_{iq} \equiv C_{ii} \hat{g}_{i,q} \quad (i=2,3,4)$$

and the coefficients $\hat{C}_{\alpha\beta}$ are determined by (III.16). A simple calculation leads to:

$$\bar{c} = ({}^1/k_B Tn)^{\frac{1}{2}}$$

$$C_{15} = {}^1/\sqrt{k_B \, n \, C_p \, (q)}$$

$$C_{25} = (\frac{1}{k_B Tn})^{\frac{1}{2}} \, c(q) \qquad \text{(III.20)}$$

$$C_{21} = \frac{1}{\sqrt{k_B n}} \left[\frac{1}{C_v(q)} - \frac{1}{C_p(q)} \right]^{\frac{1}{2}}$$

Here $C_p(q)$ is given by:

$$k_B n \, C_p(q) = \frac{1}{\Omega} \langle \, |\hat{s}_{-q} \, \hat{s}_{+q}| \, \rangle \qquad \text{(III.21)}$$

while $C_v(q)$ and $c(q)$ are <u>defined</u> by the following set of equations:

$$\frac{1}{k_B n} \left[\frac{1}{C_v(q)} - \frac{1}{C_p(q)} \right] \equiv$$

$$\frac{\Omega}{\left[\dfrac{\langle |\hat{s}_{-q} \, \hat{s}_{+q}| \rangle \langle |\hat{n}_{-q} \, \hat{n}_{+q}| \rangle}{\langle |\hat{s}_{-q} \, \hat{n}_{+q}| \rangle^2} - 1 \right] \langle |\hat{s}_{-q} \, \hat{s}_{+q}| \rangle} \qquad \text{(a)}$$

$$\frac{1}{k_B Tn} \, c^2(q) = \frac{1}{k_B n} \left[\frac{1}{C_v(q)} - \frac{1}{C_p(q)} \right] \frac{\langle |\hat{s}_{-q} \, \hat{s}_q| \rangle}{\langle |\hat{n}_{-q} \, \hat{n}_q| \rangle} \quad \begin{array}{l} \text{(b)} \\ \text{(III.22)} \end{array}$$

These definitions, apparently complicated, have the virtue that, in the long wave-length limit $q \to o$, $C_p(q)$, $C_v(q)$ and $c(q)$ tend to the equilibrium quantities C_p, C_v (specific heats per particle at constant pressure and volume) and c (adiabatic sound velocity). For example, we have

$$\frac{1}{\Omega} \lim_{q \to o} \langle | \hat{s}_{-q} \hat{s}_{+q} | \rangle = \frac{1}{\Omega} \lim_{q \to o} \int dr \, dr' \exp. iq(r-r')$$

$$\langle (s(r)-\bar{s})(s(r')-\bar{s}) \rangle$$

$$= k_B n C_p \qquad (III.23)$$

where, in the equality, we have used a well-known result of equilibrium theory for the total entropy fluctuation.[2]

To summarize our results, we have thus constructed the local equilibrium states (III.14) with:

$$a_{1,q} = \hat{s}_q / \sqrt{k_B n C_p(q)} \qquad (a)$$

$$a_{2,q} = (1/k_B Tn)^{\frac{1}{2}} c(q) \, \hat{n}_q + (1/k_B n)^{\frac{1}{2}} \times$$

$$\times \left[1/C_v(q) - 1/C_p(q) \right]^{\frac{1}{2}} \hat{s}_q \quad (b) \quad (III.24)$$

$$a_{i,q} = (1/k_B Tn)^{\frac{1}{2}} \hat{g}_{i,q} \quad (i=x,y,z \equiv 3,4,5) \quad (c)$$

If q is the smallest parameter of the problem, these can be replaced by:

$$a_{1,q} = \hat{s}_q / \sqrt{k_B n C_p} \qquad (a)$$

$$\qquad \qquad (III.24')$$

$$a_{2,q} = (1/k_B Tn)^{\frac{1}{2}} c \, \hat{n}_q + (1/k_B n)^{\frac{1}{2}} \left[1/C_v - 1/C_p \right]^{\frac{1}{2}} \hat{s}_q$$

$$\qquad \qquad (b)$$

$$a_{i,q} = (1/k_B Tn)^{\frac{1}{2}} \hat{g}_{i,q} \quad (i=x,y,z \equiv 3,4,5) \quad (c)$$

where C_p, C_v and c are the usual thermodynamic
quantities. However, in the canonical ensemble
used here, one should be careful not to go too
fast to the $q=0$ limit in the operators themselves
(in particular with $\hat{n}_q$ because $\lim_{q \to 0} \hat{n}_q = 0$ although
$\lim_{q \to 0} \langle |\hat{n}_q \, \hat{n}_{-q}| \rangle = nk_B T \, \chi_T \neq 0$). This difficulty does
not appear in the grand canonical ensemble.

   (iii) <u>Transport equation</u>: Let us
denote by $|\tilde{j},q\rangle$ the eigenstates of the Liouville
operator:

$$\mathbf{L}_N \, |\tilde{j},q\rangle = z_{j,q} \, |\tilde{j},q\rangle \qquad\qquad (III.25)$$

in this equation, we have used the fact that the
Hamiltonian of the system is translation invariant:
this has as a consequence that the eigenstates
are characterized by the wave number q.

  Instead of looking at the general problem
(III.25), we limit ourselves to the hydrodynamical
eigenstates ($\nu \in 1,\dots 5$); these have the following
property: 1) the corresponding eigenvalues $z_{\nu,q}$ go
to zero when $q \to 0$ (because in the $q=0$ limit, the
hydrodynamical states are exactly conserved).

   2) these eigenstates are mostly composed
of the local equilibrium states $|\alpha,q\rangle$.

  This latter idea is introduced into the theory
with the help of the projector P_q (we take q discrete for simplicity):

$$P_q = 1 - \frac{1}{\Omega} \sum_{\alpha=1}^{5} |\alpha,q\rangle\langle\alpha,q| \qquad\qquad (III.26)$$

which removes the local equilibrium part of a
given state.

Multiplying (III.25) by $\langle \alpha, q |$, we write:

$$z_{\mathbf{v},q} \langle \alpha, q | \tilde{\mathbf{v}}, q \rangle = \langle \alpha, q | L_N | \tilde{\mathbf{v}}, q \rangle =$$

$$= \frac{1}{\Omega} \sum_{\beta} \langle \alpha, q | L_N | \beta, q \rangle \langle \beta, q | \tilde{\mathbf{v}}, q \rangle + \langle \alpha, q | L_N P_q | \tilde{\mathbf{v}}, q \rangle \tag{III.27}$$

We have also

$$z_{\mathbf{v},q} \, P_q | \tilde{\mathbf{v}}, q \rangle = P_q z_{\mathbf{v},q} | \tilde{\mathbf{v}}, q \rangle = P_q L_N | \tilde{\mathbf{v}}, q \rangle =$$

$$= P_q L_N P_q | \tilde{\mathbf{v}}, q \rangle + \frac{1}{\Omega} \sum_{\beta} P_q L_N | \beta, q \rangle \langle \beta, q | \tilde{\mathbf{v}}, q \rangle \tag{III.28}$$

or, since $P_q^2 = P_q$,

$$P_q | \tilde{\mathbf{v}}, q \rangle = \frac{1}{z_{\mathbf{v}} - P_q L_N P_q} \sum_{\beta} P_q L_N | \beta, q \rangle \langle \beta, q | \tilde{\mathbf{v}}, q \rangle / \Omega \tag{III.29}$$

Introducing this latter result into (III.27), we get

$$\sum_{\beta} \left[z_{\mathbf{v},q} \delta_{\alpha,\beta}^{Kr} - L_{\alpha\beta}(q) - V_{\alpha\beta}(q, z_{\mathbf{v},q}) \right] \langle \beta, q | \tilde{\mathbf{v}}, q \rangle = 0 \tag{III.30}$$

$$L_{\alpha\beta}(q) = \frac{1}{\Omega} \langle \alpha, q | L_N | \beta, q \rangle \tag{III.31}$$

$$V_{\alpha\beta}(q, z) = \frac{1}{\Omega} \langle \alpha, q | L_N P_q \frac{1}{z - P_q L_N P_q} L_N P_q | \beta, q \rangle \tag{III.32}$$

If the coefficients L and V were known, eq. (III.30) would give us a 5 x 5 system of equations which would allow us to determine:

a) the local equilibrium part of the hydrodynamical eigenstates.

b) the corresponding eigenmodes $z_{\mathbf{v},q}$.
It is clear that the formal definition of the coefficients $V_{\alpha\beta}(q,z)$- which still involve the complete Liouville operator - does not allow us

explicitly to fulfil this program. However, we
will assume that eq. (III.30) provides us with
the linear transport theory, provided the $L_{\alpha\beta}$ are
identified with thermodynamic derivatives and the
$V_{\alpha\beta}$ with the matrix of transport coefficients.

Let us first look at the $L_{\alpha\beta}$; we have

$$L_{\alpha\beta}(q) = \frac{1}{\Omega}\langle\alpha,q|L|\beta,q\rangle = \frac{1}{\Omega}\langle|a_{\alpha,-q}\,L_N a_{\beta,q}|\rangle.$$

$$= \frac{1}{\Omega}\langle|[a_{\alpha,-q},L_N]\,a_{\beta,q}|\rangle \quad \text{(see (III.5))}$$

$$= +\frac{q}{\Omega}\langle|j_{\alpha,-q}\,a_{\beta,q}|\rangle \quad \text{(see (III.7))}$$

Choosing q along the x axis, it is readily
seen that, for symmetry reasons, the only non-
vanishing coefficients are:

$$L_{23}(q)=L_{32}(q)=q_x\langle|j_{2,-q}\,a_{3,q}|\rangle \tag{III.34}$$

From (III.24 , 10, 12), the equilibrium correlation involved in the right-hand side of (III.34)
is readily computed; as is shown in an appendix
we have:

$$L_{23}(q) = L_{32}(q) = q_x c(q) \tag{III.35}$$

Similarly, among the 25 coefficients $V_{\alpha\beta}(q,z)$,
which, from (III.7,32) can be rewritten as:

$$V_{\alpha\beta}(q,z)= \frac{1}{\Omega}\Big[q_x\langle|j^x_{\alpha,-q}\,P_q\frac{1}{z-P_q L_N P_q}\,P_q j^x_{\beta,+q}|\rangle q_x\Big] \tag{III.36}$$

symmetry arguments together with $P_q g_x \equiv 0$ lead to
only six non-vanishing coefficients. After some
simple algebra, which will not be reproduced here,

we obtain:

$$V_{11}(q,z) = \frac{-i \, q^2 \, K(q,z)}{n \, C_p(q)} \, . \tag{a}$$

$$V_{22}(q,z) = \frac{-i \, q^2 K(q,z)}{n} \left[\frac{1}{C_v(q)} - \frac{1}{C_p(q)} \right] \, . \tag{b}$$

$$V_{21}(q,z) = V_{12}(q,z)$$

$$= \frac{-i q^2 \, K(q,z)}{n} \left[\frac{1}{C_p(q)} \left(\frac{1}{C_v(q)} - \frac{1}{C_p(q)} \right) \right]^{\frac{1}{2}} . \tag{c}$$

$$V_{33}(q,z) = -i \, \frac{q^2 (\zeta(q,z) + \frac{4}{3} \eta(q,z))}{n} \, . \tag{d}$$

$$V_{44}(q,z) = V_{55}(q,z) = \frac{-i \, q^2 \, \eta(q,z)}{n} \, . \tag{e}$$

$$\tag{III.37}$$

where we have introduced:

$$q^2 K(q,z) = - \frac{1}{k_B \Omega} \langle | \hat{s}_{-q} \, L_N P_q \, \frac{1}{i(-P_q L_N P_q + z)} \, P_q L_N \hat{s}_q | \rangle \tag{a}$$

$$q^2 \eta(q,z) = - \frac{1}{k_B T \Omega} \langle | \hat{g}_{y,-q} L_N P_q \, \frac{1}{i(-P_q L_N P_q + z)} \, P_q L_N \hat{g}_{y,q} | \rangle \tag{b}$$

$$q^2 \left[\frac{4}{3} \eta(q,z) + \zeta(q,z) \right] =$$

$$= \frac{-1}{k_B T \Omega} \langle | \hat{g}_{x,-q} L_N P_q \, \frac{1}{i(-P_q L_N P_q + z)} \, P_q L_N \hat{g}_{x,q} | \rangle \tag{c}$$

$$\tag{III.38}$$

The matrix (III.30) takes thus the following form:

$$
\begin{bmatrix}
z + \dfrac{iq^2 K}{nC_p} & \dfrac{iq^2 K}{n}\left[\dfrac{1}{C_p}\left(\dfrac{1}{C_v}-\dfrac{1}{C_p}\right)\right]^{1/2} & 0 & 0 & 0 \\[2ex]
\dfrac{iq^2 K}{n}\left[\dfrac{1}{C_p}\left(\dfrac{1}{C_v}-\dfrac{1}{C_p}\right)\right]^{1/2} & z + \dfrac{iq^2 K}{n}\left(\dfrac{1}{C_v}-\dfrac{1}{C_p}\right) & -cq_x & 0 & 0 \\[2ex]
0 & cq_x & \dfrac{iq^2(\zeta + \tfrac{4}{3}\eta)}{n} & 0 & 0 \\[2ex]
0 & 0 & 0 & z + \dfrac{iq^2 \eta}{n} & 0 \\[2ex]
0 & 0 & 0 & 0 & z + \dfrac{iq^2 \eta}{n}
\end{bmatrix}
$$

$$(III.39)$$

where we have dropped the q- and z-dependence of the coefficients C_p, C_v, c and η, ζ, K. It is thus almost in diagonal form, especially if we notice that the off-diagonal coefficients $V_{12} = V_{21}$, of order q^2 can be neglected in the small q-limit (they give corrections of order at least q^4 to the eigenvalues).

We find thus easily the following relations for the eigenvalues and for the corresponding ei-genfunctions:

$$z_1 = -\frac{iq^2 K(q,z)}{nC_p(q)} , \qquad |\tilde{1},q\rangle = a_{1,q}|\rangle$$

$$z_i = \pm c(q)q_x - iq^2 \Gamma(q,z_i), \quad |\tilde{i},q\rangle = \frac{a_{2,q} \pm a_{3,q}}{\sqrt{2}}|\rangle \quad (i=2,3)$$

$$z_j = - \frac{iq^2 \eta(q,z_j)}{n}, \qquad |\tilde{j},q\rangle = a_{j,q}|\rangle \quad (j=4,5)$$

$$\text{(III.40)}$$

where we have introduced the coefficient $\Gamma(q,z)$:

$$\Gamma(q,z) = \frac{1}{2n}\left[4\eta(q,z)/3 + \zeta(q,z) + (\frac{1}{C_v(q)} - \frac{1}{C_p(q)})K(q,z)\right]$$

$$\text{(III.41)}$$

The structure of these results strongly suggests that we have obtained here an N-body representation of the hydrodynamic modes, first introduced in macroscopic terms in Chapter I and then constructed in Chapter II starting from kinetic theory. Indeed, we see immediately the analogy between the eigen-values in (III.40) and the corresponding expressions in (II.50); moreover the microscopic definitions (III.38) for the frequency and wave number depen-dent transport coefficients are closely connected to the Laplace transform representation of the Green-Kubo formula (I.5). Indeed, we can formally write this latter formula as

$$X = \lim_{z \to i\varepsilon} \lim_{\Omega} \frac{-1}{k_B T \Omega} \langle \hat{J}^x \frac{1}{i(-\hat{L}_N + z)} \hat{J}^x \rangle \qquad \text{(III.42)}$$

and, as we shall see shortly, eqs. (III.38) and (III.42) become indeed identical when $q \to 0$, $z \to i\varepsilon$.

From (III.38), we see that a discussion of the general case requires an analysis of the operator

$(P_q \frac{L}{N} P_q - z)^{-1}$ which is a complicated object, and we will restrict ourselves here to the limit $q \to o$. In this case, we get the following eigenvalues:

$$z_1 = \frac{-iq^2 \, K(o,o)}{nC_p} + O\ (q^3) \qquad\qquad (a)$$

$$z_i = \ \pm cq_x - iq^2 \, \Gamma(o,o) + O(q^3) \ (i=2,3) \ (b) \qquad (III.43)$$

$$z_j = -\ iq^2 \, \eta(o,o)/n + O(q^3) \qquad (j=3,4) \ (c)$$

III.b) Long time behaviour of the Green-Kubo integrands:

For definiteness, we consider the shear viscosity coefficient (III.38b) in the limit $q \to o$. We first establish its connection with the Green-Kubo formula (III.42). From (III.9), it is easy to verify that:

$$L_N \hat{g}_{y,q} \vert \rangle = -\ q_x \ \hat{\tau}_{xy,q} \vert \rangle \qquad\qquad (III.44)$$

Moreover, (III.26) and (III.40) allow us to show that:

$$\lim_{q \to o} P_q \ \hat{\tau}_{xy,q} = \hat{\tau}_{xy,o} \equiv \hat{J}^\eta \qquad\qquad (III.45)$$

where we have used the definitions (I.7,8). We thus obtain

$$\eta(o,z) \equiv \lim_{q \to o} \eta(q,z) = \frac{-1}{k_B T \Omega} \langle \hat{J}^\eta \frac{1}{i(z - P_o L_N P_o)} \hat{J}^\eta \rangle$$
$$(III.46)$$

where $P_o \equiv \lim_{q \to o} P_q$. This formula can be further simplified if we notice that:

$$\lim_{q \to 0} P_q L_N P_q = \lim_{q \to 0} \left[L_N - \sum_{\alpha} |\alpha, q \rangle \langle \alpha, q | L_N | \alpha, q \rangle \langle \alpha, q | \right]$$

$$= L_N \qquad \text{(III.47)}$$

because, in the $q \to 0$ limit, the local equilibrium states are exactly conserved.

Thus:

$$\eta(o,z) = \frac{-1}{k_B T \Omega} \left\langle \hat{J}^\eta \frac{1}{i(z - L_N)} \hat{J}^\eta \right\rangle \qquad \text{(III.48)}$$

which is the finite frequency generalization of (III.42); this generalization is nothing else than the Laplace transform of the Green-Kubo integrand of (I.5); we have indeed:

$$\eta(o,z) = \frac{1}{\Omega k_B T} \int_0^\infty d\tau \ \exp iz\tau \left\langle \hat{J}^\eta(\tau) \ \hat{J}^\eta \right\rangle \qquad \text{(III.49)}$$

$$\frac{1}{\Omega k_B T} \left\langle \hat{J}^\eta(t) \hat{J}^\eta \right\rangle = \frac{+1}{2\pi} \oint_C dz \ \exp{-izt} \ \eta(o,z) \qquad \text{(III.50)}$$

where the contour C is parallel to the real axis.

It was by computer experiments that Alder and Wainwright [19] first discovered that these Green-Kubo integrands decay very slowly for long times; more precisely, they observed a t^{-1} decay, in two dimensions, for the Green-Kubo integrand of self-diffusion. It has since been shown that this situation is general [5] and we want to illustrate this point here for the shear viscosity in three dimensions, using the Kadanoff-Swift formalism.

Remembering that the long time behaviour is dominated by small z values in the Laplace transform, we have to investigate (III.48) for z small. The central question in this problem is a knowledge of the inverse operator:

$$1/(z-L_N) \qquad\qquad (III.51)$$

Of course, we can write formally:

$$\frac{1}{(L_N-z)} = \sum_{\text{all } j,q} |\tilde{j},q\rangle \frac{1}{z_{j,q}-z} \langle \tilde{j},q| \qquad (III.52)$$

with the help of (III.25), but this representation
is of little help as long as we have no explicit
information about the spectrum of L_N. This problem
is really the heart of non-equilibrium statistical
mechanics and one has no general answer to it.

Qualitatively however, we expect to find in the
spectrum of L_N:

1) the hydrodynamical modes $|\tilde{\nu},q\rangle$ ($\nu \in (1..5)$):
these are of no help here because we have a total
wave number $q \to o$, in which case the flow operator
J^{η} is orthogonal to these modes (see (III.45)).

2) the so-called "relaxation" modes. Although
these are difficult to define rigorously, they
correspond physically to the fast decay of an
arbitrary deviation from equilibrium. The corresponding eigenvalues are all finite and, <u>although
these modes are expected to give the dominant
contribution to the zero frequency transport coef-
ficients</u>, they should give roughly an exponential
decay to the Kubo-integrand: they should not
contribute significantly to the long time limit of
(III.50).

3) nonetheless, we can find a good candidate
for a slow decay in modes built up by the <u>super-</u>

position of two (or more) independent hydrodynamical modes, with total wave number zero.

We suppose here that we have:

$$L_N \tilde{a}_{\nu,q'} \tilde{a}_{\nu',-q'} |\rangle = (z_{\nu,q'} + z_{\nu',-q'}) \tilde{a}_{\nu,q'} \tilde{a}_{\nu',-q'} |\rangle$$

$$(\nu, \nu' \in (1..5)) \qquad (III.53)$$

where $\tilde{a}_{\nu,q}$ denote the proper linear combinations of the $a_{\nu,q}$ defining the eigenmodes (III.40).

The following remarks can be made about eq. (III.53):

1) To the best of our knowledge, there exists no proof that this superposition of independent modes really furnishes proper eigenfunctions of the Liouville operator L_N. This is however a very natural assumption because, in a large system, it should be possible to build two independent hydrodynamical modes involving different particles, which, up to order $1/N$, completely ignore each other.

2) It can be shown, again in the limit of a large system, that these modes are orthogonal to the single hydrodynamical modes (III.40). Similarly, we could construct orthogonal eigenfunctions involving three, four... hydrodynamical modes. However, this basis is obviously not complete because the relaxation modes are not included in this mode description.

3) The interesting property of the eigenvalues (III.53) is that, for small enough q', they go to zero: the corresponding singularities in (III.50) are thus as close as we wish to the real axis and

can thus give rise to a very slow decay of the
Kubo integrand.

Where acting on q=o states on the left and on
the right, the operator (III.51) has the following
representation:

$$\frac{1}{L_N^{-z}} = \frac{1}{2!\Omega^2} \sum_{q'} \sum_{\boldsymbol{\nu},\boldsymbol{\nu}'=1}^{5} |\tilde{\boldsymbol{\nu}}q';\tilde{\boldsymbol{\nu}}',-q') \frac{1}{z_{\boldsymbol{\nu},q'}+z_{\boldsymbol{\nu}',-q'}-z}$$

$$\langle \tilde{\boldsymbol{\nu}}q';\tilde{\boldsymbol{\nu}}',-q'| \qquad \text{(III.54)}$$

+(three and more modes) + (relaxation modes).
Note the factor $1/2!$ in this equation, necessary to
avoid overcounting; similarly, the factor Ω^{-2} is
a consequence of the normalization condition
(III.16). Moreover, we have not written explicitly
the terms involving three or more modes because
they can be shown not to contribute to the dominant asymptotic decay of (III.50).

With (III.54), the asymptotic evaluation of
(III.50) becomes almost trivial. We insert (III.54)
into (III.48) and obtain from (III.50):

$$\lim_{t\to\infty} \frac{1}{\Omega k_B T} \langle \hat{J}^{\eta}(t)\hat{J}^{\eta}\rangle = \lim_{\varepsilon\to o} \frac{1}{2\pi i} \oint_C \exp{-izt}\, dz$$

$$\frac{1}{8\pi^3} \sum_{\boldsymbol{\nu},\boldsymbol{\nu}'=1}^{5} \int d^3q' \frac{1}{2!} \frac{1}{\Omega^2 k_B T} \langle |J^{\eta}| \tilde{\boldsymbol{\nu}}q';\tilde{\boldsymbol{\nu}}',-q'\rangle$$

$$\frac{1}{z_{\boldsymbol{\nu},q'}+z_{\boldsymbol{\nu}',-q'}-z} \langle \tilde{\boldsymbol{\nu}}q';\tilde{\boldsymbol{\nu}}',-q'|\hat{J}^{\eta}|\rangle \qquad \text{(III.55)}$$

Assuming that the q' and z integrals may be
commuted, we get:

$$\lim_{t\to\infty} \frac{1}{\Omega k_B T} \langle \hat{J}^{\eta}(t)\, \hat{J}^{\eta} \rangle =$$

$$= \lim_{t\to\infty} \frac{1}{2\Omega^2 k_B T} \frac{1}{8\pi^3} \sum_{\nu,\nu'=1}^{5} \int d^3q'$$

$$\langle \hat{J}^{\eta} | \tilde{\psi}_{q'}\, ; \tilde{\nu}', -q \rangle \langle \tilde{\psi}_{q'}\, ; \tilde{\nu}', -q | \hat{J}^{\eta} \rangle \exp{-i}\left[z_{\nu,q'} + z_{\nu',-q} \right] t$$

$$(III.56)$$

We have, in principle, to sum over 5 x 5 possible
combinations of the modes, but a considerable simplification occurs when we take into account the various symmetries of the problem. Before doing this,
let us point out that the modes (III.40) correspond
to q aligned along the x-axis; obviously, in
(III.56) we need these modes for an arbitrary direction of q. From simple geometrical arguments,
it is readily seen that it is sufficient to rede-
fine the momentum operators $a_{i,q}$ (i=3,4,5) according
to:

$$a_{3,q} = \left(^1/k_B T n\right)^{\frac{1}{2}} \left[\hat{\underline{g}}_q \ \underline{1}q \right] \quad \text{(longitudinal momentum)(a)}$$

$$a_{4,q} = \left(^1/k_B T n\right)^{\frac{1}{2}} \left\{ -\hat{g}_{x,q}\left[(1q)_x(1q)_y\right] + \hat{g}_{y,q} \times \right.$$

$$\times \left[(1q)_x^2 + (1q)_z^2 \right] - \hat{g}_{z,q}\left[(1q)_z(1q)_y\right] \right\} \Big/ \left((1q)_x^2 + (1q)_z^2 \right)$$

$$\text{(transverse momentum (b)}$$

$$a_{5,q} = \left(^1/k_B T n\right)^{\frac{1}{2}} \left\{ \hat{g}_{z,q}(1q)_x - \hat{g}_{x,q}(1q)_z \right\} \Big/ \sqrt{(1q)_x^2 + (1q)_z^2}$$

$$\text{(transverse momentum (c)}$$

$$(III.57)$$

in order to maintain the validity of (III.40) for

arbitrary direction of the unit vector $\mathbf{1}_q$.

With this remark in mind, we notice that the quantity $\langle \hat{J}^{\eta} | \tilde{\mathbf{v}}_{q'} ; \tilde{\mathbf{v}}'_{-q'} \rangle$ involves an equilibrium average over $\hat{J}^{\eta}$, which, from (I.8) is the x-y component of a tensor. From (III.40,57), we also see that in the small q' limit (which is legitimate to consider because, for long times, small q' dominates the integral (III.56)), the only tensor components of this kind which can be built with the mode $| \tilde{\mathbf{v}}_{q'} , \tilde{\mathbf{v}}'_{-q'} \rangle$ are either 1) a shear-shear mode $| \tilde{\mathbf{v}}=\tilde{4},\tilde{5},q' ; \tilde{\mathbf{v}}'=\tilde{4},\tilde{5};-q' \rangle$ or   2) a sound-sound mode $| \tilde{\mathbf{v}}=\tilde{2},\tilde{3},q' ; \tilde{\mathbf{v}}'=\tilde{3},\tilde{2};-q' \rangle$. Moreover, the tensors involved in these modes depend on the momentum variables alone. This has as a consequence that when we perform the equilibrium average implied by $\langle \hat{J}^{\eta} | \tilde{\mathbf{v}}_{q'} ; \tilde{\mathbf{v}}', -q' \rangle$ only the kinetic part of the momentum flow $\hat{J}^{\eta}$ gives a non-vanishing contribution.

More precisely, using (III.40,57) and (I.8), we get as the only non-vanishing averages $\langle \hat{J}^{\eta} | \tilde{\mathbf{v}}_{q'} ; \tilde{\mathbf{v}}', -q' \rangle$:

$$\lim_{q' \to 0} \langle \hat{J}^{\eta} | \tilde{\mathbf{v}}=\tilde{4},q' ; \tilde{\mathbf{v}}' = \tilde{4},-q' \rangle =$$

$$= \left\{ \iint dr^N dp^N \sum_{a=1}^{N} p_{a,x} p_{a,y} \sum_{b=1}^{N} p_{b,x} \sum_{c=1}^{N} p_{c,y} \right.$$

$$\left. \times \frac{e^{-\beta H_N}}{Z_N} \right\} \frac{(-2 (\mathbf{1}_q)_x (\mathbf{1}_{q'})_y)}{(n k_B T)} = \Omega(k_B T)(-2 (\mathbf{1}_{q'})_x (\mathbf{1}_{q'})_y).$$

$$\text{(III.58)}$$

$$\lim_{q \to 0} \langle \hat{J}^{\eta} | \tilde{\mathbf{v}}=\tilde{4},q' ; \tilde{\mathbf{v}}=\tilde{5},-q' \rangle = \lim_{q \to 0} \langle \hat{J}^{\eta} | \tilde{\mathbf{v}}=\tilde{5},q' ; \tilde{\mathbf{v}}=\tilde{4},-q' \rangle =$$

$$\Omega(k_B T)(-(\mathbf{1}_{q'})_z) \qquad \text{(III.59)}$$

$$\lim_{q \to o} \langle \hat{J}^{\eta} | \tilde{\nu} = \tilde{2}, q' ; \tilde{\nu}' = \tilde{3}, -q' \rangle = \lim_{q \to o} \langle \hat{J}^{\eta} | \tilde{\nu} = \tilde{3}, q' ; \tilde{\nu}' = \tilde{2}, -q' \rangle$$

$$= \Omega(k_B T) \, 1_{q_x'^2} \, 1_{q_y'^2} \qquad (*) \text{(III.60)}$$

The same formulas are valid for the "conjugate" expressions $\langle \tilde{\nu}, q' ; \tilde{\nu}', -q' | \hat{J}^{\eta} \rangle$. Inserting these results into (III.56), we get, with the help of (III.43):

$$\lim_{t \to \infty} \langle \hat{J}^{\eta}(t) \hat{J}^{\eta} \rangle = \lim_{t \to \infty} \left[\frac{k_B T}{2} \frac{4\pi}{8\pi^3} \int_0^\infty q'^2 dq' \left\{ j_1 \cdot \right. \right.$$

$$\left. \left. \cdot \text{ ext} - 2\eta q'^2 t + j_2 \cdot \exp{-2\Gamma q'^2 t} \right\} \right] \qquad \text{(III.61)}$$

where

$$j_1 = \frac{\int d\bar{\Omega} [4 \, (1_{q_x}^2 \, 1_{q_y}^2) + 2 \, (1_{q_z}^2)]}{\int d\bar{\Omega}} = \frac{14}{15} \qquad \text{(III.62)}$$

and

$$j_2 = \frac{\int d\bar{\Omega} \, (1_{q_x})^2 \cdot (1_{q_y})^2}{\int d\bar{\Omega}} = \frac{2}{15} \qquad \text{(III.63)}$$

The remaining q'-integral is trivial to perform and leads to:

$$\lim_{t \to \infty} \frac{1}{\Omega k_B T} \langle \hat{J}^{\eta}(t) \, \hat{J}^{\eta} \rangle =$$

$$k_B T \left[\frac{7}{15} \frac{1}{(8\pi \eta t)^{\frac{3}{2}}} + \frac{1}{15} \frac{1}{(8\pi \Gamma t)^{\frac{3}{2}}} \right] \qquad \text{(III.64)}$$

which shows indeed a slow $t^{-\frac{3}{2}}$ decay.

Let us close this chapter by a few remarks of general interest:

1) The simplifying feature of shear viscosity

$(*)$The combinations $\tilde{\nu} = \tilde{\nu}' = \tilde{2}$ and $\tilde{\nu} = \tilde{\nu}' = \tilde{3}$ give contributions which are smaller than $t^{-3/2}$ in the $t \to \infty$ limit of (III.56), as also do $\tilde{\nu} = \tilde{2}, \tilde{3}, \tilde{\nu}' = \tilde{4}, \tilde{5}$.

is that the coefficients in front of (III.62) only
depend on temperature. This is connected with the
absence of potential contributions noted above.
The other transport coefficients can be analyzed
in a similar way, but there we have to evaluate
complicated equilibrium correlations instead of
the simple integral involved in (III.58). We shall
not display the corresponding formulas here [20].

2) If the same calculation is performed in
2 dimensions, one finds:

$$\langle \hat{J}^{\eta}(t)\, \hat{J}^{\eta}(o) \rangle_{2d} \propto \frac{1}{t} \qquad\qquad (III.65)$$

But then, we get:

$$\eta_{2d} \propto \int_0^\infty \langle \hat{J}^{\eta}(t)\hat{J}^{\eta}(o) \rangle_{2d} = \infty\ ! \qquad (III.66)$$

This indicates a serious difficulty, because
the method itself is based on the very existence
of these transport coefficients! Although there
exist in the literature some attempts to overcome
this difficulty [21], we are not aware of any method
that solves the contradiction implied by (III.66)
on the sole basis of the traditional Green-Kubo
formulas.

3) Although very plausible, the Kadanoff-Swift
formalism presented here (and the equivalent methods
due to Hauge et al., Pomeau etc.[5]) is largely
based on assumptions of a semi-macroscopic nature.
It is thus very comforting to know that a fully
microscopic theory - based on the kinetic equation
formalism - leads to identical results [22]. Although
the published proof is limited to moderate densities,

the same result can be proved in full generality[23].
As we shall present an analysis of this kinetic
method in the next chapter for the van der Waals
fluid, we shall not discuss it for our present
problem.

4) A remarkable feature of (III.64) is that it
is completely model independent, involving solely
the transport coefficients of the fluid .

5) The usefulness of this result is still open
to question. Indeed, we have only derived an asymptotic formula, valid for long times, and it is very
difficult to assert the time scale after which it
is valid. Theoretically, this would require an
evaluation of the corrections indicated in (III.54)
and, in general, this is beyond our present theoretical abilities. Yet, from the results of computer
experiments, it seems that (III.64) should only be
valid for very long times in three dimensions at
fluid densities [24].

IIIc) <u>Critical dynamics</u>

Hydrodynamical concepts also play an essential role in this problem; indeed, we can expect
a <u>priori</u> that long-wave-length phenomena are
important because the main characteristic length
of the problem is the correlation length which
becomes infinite at the critical point.

The same type of formula as we have established in Section IIIb can be used (with,

however, the non-trivial difficulty that we
cannot go directly to the $q \to 0$, $z \to 0$ limit
for the transport coefficients $X(q,z)$) to
predict the singularities of critical transport
coefficients. The equilibrium quantities which
appear in these formulas can be evaluated with
the help of the idea of static scaling[31,32]. We
then obtain a set of self-consistent equations
which allow us to express the singular part of
the transport coefficients in terms of the static
singularities. For lack of space, we shall not
discuss this in detail here and we refer the
reader to the original papers for this remarkably
successful theory[6,30].

IV.   MODE COUPLING (II) AND THE TRANSPORT COEFFI-
CIENTS OF THE VAN DER WAALS FLUID:

IV.a) Introduction.

The number of soluble models for three-
dimensional fluids is very limited. Essentially,
we have:

1) <u>The dilute or moderately dilute gas</u>: the
equilibrium theory is given by the well-known
virial expansion while the non-equilibrium theory
is based on the Boltzmann equation (see Chapter II,
Section a). We also have generalizations of this
Boltzmann equation for higher densities although

one is rapidly confronted with divergence difficulties, giving rise to logarithmic terms in the density expansion of transport coefficients [25].

2) <u>The dilute or hot plasma</u>: The equilibrium
description of this system is provided by the
Debye-Hückel theory; the non-equilibrium analogue
is furnished by the Balescu-Lenard-Guernsey kinetic equation [26].

3) <u>The heavy particle</u>: as we are concerned
here with one single heavy particle, the equilibrium
properties are trivially described by the Maxwell-
Boltzmann distribution; the transport properties
of this Brownian particle are given by the well-
known Fokker-Planck equation [27].

However, at equilibrium there exists still
another model which turns out to be very convenient
in the qualitative description of the thermodynamic
properties of a fluid: it is the so-called van der
Waals fluid.[28] Formulated in our modern language,
this model is based on the splitting of the total
interaction potential into two parts:

$$V(r) = V^S(r) + \gamma^3 V^L(\gamma r) \tag{IV.1}$$

Here $V^S(r)$ describes the short-range part of the
potential (mostly repulsive), while $\gamma^3 V^L(\gamma r)$
represents the long-range (attractive) part; γ is
a measure of the inverse range of the potential and
is supposed to be small: the attractive part is
thus assumed to be weak and long range, in such a
way that the total potential energy

$$\langle V^L \rangle = \int d^3r \, \gamma^3 V^L(\gamma r) = \int d^3x \, V^L(x) \tag{IV.2}$$

remains finite.

Physically, the simplicity of this model comes from the fact that the long-range potential involves particles which are mostly very far from each other: in fact, they are so far apart, that the short range part of the potential, $V^S(r)$, is essentially unable to correlate them: as a consequence, the potential energy due to V^L can be calculated as if the particles were free. By this kind of argument, one gets for example for the equation of state:

$$p(n,T) = p^S(n,T) + \frac{n^2}{2} \tilde{V}_0{}^L + O(\gamma^3) \qquad (IV.3)$$

where

$$\tilde{V}_y{}^L = \int dx \, e^{iyx} \, v^L(x) \qquad (IV.4)$$

We want to show here how it is possible to extend this van der Waals theory to the calculation of transport coefficients. In order to do this, we will use kinetic theory; however, as the explicit calculations rapidly become very involved, we will only sketch the main features of the theory, refering the reader to the original papers for the technical details [7].

We again start from the Green-Kubo formula (I.5), and first indicate how an explicit evaluation of this expression can be reduced to the solution of a kinetic equation. To simplify, we will consider here the kinetic part of η, in an equilibrium ensemble depending only on the kinetic energy H_0; we thus look for:

$$\eta_o^{\kappa\kappa} = \lim_{t\to\infty} \frac{1}{\Omega\, k_B T} \int_0^t d\tau \left\langle J^{\eta,\kappa}\, e^{-iL_N\tau}\, J^{\eta,\kappa} \right\rangle_0 \tag{IV.5}$$

where:

$$J^{\eta,\kappa} = \sum_{a=1}^{N} p_{a,x} p_{a,y} \tag{IV.6}$$

and

$$\langle \cdots \rangle_0 = \frac{\int dr^N dp^N \ldots \exp{-\beta H_0}}{Z_0} \tag{IV.7}$$

We retain, of course, the interactions in the Liouville operator L_N. The calculation of the complete
η runs along similar lines but the technique is
much more involved.

Using the symmetric role played by all the
particles in the system, we can rewrite (IV.5) as

$$\eta_o^{\kappa\kappa} = \lim_{t\to\infty} \frac{n}{k_B T} \int_0^t d\tau \int dr^N dp^N p_{1,x} p_{1,y}\, \delta f_N(r^N,p^N;\tau) \tag{IV.8}$$

where

$$\delta f_N(r^N,p^N;\tau) = e^{-iL_N\tau}\, \delta f_N(r^N,p^N;o) \tag{IV.9}$$

$$\delta f_N(r^N,p^N;o) = \sum_{a=1}^{N} p_{a,x}\, p_{a,y}\, \frac{e^{-\beta H_0}}{Z_0} \tag{IV.10}$$

From (IV.9), we immediately see that $\delta f_N(r^N,p^N;\tau)$
satisfies the Liouville equation:

$$i\partial_\tau\, \delta f_N(r^N,p^N;\tau) = L_N \cdot \delta f_N(r^N,p^N;\tau) \tag{IV.11}$$

The possibility of deriving a kinetic theory
for η^κ is based on the fact that, in order to
evaluate (IV.8), we do not need to know the time
development of the complete N-body distribution
δf_N. Indeed, we can rewrite (IV.8) as:

$$\eta_0^{\kappa\kappa} = \frac{n}{k_B T} \lim_{t \to \infty} \int_0^t d\tau \int dp_1 p_1^x p_1^y \, \delta\varphi_1(p_1;\tau) \qquad (IV.12)$$

where:

$$\delta\varphi_1(p_1;\tau) = \int dr^N dp^{N-1} \delta\rho_N(r^N, p^N;\tau) \qquad (IV.13)$$

and:

$$\delta\varphi_1(p_1;o) = p_{1,x} p_{1,y} \, \varphi_1^{eq}(p_1) \qquad (IV.14)$$

Thus, we need only the <u>one-particle</u> velocity distribution function $\delta\varphi_1(p_1;t)$ instead of $\delta\rho_N(r^N, p^N;\tau)$. The derivation of a kinetic equation for $\delta\varphi_1(p_1;t)$ starting from the Liouville equation (IV.9) is a central problem of irreversible statistical mechanics; yet, as in Chapter II, we shall not present this calculation here but shall simply give the result, which can be obtained from (II.37) by setting q=o. We have indeed:[*]

$$\partial_t \delta\varphi(p_1;t) = C_0^\ell \, \delta\varphi(p_1;t) \qquad (IV.15')$$

where the explicit form of C_0^ℓ will not be needed here.

Inserting the formal solution of (IV.15') into (IV.12), we get finally:

$$\eta_0^{\kappa\kappa} = -\frac{n}{k_B T} \int dp_1 \, p_{1,x} \, p_{1,y} \, \frac{1}{C_0^\ell} \, p_{1,x} p_{1,y} \, \varphi_1^{eq}(p_1)$$

$$(IV.15)$$

If we want to analyse $\eta_0^{\kappa\kappa}$ for a van der Waals potential (IV.1), we thus have to study the

---

[*] More precisely, it can be shown[13] that the operator C_0^ℓ appearing in (IV.15) is obtained from (II.38) by setting q=o <u>and</u> putting formally $\Omega_q=1$. This point will play no role in the present qualitative discussion and we shall ignore it.

linearized collision operator in this case; this
is the object of the next section.

    IV.b) <u>The collision operator of the van der
Waals fluid</u>:

    Because the attractive potential V^L is,
in some sense, weak, it is tempting to try a naïve
perturbation expansion of C_o^{ℓ} in powers of V^L:

$$C_o^{\ell} = C^S + C^{V_L} + C^{V_L^2} + \ldots \qquad (IV.16)$$

where C^S denotes the pure short range operator,
and $C^{V_L^n}$ the correction of nth order in V^L. However, such an expansion does not converge!

    Instead of showing this in detail, let us
illustrate the difficulties connected with the use
of (IV.16) on a very simple contribution to this
expansion: we consider the Born approximation to
C^{V_L}. We thus take a process of second order in the
interactions, involving one short- and one long-
range interaction. This is depicted in Fig. 3,
where we schematically indicate the trajectory of
the two colliding particles and where the white
and black rectangles respectively represent the
short range and long range interaction.

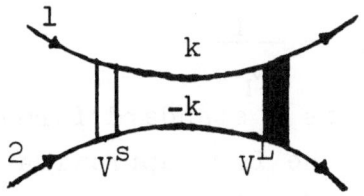

<u>Fig. 3</u>: Schematic description of a collision
process in the Born approximation.

Although it is possible to formulate precise
rules which allow us to associate uniquely a given
analytical contribution to C_o^ℓ with a graph of the
type indicated in Fig. 3, we shall not do so here
and we shall merely indicate the structure of this
contribution; this can easily be understood by
analogy with usual scattering theory: it is [12)33)]

$$\frac{1}{\Omega} \int_0^\infty d\tau \left\{ \frac{1}{8\pi^3} \int d^3k \; kV_k^{I,} \frac{\partial}{\partial v_{12}} \right.$$

$$\left. \tilde{G}_k^o (1;\tau) \tilde{G}_{-k}^o (2;\tau) \; kV_k^S \; \frac{\partial}{\partial v_{12}} \right\} \qquad (IV.17)$$

Here we have an integration over the wave-number
exchanged in the collision process, a factor
$\delta L_{12} = k \; V_k \frac{\partial}{\partial v_{12}}$ for each interaction and a
propagator $G_k^o (1; \tau)$ representing the free
motion of each particle during the time interval
τ between the two successive interactions; finally,
we have an integral over this time interval. Let
us also point out the explicit form of the propagator:

$$\tilde{G}_k^o(1;\tau) = \exp \; ikv_1\tau = \frac{-1}{2\pi i} \oint_C \exp-iz\tau \; G_k^o(1,z) \quad (IV.18)$$

where

$$G_k^o (1;z) = \frac{1}{kv_1 - z} \qquad (IV.19)$$

while eq. (IV.17) is a standard formula for classical scattering in the Born approximation, the long
range character of V^L gives it a unique, and un-
pleasant, feature: indeed, the Fourier transform
V_k^L is sharply peaked (with a width $\approx \gamma$) around

k=o and the wave-numbers which are integrated over
in (IV.17) are therefore limited by the condition
k $\leqslant \gamma$ (see Fig. 4). For these small k-values,
$G_k^o(j;z)$ describes the free propagation of particle
j over large distances, of the order γ^{-1}. As we are
considering a dense fluid, such a picture is totally
unrealistic:

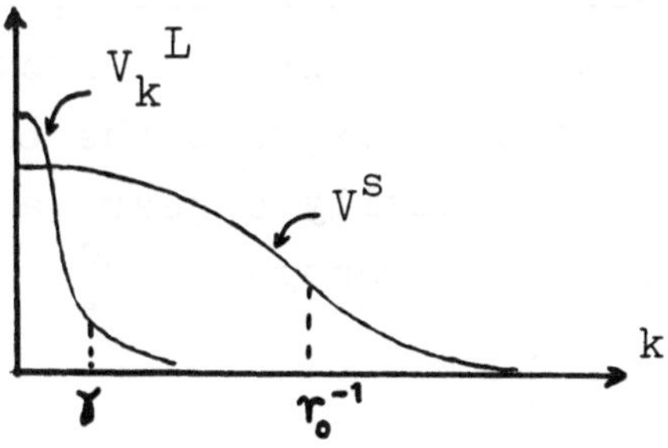

Fig. 4: <u>Fourier transforms of the long range and
short range potential</u>.

Over such distances, particle j feels forces (both
short and long range) due to all the others. This
is the physical origin of the divergence of the
naïve expansion (IV.16). Even in the simplest
approximation, one has to replace the free propagation by the exact motion in the presence of the
other particles. This is schematically depicted in
Fig. 5; mathematically, this amounts to

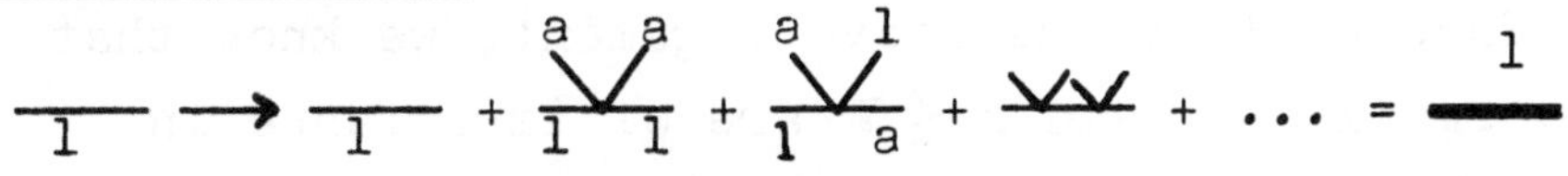

$\bigvee$ = collision with a fluid particle.

Fig. 5: <u>Schematic description of the propagator
renormalization</u>.

"renormalizing" the propagator and to making the
replacement:

$$G_k^o(1;z) \longrightarrow G_k(1;z) = \frac{1}{kv_1 + iC_k^\ell(z) - z} \qquad (IV.20)$$

where $C_k^\ell(z)$ is the generalization, to finite wave number and frequency, of the linearized collision operator C_o^ℓ; it involves the effect of both short and long range forces.

To summarize this discussion, we have arrived at the idea that, in a system involving long range forces, one should reexpress the collision operator C_o^ℓ, which initially appears as a functional of the unperturbed propagator G_k^o:

$$C_o^\ell = C(\{G_k^o\}) \qquad (IV.21)$$

as a functional of the renormalized propagator (IV.20):

$$C_o^\ell = \tilde{C}(\{G_k\}) \qquad (IV.22)$$

We can expect in this case serious mathematical difficulties because G_k is not a simple function like G_k^o (see (IV.19)), but is in general a complicated inverse operator (see (IV.20)). The theory of hydrodynamical modes developed in chapter II will, however, help us considerably in tackling this problem.

Indeed, from the above argument, we know that only small values of k $\lesssim \gamma$ are of importance in describing the effect of the long range forces. Let us first suppose the stronger condition k≪γ; thus k is much smaller than all the molecular parameters of the problems and we are in the hydrodynamic regime. In this case (using the fact that

the relevant values of z are also small) we can
use eq. (II.47) to express the dominant part of
the propagator, namely:(*)

$$G_k(1;z) = \sum_{\mu=1}^{5} |\psi_\mu^{1k}\rangle \frac{1}{i\,\lambda_\mu^k - z} \langle \bar{\psi}_\mu^{1k}| \qquad (IV.23)$$

$$k \to o$$
$$z \to o$$

where the λ_μ^k are numbers, expressed solely in
terms of the thermodynamic and transport coefficients of the van der Waals fluid, while the eigenfunctions are linear combinations of the collision invariants $1, \underline{v}, v^2$. For example:

$$|\psi_\mu^{1k}\rangle = \sum_{\alpha=1}^{5} C_{\mu\alpha} |\alpha\rangle \qquad (IV.24)$$

where the coefficients $C_{\mu\alpha}$ depend on only the
thermodynamic properties of the fluid.

We are, however, confronted here with a more
delicate problem because k, albeit small, can be
of order γ . We are thus not strictly in the hydrodynamic regime and we have to consider the asymptotic behaviour of the propagator in the double
limit:

$$k \to o$$

$$k/\gamma = y = \text{ finite quantity.} \qquad (IV.25)$$

The analysis of (IV.23) in this limit is very
delicate and will not be reproduced here; it is
essentially of a selfconsistent nature because the
<u>eigenvalues which</u> generalize the λ_μ^K involve the

(*) As in chapter II, section b, we have to take
into account that, even for $k \to o$, the modes $|\psi_\mu^{1k}\rangle$
depend on the direction of $\underline{1}_k$.

transport coefficients of the van der Waals fluid;
these are precisely the quantities we are trying
to determine. The result is, however, very simple[7]:
the structure (IV.23) remains valid but the eigenfunctions and eigenvalues now depend on y:

$$G_k (1,z) \quad = \sum_{\mu=1}^{5} | \psi_\mu^{1k}(y) \rangle \frac{1}{i \lambda_\mu^k(y)-z} \langle \psi_\mu^{1k}(y) | \quad (IV.26)$$
$$k \to o$$
$$k/\gamma = y \text{ finite}$$
$$z \to o$$

where the linear relationship (IV.24) still holds
true (with $C_{\mu\alpha}(y)$) while the eigenvalues are now
(compare with (II.50)):

$$\lambda_{1,2}^k (y) = \pm ic(y)k - \frac{1}{2}\frac{k^2}{n}\left[\frac{4}{3}\eta^s + (\frac{1}{C_p(y)} - \frac{1}{C_v^s})\kappa^s + \zeta^s\right] \quad (a)$$

$$\lambda_{3,4}^k = -\eta^s k^2 \quad (b)$$

$$\lambda_5^k = -\frac{k^2 \kappa^s}{nC_p(y)} \quad (c) \quad (IV.27)$$

Here all quantities indexed by the superscript
s refer to the purely short range reference system,
while the y-dependent quantities denote <u>generalized</u>
thermodynamic properties (*). We have indeed:

$$c^2(y) = c^{s2} + n \widetilde{V}_y^L \quad (IV.28)$$

$$C_p(y) = C_p^s + \frac{T}{n}\left[(\frac{\partial p}{\partial T})_n^s\right]^2\left[\chi_T(y) - \chi_T^s\right] \quad (IV.29)$$

where the generalized compressibility $\chi_T(y)$ is

(*) As a matter of fact, we have here a concrete
example, for a van der Waals fluid, of the q-dependent thermodynamic quantities introduced in
the Kadanoff-Swift formalism; see eqs. (III.21,22).

defined by:

$$\chi_T(y) = \frac{1}{n\left[(\frac{\partial p}{\partial n})^S_T + n\, \tilde{V}^L_y\right]} \tag{IV.30}$$

These same generalized quantities appear in the coefficients $c_{\mu\beta}(y)$ (see (IV.29) and (II.49)). It is a simple exercise to check that, when $y \to o$, eq. (IV.28-30) reduce indeed to the corresponding equilibrium properties of the van der Waals fluid.

As already mentioned, the proof of these results on the basis of kinetic theory is not trivial. Yet the answer is easy to understand on a macroscopic basis: it is indeed easy to show that eqs. (IV.27) give the eigenvalues of the macroscopic equation of motion of a van der Waals fluid[30]. These are obtained from (I.15) by assuming that the only effect of the long range interaction is to introduce an average force term in the Navier-Stokes equations. Thus, instead of (I.15), we write:

$$-i\omega n_{\underline{q},\omega} + i\underline{q}n\underline{v}_{\underline{q},\omega} = n_{\underline{q}}(t=o) \tag{a}$$

$$-i\omega \underline{v}_{\underline{q},\omega} + i\underline{q}\frac{1}{n}(\frac{\partial p}{\partial n})^S_T n_{\underline{q},\omega} + i\underline{q}\frac{1}{n}\,n\,\tilde{V}^L_y\, n_{\underline{q},\omega}$$

$$+ i\underline{q}\frac{1}{n}(\frac{\partial p}{\partial T})^S_n T_{\underline{q},\omega}$$

$$+\left\{\frac{\eta^S}{n}q^2\underline{v}_{\underline{q},\omega} + \frac{(\zeta^S + \frac{1}{3}\eta^S)}{n}\underline{q}\left[\underline{q}\underline{v}_{\underline{q},\omega}\right]\right\} = \underline{v}_{\underline{q}}(t=o) \tag{b}$$

$$-i\omega T_{\underline{q},\omega} + i\underline{q}\frac{T}{nC_v}(\frac{\partial p}{\partial T})^S_n \underline{v}_{\underline{q},\omega} + \frac{\kappa^S}{nC^S_v}q^2 T_{\underline{q},\omega} = T_{\underline{q}}(t=0) \tag{c}$$

$$\tag{IV.31}$$

where we have underlined the supplementary long
range term. It is easy to verify that the eigenvalues associated with (IV.31) are precisely
given by (IV.27).

Once we know how to work with the renormalized
propagator, the analysis of the γ dependence of
(IV.22) can be made straightforwardly although
this proof involves a detailed analysis of the
many-body structure of C_0^ℓ. Such a study would lead
us too far astray and we shall merely give the
main results:

1) The collision operator can be expanded
in powers of the parameter γ:

$$C_0^\ell = C^s + \gamma C^{(1)} + \ldots \qquad\qquad (IV.32)$$

2) The first correction $C^{(1)}$ is of the type
depicted in Fig. 6, where the bubble represents
the most general hard core collision process,
involving an arbitrary number of particles. Moreover the heavy straight line represents the

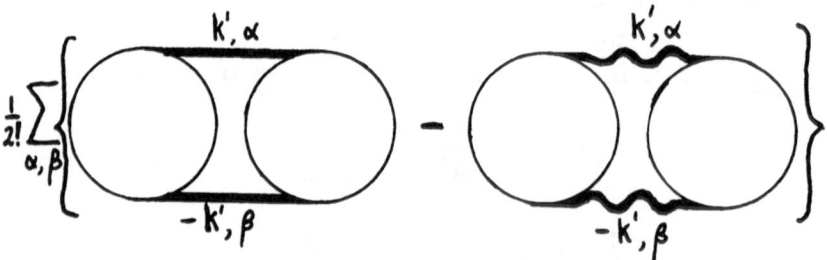

Fig. 6: Schematic description of the first
correction $C^{(1)}$.

"van der Waals propagator" (IV.26), while the
heavy wavy line represents the corresponding

propagator for the reference system, obtained from
(IV.26) by formally setting $V_y^L \equiv 0$.

Denoting by $\widetilde{C}^s$ the bubble operator in Fig. 6,
(which is assumed to have a finite limit when
$k \rightarrow 0$), we can readily check that the graph of Fig.6
indeed gives a correction of order γ; we have:

$$\gamma C^{(1)} = \frac{1}{2!8\pi^3 n} \sum_{\beta=1}^{5} \int d^3k' \int dv_2 \left[\widetilde{C}^s \right.$$

$$\left| \psi_\alpha^{1k'}(y) \right\rangle \left| \psi_\beta^{-1k'}(y) \right\rangle \frac{1}{i\lambda_\alpha^{k'}(y)+i\lambda_\beta^{k'}(y)} \left\langle \psi_\alpha^{1k'}(y) \right| \left\langle \psi_\beta^{-1k'}(y) \right| \widetilde{C}^s$$

$$\left. - \text{ hard core} \right] \qquad\qquad (IV.33)$$

or

$$\gamma C^{(1)} \sim \int_{k<\gamma} \frac{k'^2 dk'}{k'^2} \sim \gamma \qquad\qquad (IV.34)$$

Let us stress that, in this estimate, we have
taken into account that for $k' > \gamma$, the first term in
the bracket of (IV.33) exactly reduces to the hard
core term (because $\widetilde{V}_{k'}^L / \gamma = 0$ for $k' > \gamma$)

Before closing this section, let us point out
that both Fig. 6 and eq. (IV.33) clearly indicate
that the purely hard core term gives a non-negli-
gible contribution coming from the hydrodynamical
modes: this effect is closely related to the $t^{-3/2}$
behaviour of the Green-Kubo integrands discussed
in Chapter III. This point will be further discussed in interpreting the final results of section
IV.c.

IV.c) <u>Corrections to the transport coeffi-</u>
<u>cients of the van der Waals fluid:</u>
Inserting (IV.32) into (IV.15), we
get the expansion

$$\eta_0^{KK} = (\eta_0^{KK})^S + \gamma(\eta_0^{KK})^{(1)} + \dots \qquad (IV.35)$$

with

$$(\eta_0^{KK})^{(1)} = \frac{n}{k_B T} \int dp_1 p_{1,x} p_{1,y} \left[\frac{1}{c^s} c^{(1)} \frac{1}{c^s} \right] p_{1,x} p_{1,y} \varphi_1^{eq}(p \qquad (IV.36)$$

and, from the structure of the bracketed term in
this last equation - which involves both the short
range operators c^s and $\widetilde{\widetilde{c}}^s$ (see (IV.33))- we of
course expect that some model for the short range
fluid is required before a closed expression for
$(\eta_0^{KK})^{(1)}$ can be obtained.

There is however a remarkable property of the
operators c^s and $\widetilde{\widetilde{c}}^s$ which makes possible a model-
independent evaluation of (IV.36). This property
can be expressed in the following way:

$$\frac{1}{c^s(v_1)} \int dv_2 \, \widetilde{\widetilde{c}}^s(v_1,v_2) \Big[\langle v_1 | \psi_\alpha^{1k'}(y) \rangle \langle v_2 | \psi_\beta^{-1k'}(y) \rangle$$

$$+ \langle v_2 | \psi_\alpha^{1k'}(y) \rangle \langle v_1 | \psi_\beta^{-1k'}(y) \rangle \Big] =$$

$$= - \langle v_1 | \psi_\alpha^{1k'} * \psi_\beta^{-1k'}(y) \rangle \qquad (IV.37)$$

where the right hand side is a convenient notation
for the following function:

$$\langle v_1 | \psi_\alpha^{1k'} * \psi_\beta^{-1k'}(y) \rangle =$$

$$= \langle v_1 | \psi_\alpha^{1k'}(y) \rangle \langle v_1 | \psi_\beta^{-1k'}(y) \rangle / \varphi_{(v_1)}^{eq} \qquad (IV.38)$$

This property is remarkable indeed because, as is stressed by the notation, $C^S(v_1)$ is a one-particle operator (see for example (II.3) for the dilute gas limit) while $\tilde{\tilde{C}}^S(v_1,v_2)$ is a two-particle operator acting on two simultaneous deviations from equilibrium: this appears clearly from Fig. 6 and eq. (IV.37). The central point in the derivation of (IV.37) is that this operator $\tilde{\tilde{C}}^S$ does not act on arbitrary functions but, by (IV.24), only on linear combinations of the collision invariants. We shall not prove eq. (IV.37) here but, in an appendix, we show how this result can be simply proved in the dilute gas limit.

Inserting (IV.37) into (IV.36), together with the analogous result for the conjugate quantity $(\tilde{\tilde{C}}^S \frac{1}{C^S})$, we get simply:

$$\eta_o^{\kappa\kappa(1)} = \frac{1}{2!k_B T \, 8n^3} \sum_{\alpha,\beta=1}^{5} \int d^3k'$$

$$\left\{ \left[\int dp_1 p_{1,x} p_{1,y} \left\langle v_1 \middle| \psi_\alpha^{1k'} * \psi_\beta^{-1k'}(1,y) \right\rangle \right] \frac{1}{i[\lambda_\alpha^{k'}(y)+\lambda_\beta^{-k'}(y)]} \right.$$

$$\left. \left[\int dp_1' \left\langle \bar{\psi}_\alpha^{1k'} * \bar{\psi}_\beta^{-1k'}(1,y) \middle| v_1 \right\rangle p_{1,x}' \, p_{1,y}' \right] - \text{hard core} \right\}$$

$$(IV.39)$$

The integrals

$$\int dp_1 \, p_{1,x} \, p_{1,y} \left\langle v_1 \middle| \psi^{1k} * \psi^{-1k'}(1,y) \right\rangle \qquad (IV.40)$$

are trivial to perform (see (IV.24)) and, as in section II.b), most of the terms in the α, β sums disappear for symmetry reasons. The correction $\eta_o^{\kappa\kappa(1)}$ can then be expressed as the sum over a few integrals over k´.

If we now ask for the general case, involving
the potential terms of η , a similar result holds.
The only difference is that the simple kinetic
averages (IV.40) are replaced by more complicated
equilibrium fluctuations. Let us simply give here
the final result; we have found:

$$\gamma \eta^{(1)} = \frac{\gamma k_B T}{30 \pi^2} \int_0^\infty dy \left\{ \left[1 + \frac{1}{c^2(y)} ny \frac{\partial V_y^L}{\partial y} \right]^2 \frac{1}{\Gamma(y)} - \frac{1}{\Gamma^s} \right.$$

$$\left. + (n \gamma_T(y) \frac{\gamma(y)-1}{\gamma(y)} n y \frac{\partial V^L}{\partial y})^2 \frac{nC_p(y)}{2 \kappa^s} \right\} \qquad \text{(IV.41)}$$

where

$$\Gamma(y) = \frac{1}{2n} \left[\frac{4}{3} \eta^s + \zeta^s + \left(\frac{1}{C_v^s} - \frac{1}{C_p(y)} \right) \kappa^s \right] \qquad \text{(IV.42)}$$

$$\gamma(y) = C_p(y)/C_v^s \qquad \text{(IV.43)}$$

We have also obtained the corresponding formulas for thermal conductivity and bulk viscosity, but
we shall not display them here.

The remarkable feature of eq. (IV.41) is, of
course, its model independent nature: $\eta^{(1)}$ is entirely determined in terms of the long range potential $\tilde{V}_y^L$ and of the equilibrium and transport properties of the reference short range fluid. Although
it is very pleasant to have a kinetic, microscopic
derivation of this result, we can guess that eq.
(IV.41) can be derived by purely macroscopic arguments. As a matter of fact, the attentive reader

will have already noticed the analogy of eq.
(IV.39) with the zero frequency Laplace transform
of the Green-Kubo integrand for the shear viscosity
(see (III.55)). Indeed, it can be shown that eq.
(IV.41) can be obtained, macroscopically, by the
same method which we used in deriving the $t^{-\frac{3}{2}}$ behaviour; the only difficult point is that one has
to evaluate the averages $\langle \hat{j}^{\eta} | \tilde{v} \ q' ; \tilde{v}', -q' \rangle$ in the
$q' \to o$, $y = q'/\gamma$ finite, limit. The dominant correction
(IV.41) then appears as the time-integral of the
difference between the long time behaviour of the
Green-Kubo integrands in the van der Waals fluid -
with eigenmodes determined by (IV.31) - and in the
hard core reference fluid - with eigenmodes fixed
by (III.42).

   At present, we have no idea of the usefulness
of eq. (IV.41) in interpreting the properties of
realistic fluids.

APPENDIX A

The Equations of Linearized Hydrodynamics

The macroscopic conserved quantities in a fluid are the particle density $n(r,t)$, the momentum density $g(r,t)$ and the energy density $\bar{\varepsilon}(r,t)$ (this involves both the internal energy $\varepsilon(r,t)$ and the kinetic energy $\frac{n}{2}v^2$). The corresponding conservation equations are (recall that m=1):

$$\partial_t \, n(r,t) + \nabla g(r,t) = 0$$
$$\partial_t \, g(r,t) + \nabla\tau(r,t) = 0 \qquad\qquad (A.1)$$
$$\partial_t \, \bar{\varepsilon}(r,t) + \nabla j^{\,\varepsilon}(r,t) = 0$$

where $\tau(r,t)$ is the stress tensor and j^{ε} the energy flow; to simplify, we have not written down explicitly the (obvious) tensorial character of the various quantities appearing in this equation.

In order to obtain from (A.1) a closed system of equations, we use:

1) the phenomenological assumptions for the currents τ and j^{ε} in real fluids [2b)

2) the assumptions that thermodynamic quantities are, in the hydrodynamical regime, interrelated in the same way as at equilibrium.

3) a linearization of the resulting equations around absolute equilibrium.

Using the following definition for the velocity field $v(r,t)$:

$$g(r,t) = n(r,t) \times v(r,t) , \qquad\qquad (A.2)$$

we get from assumptions 1) and 3):

$$g(r,t) = n \, v(r,t) \qquad\qquad (A.3)$$

$$\tau_{ij}(r,t)=p(r,t)\delta_{ij}^{Kr}- \eta(\frac{\partial v_i(r,t)}{\partial r_j}) + \frac{\partial v_j(r,t)}{\partial r_i})$$

$$-\delta_{ij}^{Kr}(\zeta- \tfrac{2}{3} \eta)\nabla v(r,t) \qquad (A.4)$$

$$(\text{Stokes–Navier hypothesis})$$

$$\bar{\epsilon}(r,t) = \epsilon(r,t) \qquad (A.5)$$

$$j^{\epsilon}(r,t) = h v (r,t)- \kappa \nabla T(r,t)$$

where n and h respectively denote the equilibrium
particle and enthalpy densities, while $T(r,t)$ is
the local temperature. More generally, we shall
use the convention that where the (r,t) dependence
of a given variable is not indicated, its absolute
equilibrium value should be taken.

We also write:

$$n(r,t) = n + \delta n(r,t) \qquad (A.6)$$

$$\epsilon(r,t) = \epsilon + \delta\epsilon(r,t) \qquad (A.7)$$

etc..., and we notice that the second law of thermodynamics can be written as:

$$dS = \frac{dQ}{T} = \tfrac{1}{T} (d E + pd\Omega)$$

$$= \tfrac{1}{T} (d(\epsilon\Omega) + pd\Omega)$$

$$= \tfrac{1}{T} (\Omega d\epsilon + (p+ \epsilon)d\Omega) \qquad (A.8)$$

Thus, the entropy density fluctuation $ds = dS/\Omega$
can be written:

$$ds = \tfrac{1}{T} (d\epsilon - \tfrac{h}{n} dn) \qquad (A.9)$$

According to assumption 2), this equation may be
used to connect the non-equilibrium quantities
$\delta s(r,t)$, $\delta\epsilon(r,t)$ and $\delta n(r,t)$.

Collecting the results, we obtain from (A.1):

$$\partial_t \delta n(r,t) + n \nabla v(r,t)= o$$

$$\partial_t v(r,t) + \frac{1}{n}\nabla\delta p(r,t) - \frac{\eta}{n}\nabla^2 v(r,t) - \frac{\zeta + \eta/3}{n}\nabla(\nabla v(r,t)) = 0$$

$$\partial_t \,\delta s(r,t) - \frac{\kappa}{T}\nabla^2\,\delta T(r,t) = 0 \qquad\qquad (A.10)$$

These equations will form a closed system provided we use the thermodynamic relation:

$$\delta p(r,t) = \left(\frac{\partial p}{\partial n}\right)_T \delta n(r,t) + \left(\frac{\partial p}{\partial T}\right)_n \delta T(r,t) \qquad (A.11)$$

$$\delta s(r,t) = \frac{1}{\Omega}\left(\frac{\partial S}{\partial n}\right)_T \delta n(r,t) + \frac{1}{\Omega}\left(\frac{\partial S}{\partial T}\right)_n \delta T(r,t)$$

With $\left(\frac{\partial S}{\partial T}\right)_n = N\,c_v/T$ (where c_v is the specific heat per particle at constant volume) and $\left(\frac{\partial S}{\partial n}\right)_T =$
$= -\Omega\frac{1}{n}\left(\frac{\partial p}{\partial T}\right)_n$, we get readily:

$$\partial_t\,\delta n(r,t) + n\nabla v(r,t) = 0$$

$$\partial_t\,v(r,t) + \frac{1}{n}\left(\frac{\partial p}{\partial n}\right)_T\nabla\delta n(r,t) + \frac{1}{n}\left(\frac{\partial p}{\partial T}\right)_n\nabla\delta T(r,t)$$

$$-\frac{\eta}{n}\nabla^2 v(r,t) - \left(\frac{\zeta + \frac{\eta}{3}}{n}\right)\nabla(\nabla v(r,t)) = 0$$

$$\partial_t\,\delta T(r,t) + \frac{T}{nc_v}\left(\frac{\partial p}{\partial T}\right)_n\nabla v(r,t) - \frac{\kappa}{nc_v}\nabla^2\delta T(r,t) = 0$$

$$(A.12)$$

By Fourier–Laplace transform of eq. (A.12), we get eq. (I.15) of the text.

APPENDIX   B

Calculation of L_{23} (q):

Let us denote by $\langle \cdot\cdot \rangle_p$ the average over the momenta in an equilibrium system; we have

$$\langle \cdot\cdot \rangle_p = \frac{\int dp^N \ldots \exp{-\beta H_o}}{(2\pi k_B T)^{3N/2}} \tag{B.1}$$

It is immediate to verify the following relations (see III.8, 10):

$$\langle \hat{g}_i(r)\hat{j}_k(r') \rangle_p = k_B T\, \delta(r-r')\langle \hat{n}(r) \rangle_p\, \delta_{i,k}^{Kr}$$

$$(i,k \in (x,y,z)) \tag{B.2}$$

and

$$\langle \hat{g}_i(r)\ \hat{j}_k^\epsilon(r') \rangle_p = k_B T\delta(r-r')\Big[\langle \hat{\epsilon}(r) \rangle_p + \langle \hat{p}(r') \rangle_p\Big]\delta_{i,k}^{Kr} \tag{B.3}$$

where $\hat{p}$ denotes the pressure operator.

From (III.12), we get thus:

$$\langle \hat{g}_i(r)\hat{j}_k^{\ s}(r') \rangle_p = \delta(r-r')\Big[\langle \hat{\epsilon}(r) \rangle_p + \langle \hat{p}(r) \rangle_p$$

$$-\frac{h}{n}\langle \hat{n}(r) \rangle_p\Big]\,\delta_{i,k}^{Kr} \tag{B.4}$$

Taking now the total average of this expression we obtain:

$$\langle |\hat{g}_i(r)\ \hat{j}_k^{\ s}(r')| \rangle = 0 \tag{B.5}$$

We now take the definition (III.34) of L_{23}(q) and insert (III.24), (III.8) and (III.12) into it; we get:

$$L_{23}(q) = \frac{q_x}{\Omega}\langle |\Big[(\frac{1}{k_B Tn})^{\frac{1}{2}}\ c(q)\ \hat{j}_{-q,x} + (\frac{1}{k_B n})^{\frac{1}{2}}\times$$

$$\times \Big[\frac{1}{C_v(q)} - \frac{1}{C_p(q)}\Big]^{\frac{1}{2}}\ \hat{j}_{-q,x}^{\ s}\Big](\frac{1}{k_B Tn})^{\frac{1}{2}}\hat{g}_{q,x}| \rangle \tag{B.6}$$

From (B.5),   this is simply:

$$L_{23}(q) = \frac{q_x}{\Omega} \; \frac{1}{nk_B T} \; c(q) \; \langle | \; \hat{j}_{-q,x} \; \hat{g}_{q,x} \; | \rangle$$

and from (B.2):

$$L_{23}(q) = q_x \; c(q)$$

which is the required result III.35.

APPENDIX C

Equation (IV.37) in the Boltzmann limit

We have already considered in eq. (II.3) the linearized Boltzmann collision operator. For our present purpose, it is convenient to rewrite it in the following form:

$$c^{\ell B}(v_1)f_1(v_1) \equiv \int dv_2 \left[c^B(v_1,v_2|f_1(v_1)\varphi^{eq}(v_2)) \right.$$

$$\left. + c^B(v_1,v_2|\varphi^{eq}(v_1) f_1 (v_2)) \right] \qquad (C.1)$$

where the two-body operator $c^B(v_1,v_2|f_1(v_1)\bar{f}_1(v_2))$ is defined by:

$$c^B(v_1,v_2|f_1(v_1)\bar{f}_1(v_2)) = n \int d\Omega\, \sigma(\Omega,|v_{12}|)|v_{12}| \times$$

$$\left[f_1(v_1') \bar{f}_1(v_2') - f_1(v_1)\bar{f}_1(v_2) \right] . \qquad (C.2)$$

Except for an integral over v_2 which is lacking, this is simply the full (non-linear) Boltzmann operator.

One remarkable property of this operator is that it conserves the collision invariants; indeed, it is easy to show that:

$$c^B(v_1,v_2|(\alpha(v_1)+\alpha(v_2)) f_1(v_1)\bar{f}_1(v_2))$$

$$=(\alpha(v_1)+\alpha(v_2)) c^B(v_1,v_2|f_1(v_1)\bar{f}_1(v_2)) \qquad (C.3)$$

where f and $\bar{f}$ are arbitrary one particle functions while $\alpha(v)$ is any of the five quantities:

$$\alpha(v) \in 1, \underline{v}, v^2/2 \qquad (C.4)$$

Moreover c^B gives zero when acting on a product of Maxwellians:

$$c^B(v_1,v_2|\varphi^{eq}(v_1) \varphi^{eq}(v_2)) = 0 \qquad (C.5)$$

After these preliminaries, let us come back
to the proof of eq. (IV.37). Because of the linear
relation (IV.24), the validity of this equation
will be established if we can show that:

$$\frac{1}{c^S(v_1)}\int dv_2 \tilde{\tilde{c}}^S(v_1,v_2)\left[\langle v_1|\alpha\rangle\langle v_2|\beta\rangle + \langle v_1|\beta\rangle\langle v_2|\alpha\rangle\right]$$

$$= \langle v_1|\alpha * \beta\rangle \qquad\qquad (C.6)$$

In the Boltzmann limit, with our notation (C.2),
this becomes simply

$$\frac{1}{c^S(v_1)}\, I = -\langle v_1|\alpha * \beta\rangle \qquad\qquad (C.7)$$

where we have introduced the quantity I defined by
by: (see also (II.13′)):

$$I=\int dv_2 c^B(v_1,v_2|(\alpha(v_1)\beta(v_2)+\alpha(v_2)\beta(v_1))\varphi^{eq}(v_1)\varphi^{eq}(v_2)$$

$$(C.8)$$

We transform I by adding and subtracting the same
term:

$$I =\int dv_2 c^B(v_1,v_2|(\alpha(v_1)+\alpha(v_2))(\beta(v_1)+\beta(v_2))$$

$$\varphi^{eq}(v_1)\varphi^{eq}(v_2) -\int dv_2\ c^B(v_1,v_2|(\alpha(v_1)\beta(v_1)+\alpha(v_2)\beta(v_2))$$

$$\varphi^{eq}(v_1)\varphi^{eq}(v_2) \qquad\qquad (C.9)$$

Using twice (C.3) and then (C.5), we see that the
first term in the right hand side of (C.9) vanishes.
From (C.1), we are left with:

$$I=-\int dv_2\left[c^B(v_1,v_2|\alpha(v_1)\beta(v_1)\varphi_1^{eq}(v_1)\varphi_1^{eq}(v_2))\right.$$

$$+ \left. c^B(v_1,v_2|\varphi_1^{eq}(v_1)\alpha(v_2)\beta(v_2)\varphi_1^{eq}(v_2))\right]$$

$$= - c^{0B}(v_1)\alpha(v_1)\beta(v_1)\varphi_1^{eq}(v_1) \qquad\qquad (C.10)$$

Inserting this result into (C.7) we obtain the
required result, (C.6). The proof of (IV.37) for
a dense fluid is based on the same principles.

REFERENCES

1) See for example P.C. Martin in "Many Body Physics", eds. de Witt and Balian (Gordon and Breach, New York, 1968).

2) L. LANDAU and E. LIFSHITZ: a) "Fluid Mechanics", (Pergamon Press), 1958 b) "Statistical Physics", (Pergamon Press) 1958

3) L. KADANOFF and P.C. MARTIN: Ann. Phys. 24, 419 (1963).

4) B. ALDER and T. WAINWRIGHT: Phys. Rev. A1, 18 (1970)

5) M. ERNST, E. HAUGE and J. VAN LEEUWEN: Phys. Rev. Letters, 25, 1254 (1970); Phys. Rev. B5, 2055 (1971)
   J. DORFMAN and E. COHEN: Phys. Rev. Letters, 25, 1257 (1970) and to appear.
   Y. POMEAU: Phys. Rev. (to appear)

6) L. KADANOFF and J. SWIFT: Phys. Rev. 166, 89 (1968); see also
   K. KAWASAKI, Prog. Th. Phys. Jap., 39, 1133 (1968), 40, 11(1968), 40, 4(1968),40,5,930(1968).

7) P. RESIBOIS, J. PIASECKI and Y. POMEAU: Phys. Rev. Letters, 39A, 33 (1972).
   P. RESIBOIS and J. PIASECKI: J. Math. Phys. (to appear)
   P. RESIBOIS, Y. POMEAU and J. PIASECKI: (to appear)

8) See for example G. UHLENBECK and G. FORD: "Lectures in Statistical Mechanics" (American Mathematical Society, Providence, 1963)

9) See for example P. DENNERY and A. KRZYWICKI: "Mathematics for Physicists" (Harper and Row, 1967)

10) See for example D. BOHM: "Quantum Theory", Constable, London (1954).

11) P. RESIBOIS: J. Stat. Phys., 2, 21 (1970).

12) P. RESIBOIS: Bull. Ac. Roy. Belgique,Cl.Sci. 56, 160 (1970).

13) See for example: I. PRIGOGINE, "Non Equilibrium Statistical Mechanics";(Interscience Pub., 1962)

P. RESIBOIS in "Many-Particle Physics", E. Meeron Ed. (Gordon and Breach, New York, 1967).

14) P. RESIBOIS: Physica, 49, 591 (1970)

15) P. RESIBOIS: Physica (to appear)

16) P. RESIBOIS: J. Chem. Phys. 41, 2979 (1964)

17) G. SEVERNE: Physica, 31, 877 (1966)

18) L. KADANOFF and J. SWIFT: Loc. cit.

19) B. ALDER and T. WAINWRIGHT: Loc. cit.

20) Y. POMEAU: Loc. cit.

21) See for example R. ZWANZIG in "Proceeding of the 1971 IUPAP Conference on Statistical Mechanics" (Chicago Univ. Press, Chicago, to appear 1972)

22) J. DORFMAN and L. COHEN: Loc. cit.

23) P. RESIBOIS and Y. POMEAU: unpublished.

24) L. VERLET (private communication)

25) See for example E. COHEN: "Fundamental Problems in Statistical Mechanics", E. Cohen, Ed.(North Holland Pub., Amsterdam, 1968)

26) See for example R. BALESCU: "Statistical Mechanics of charged Particles";(Interscience, 1963)

27) See for example P. RESIBOIS: "Electrolyte Theory" (Harper and Row, New York, 1968).

28) J. VAN DER WAALS: Dissertation, Leiden, 1873.

29) See: M. KAC, G. UHLENBECK and P. HEMMER: J. Math. Phys. 4, 416 (1961); ibid 4, 229 (1963); ibid 5, 60 (1964).

   N. VAN KAMPEN: Physica 135, 362 (1964)
   J. LEBOWITZ and O. PENROSE: J. Math. Phys. 7, 98 (1966).
   P. HEMMER: J. Math. Phys. 5, 75 (1964)
   J. LEBOWITZ, G. STELL and S. BAER: J. Math. Phys. 6, 1282 (1966)
   J. JALICHEE, A. SIEGERT and D. VEZETTI: J. Math. Phys. 10, 1442 (1969), ibid 11, 3168 (1970).

30) K. KAWASAKI: Prog. Th. Phys. Jap. 41,1190 (1969) and references quoted therein.

31) See for example: M. FISHER: Rep. Prog. Physics 30, 615 (1967)
    L. KADANOFF and al.: Rev. Mod. Phys. 39, 395 (1967).
    E. STANLEY: "Phase transitions and Critical Phenomena" (Clarendon Press, Oxford, 1971).

32) The recent developments can be found in K. WILSON: Phys. Rev. 4B, 3174, 3184 (1971) and references quoted therein.

33) See also BALESCU's lectures in this volume.

FLUCTUATIONS

N. G. van Kampen

University of Utrecht

Utrecht, Holland

1.   Introduction

Originally thermodynamics was only concerned with the macroscopically observed values of physical quantities, which in statistical mechanics are given by averages. However, Brownian motion showed that thermodynamics is not the whole story. In fact, Gibbs' statistical treatment led to fluctuations about these averages. (Subsequently Einstein noticed that the relation $S = k \log W$ enables one to treat these fluctuations thermodynamically as well). But these are fluctuations in the instantaneous values of the physical quantities, and their theory is well known. We are here concerned with the behaviour of the fluctuations as functions of time. Their study is much harder, because it involves not only the equilibrium distribution, but also the evolution in time, which in principle requires the solution of the equations of motion of the physical system considered.

There are several reasons why interest in fluctuations has grown in recent times. First, they form an obstacle in precise measurements and delicate experiments, and techniques to eliminate their effect are of practical importance. Secondly they may be used as a source of information concerning the system; the most striking example being the fluctuations of the electromagnetic field usually referred to as the $3^{\circ}K$ radiation. Thirdly they may give rise to

macroscopic effects such as the Van-der-Waals
force.

The proper framework for describing time-
dependent fluctuations is the theory of stochastic
processes; they are treated in sections 2 - 4,
and the master equation for Markov processes is
derived. Subsequently, in sections 5 - 8 an expansion of the master equation is given, which
makes its application to practical cases possible.
The final sections contain an introduction to the
theory of stochastic differential equations.

2. Stochastic processes

A stochastic variable X is defined by the
set of its values x (= set of "states" =
"range"), and a probability distribution over
this set. Once such a stochastic variable is
given, any related quantity $Y = f(X)$ is also
a stochastic variable. Y may be any kind of
mathematical object, in particular a function of
an auxiliary variable t, that is, $Y(t) = f(X,t)$.
Such Y(t) is called a stochastic process. It
may be regarded as a collection of "sample functions" or "realizations" $y(t) = f(x,t)$, each of
which is obtained by assigning to X one of its
possible values x.

Consider a classical dynamical system; its
instantaneous state is given by a point
$(q_1, q_2, \ldots, q_f, p_1, p_2, \ldots, p_f)$ in its phase space.
In statistical mechanics one replaces the

individual system by an <u>ensemble</u>. This amounts
to defining a stochastic variable X, whose range
consists of all points x = (q,p) in phase space
and whose probability distribution is specified
by the ensemble. The ensemble may be chosen
freely at one particular time. t = 0 and the
equations of motion then determine the probability
of finding a particular state at a later (or
earlier) time t.

A <u>dynamical quantity</u> is a function of the
state (q,p). Its value at any time t is determined by the equations of motion together with
the initial state x at t = 0, and may be
written f(x,t). By the introduction of an
ensemble, each dynamical variable is turned into
a stochastic process Y(t) = f(X,t). In this way
stochastic processes enter into physics. So far
this is nothing but a reformulation, which does
not bring us any nearer to solving the equations
of motion. However, it will appear that the
theory of stochastic processes suggests certain
assumptions, hopefully called "approximations",
which make it possible to obtain results without
having to solve the equations of motion. Some of
these results, for instance the theory of Brownian
motion, have proved very successful.

In quantum statistical mechanics the possible
values of X are the unit vectors[*] in the Hilbert

---

[*] More precisely all "rays" in Hilbert space,
that is, all unit vectors disregarding a phase
factor.

space of the system, although for describing
equilibrium it is sufficient to take only the
eigenstates of the Hamilton operator. The probability distribution is given by means of a
density matrix. The functions $Y(t)$ are the
expectation values of the various operators. The
theory of stochastic processes does not ask
whether the system is classical or quantum mechanical, but, of course, the specific properties of
the observables need not be the same in both cases.

3.   The Markov assumption

        A stochastic process $Y(t)$, defined by means
of an X as above, leads to a hierarchy of probability densities. We write

$$P_n(y_1,t_1;\ y_2,t_2;\ \ldots;\ y_n,t_n)dy_1 dy_2 \ldots dy_n$$

for the probability that $Y(t_1)$ lies between y_1
and $y_1 + dy_1$, and that also $Y(t_2)$ lies between
y_2 and $y_2 + dy_2$, etc. The P_n are defined for
$n = 1,2,3,\ldots$ and only for unequal times. Clearly one has

(i) $P_n \geqslant 0$;
(ii) P_n invariant for interchange of the pair
        x_i,t_i with the pair x_j,t_j for any i,j;
(iii) $\int P_n\ dy_n = P_{n-1}$, and $\int P_1\ dy_1 = 1.$

Kolmogorov[1)] has proved that the reverse is also
true. Any set of functions P_n obeying these

three restrictions defines a stochastic process. For future use we note that the process may also be characterized by the hierarchy of its moments, defined by

$$\mu_n(t_1, t_2, \ldots, t_n) = \langle Y(t_1)Y(t_2)\ldots Y(t_n)\rangle$$

$$= \int y_1 y_2 \ldots y_n P_n(y_1, t_1; \ldots; y_n, t_n) dy_1 dy_2 \ldots dy_n.$$

The conditional probability density $P_{n|\nu}$ for finding y_1 at t_1, and y_2 at t_2 etc., <u>given</u> that one has $Y(\tau_1) = \eta_1$, $Y(\tau_2) = \eta_2, \ldots$, is determined by

$$P_{n|\nu}(y_1, t_1; \ldots; y_n, t_n | \eta_1, \tau_1; \ldots; \eta_\nu, \tau_\nu) P(\eta_1, \tau_1; \ldots;$$
$$; \eta_\nu, \tau_\nu) =$$
$$= P_{n+\nu}(y_1, t_1; \ldots; y_n, t_n; \eta_1, \tau_1; \ldots; \eta_\nu, \tau_\nu).$$

Clearly $P_{n|\nu}$ is symmetric in the pairs y_i, t_i and also in the pairs η_j, τ_j. It is therefore no restriction to write the τ_j in chronological order, $\tau_1 < \tau_2 < \ldots < \tau_\nu$. Suppose that all t_i are later than τ_ν; a Markov process is defined as a process for which $P_{n|\nu}$ does not depend on the earlier data $\eta_1, \ldots, \eta_{\nu-1}$ but only on η_ν. More explicitly

$$P_{n|\nu}(y_1, t_1; \ldots; y_n, t_n | \eta_1, \tau_1; \ldots; \eta_\nu, \tau_\nu) =$$
$$= P_{n|1}(y_1, t_1; \ldots, y_n, t_n | \eta_\nu, \tau_\nu).$$

It follows that, if $t_1 < t_2 < \ldots < t_n$,

$$P_n(y_1, t_1; \ldots; y_n, t_n) =$$

$$= P_1(y_1, t_1) P_{1|1}(y_2, t_2 | y_1, t_1) \ldots \times$$

$$\times \ldots P_{1|1}(y_n, t_n | y_{n-1}, t_{n-1}).$$

This shows that, in the case of a Markov process, the whole hierarchy is determined by only its first member P_1 and the transition probability $P_{1|1}$. The Markov property is a very strong restriction, but it has the effect of making the hierarchy tractable. It often happens that one has reason to believe that a physical quantity, in whose fluctuating behaviour one is interested, is approximately a Markov process. It then becomes possible to find all stochastic properties of that quantity: the need for solving the actual equations of motion of the whole system has been sidestepped by assuming Markov character. Before demonstrating this technique we still have to develop the formalism somewhat further.

4.   The master equation

From the Markov character follows for $t_1 < t_2 < t_3$

$$P_3(y_1, t_1; y_2, t_2; y_3, t_3) =$$

$$= P_2(y_1, t_1; y_2, t_2) \, P_{1|1}(y_3, t_3 \mid y_2, t_2).$$

On integrating over y_2 and dividing by
$P_1(y_1,t_1)$ one obtains an integral equation for
the transition probability

$$P_{1|1}(y_3,t_3\ y_1,t_1) =$$

$$= \int P_{1|1}(y_3,t_3|\ y_2,t_2)\ P_{1|1}(y_2,t_2|y_1,t_1)dy_2.$$

This is the Chapman-Kolmogorov equation. One
easily verifies that for example

$$P_{1|1}(y_2,t_2|y_1,t_1) = \left[2\pi(t_2-t_1)\right]^{-\frac{1}{2}}\ \exp\left[-\frac{(y_2-y_1)^2}{2(t_2-t_1)}\right]$$

obeys this equation. Together with

$$P_1(y_1,t_1) = (2\pi t_1)^{-\frac{1}{2}}\ \exp(-y_1^2/2t_1)$$

it defines a Markov process for $t > 0$, which is
called the <u>Wiener process</u> and describes the random
displacement of a Brownian particle.

    If one takes $t_3 - t_2 = \Delta t$ small, one
expects that the probability that any transition
at all has taken place is proportional to Δt.
Accordingly we now assume

$$P_{1|1}(y_3,t_2+\Delta t|y_2,t_2) = A\ \delta(y_3-y_2) + \Delta t\ W_{t_2}(y_3|y_2).$$

Here W_{t_2} is the transition probability per unit
time at the time t_2, and conservation of probability tells us

$$A = 1 - \Delta t \int W_{t_2}(y'|y_2)\ dy' .$$

<u>Remark</u>: The Wiener process does not obey this

assumption. The reason is that it contains in
each Δt infinitely many infinitely small transitions. That roughly coincides with the definition of "continuous stochastic processes"[2]. Our
assumption applies to discontinuous processes, in
which finite jumps are made, but such that the
probability for any jump to take place in Δt is
of order Δt. Our contention is that this is the
correct description of physical processes, and
that the Wiener process can at best be regarded
as a first step in an approximating expansion
scheme (section 6).

Substitution leads to the differential form
of the Chapman-Kolmogorov equation,

$$\frac{\partial P_{1|1}(y,t|y_1,t_1)}{\partial t} = \int W_t(y|y') \, P_{1|1}(y',t|y_1,t_1)dy' - P_{1|1}(y,t|y_1,t_1)\int W_t(y'|y)dy'.$$

To make the meaning of this result more transparent,
we multiply the equation by $P_1(y_1,t_1)$ and integrate over y_1,

$$\frac{\partial P_1(y,t)}{\partial t} = \int \left\{ W_t(y|y')P_1(y',t) - W_t(y'|y)P_1(y,t) \right\} dy'$$

This is the celebrated <u>master equation</u>; it is a
linear homogeneous equation describing the evolution of the probability density of a Markov process. Its physical importance derives from the
fact that W refers to the short time behaviour
of the total system and therefore can often be
computed; the master equation then serves to

deduce the long time behaviour from this.

Although the Wiener process did not obey the assumption used in deriving the master equation, we see in retrospect that it does obey a master equation with a singular kernel W,

$$\frac{\partial P_1(y,t)}{\partial t} = \frac{1}{2} \frac{\partial^2 P_1(y,t)}{\partial y^2} .$$

This equation can be solved explicitly (Fourier); more generally, the master equation can be solved whenever it is invariant for shifts in y and t, i.e., when y ranges from $-\infty$ to $+\infty$ and $W_t(y|y')$ is a function of $y - y'$ alone. The reason is that in that case the successive jumps are statistically independent of each other. Another solvable master equation is the "linear Fokker-Planck equation", which gives rise to the Ornstein-Uhlenbeck process,

$$\frac{\partial P_1(y,t)}{\partial t} = \frac{\partial}{\partial y} y P_1(y,t) + \frac{1}{2} \frac{\partial^2 P_1(y,t)}{\partial y^2} .$$

However, for most master equations an approximation method is needed; this will now be developed.

5. Expansion of the master equation

We suppose that $W_t(y|y')$ is independent of t and rewrite it as a function $W(y'; \xi)$ of the starting point y' and the jump length ξ.

$$\frac{\partial P_1(y,t)}{\partial t} = \int \Big\{ W(y-\xi\,;\xi)\ P_1(y-\xi,t) -$$

$$- W(y;\xi)\ P_1(y,t) \Big\}\ d\xi .$$

Expanding the first term on the right in powers of ξ one obtains the "Kramers-Moyal expansion".

$$\frac{\partial P_1(y,t)}{\partial t} = \sum_{\nu=1}^{\infty} \frac{(-1)^\nu}{\nu!} \Big(\frac{\partial}{\partial y}\Big)^\nu \ \alpha_\nu(y)\ P_1(y,t),$$

where the α_ν are the moments of the jump probability,

$$\alpha_\nu(y) = \int_{-\infty}^{\infty} \xi^\nu\ W(y;\xi)\ d\xi .$$

It is tempting to break off after two terms and adopt the resulting "nonlinear Fokker-Planck equation"

$$\frac{\partial P_1}{\partial t} = - \frac{\partial}{\partial y} \alpha_1(y)\ P_1 + \frac{1}{2}\frac{\partial^2}{\partial y^2} \alpha_2(y)\ P_1$$

as an approximation to the master equation. This equation appears in most textbooks and is often called after Kolmogorov[3]. However, it is derived by means of the same unphysical assumptions on which the Wiener process is based. Actually this equation is <u>not</u> a systematic approximation to the master equation and has caused a great deal of confusion[4]. Our purpose is to provide a systematic expansion in powers of a parameter which obviates these difficulties.

To find the proper expansion parameters we
note that the size of a jump ξ is properly expressed in an <u>extensive</u> quantity, while the
variable y occurring in $W(y; \xi)$ expresses the
dependence of the probability on the over-all state
of the system and is therefore properly expressed
in terms of an <u>intensive</u> variable. This suggests
that, if Ω is the size of the system, one may
make the dependence of W on Ω explicit by setting

$$W(y, \xi) = \Phi\left(\frac{y}{\Omega}; \xi\right) ,$$

in such a way that Φ no longer depends implicitly
on Ω. This turns out to be true.

As an example, consider an ideal gas in a
virtually infinite reservoir with fixed density ϱ.
A smaller volume Ω is connected with it through
a hole, see figure. The probability that in a

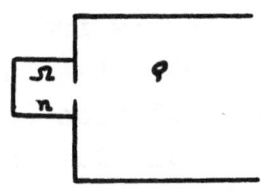

given dt the number n of molecules in Ω increases by one unit is $\alpha \varrho\, dt$. The probability
that n decreases by one unit is $\alpha \frac{n}{\Omega}\, dt$. Hence

$$W(n; \xi) = \alpha \varrho\, \delta_{\xi, 1} + \alpha \frac{n}{\Omega}\, \delta_{\xi, -1} .$$

Clearly this depends on n only through the intensive variable n/Ω and no other Ω occurs.

This turns out to be the general pattern, although minor modifications are sometimes required. For instance, additional factors Ω may appear in Φ, but they can be absorbed in the time unit. Or it may happen that Φ is not independent of Ω but consists of a power series in $1/\Omega$; this gives rise to additional terms but does not upset the general scheme. Finally it is not necessary that Ω has the physical meaning of a volume. In the case of Brownian motion, is the mass of the particle, its momentum has the role of the extensive quantity, and its velocity is the intensive quantity[*].

6.   A paradigm

Rather than formulate the expansion method in general terms we demonstrate it on a simple example, which we borrow from the theory of population statistics. Let n be the number of individuals in a population (of bacteria or other organisms). Each individual produces offspring at a rate α and has a probability β per unit time to die. This leads to

$$\dot{n} = \alpha n - \beta n \; ,$$

which is Malthus' law, assuming $\alpha > \beta$. In order to include the struggle for life, Verhulst added for each individual a death probability proportional

---

[*] No relation to the fact that Aristotle also regarded velocity as an "intensity".

to the number of individuals present. More precisely, this probability should be taken proportional to the density n/Ω , where Ω is the
size of the test tube or the amount of food. This
leads to the Malthus-Verhulst equation

$$\frac{dn}{dt} = (\alpha - \beta)n - \uparrow \frac{n^2}{\Omega} \ ,$$

which fits reality surprisingly well[5]. This is
the macroscopic, phenomenological equation or rate
equation, and ignores fluctuations.

In order to describe fluctuations one has to
introduce the probability $P(n,t)$ for having n
individuals at time t. It obeys the master equation

$$\dot{P}(n,t) = \alpha(n-1) \ P(n-1) - \alpha \ n \ P(n,t) +$$

$$+ \ \beta(n+1) \ P(n+1,t) - \beta n \ P(n,t) +$$

$$+ \ \frac{\uparrow}{\Omega} \left\{ (n+1)^2 \ P(n+1,t) - n^2 \ P(n,t) \right\} .$$

The Markov character has been introduced by assuming that each individual has fixed probabilities
for breeding or dying, regardless of age and
gestation period.

The Kramers-Moyal expansion of the master
equation is

$$\frac{\partial P(n,t)}{\partial t} = \sum_{\nu=1}^{\infty} \frac{1}{\nu!} \frac{\partial^\nu}{\partial n^\nu} \left\{ \alpha(-1)^\nu \ n + \beta n + \frac{\uparrow}{\Omega} \ n^2 \right\} P(n,t).$$

Notice that α and β occur separately, no longer in
the combination $\alpha - \beta$ alone. Thus the master
equation involves more detailed information about

the process than the macroscopic rate equations.
This is generally true and explains why the observation of fluctuations may lead to new information,
not obtainable from the measurement of the rates.
For our present purpose, however, it suffices to
take β = 0; moreover, we change the time unit
so as to have α = 1, and the unit of Ω so that
γ = 1.

7.   The Ω-expansion

It is necessary to write the powers of Ω
explicitly. We anticipate that n consists of
a macroscopic part of order Ω and fluctuations
around it of order $\Omega^{1/2}$. Accordingly we transform
from the variable n to a new variable x by
setting

$$n = \Omega\varphi(t) + \Omega^{1/2}x,$$

where $\varphi(t)$ is a function yet to be determined.
The rationale is that the probability distribution P(n,t) is represented by a sharp peak,
which slides bodily along the n-axis. Its width
is of order $\Omega^{1/2}$ at all times, and its position
$\Omega\,\varphi(t)$ is the macroscopically observed value
of n.

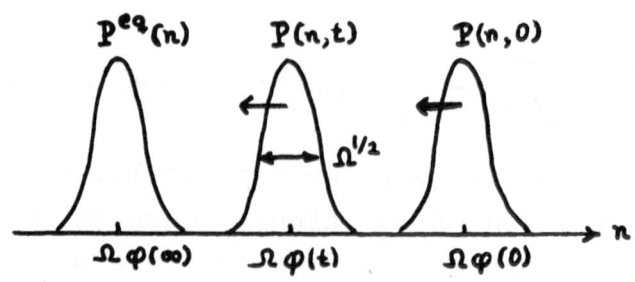

The transformation of the probability distribution reads

$$P(n,t) = P(\Omega\varphi(t) + \Omega^{1/2}x, t) = \Pi(x,t) .$$

Actually there ought to be an additional factor $\Omega^{1/2}$ on the right in order that Π be normalized, but we omit it for convenience.

We write the Kramers–Moyal expansion in the new variables, and order the terms according to the powers of Ω :

$$\frac{\partial \Pi}{\partial t} - \frac{\partial \Pi}{\partial x} \Omega^{1/2}\varphi'(t) = -\Omega^{1/2}(\varphi - \varphi^2) \frac{\partial \Pi}{\partial x} +$$

$$+(-1 + 2\varphi) \frac{\partial}{\partial x} x \Pi + \frac{1}{2}(\varphi + \varphi^2) \frac{\partial^2 \Pi}{\partial x^2} + \mathcal{O}(\Omega^{-1/2}) .$$

The largest terms are those involving $\Omega^{1/2}$. The fact that the derivative $\partial \Pi / \partial t$ is not one of them means that we are dealing with a singular perturbation problem. However, the two large terms can be caused to cancel by choosing

$$\varphi'(t) = \varphi - \varphi^2 .$$

This is the macroscopic rate equation

$$\frac{dn}{dt} = n - \frac{n^2}{\Omega} .$$

Thus we have deduced the macroscopic law from the master equation; it is simply the equation that describes how the peak $P(n,t)$ slides along the

n-axis. Note that the macroscopic equation may
well be nonlinear, although the master equation
is of course linear in P(n,t).

Having disposed of the terms of order $\Omega^{1/2}$
we collect the terms of the next order

$$\frac{\partial \Pi(x,t)}{\partial t} = (-1+2\varphi) \frac{\partial}{\partial x} x\Pi + \frac{\varphi+\varphi^2}{2} \frac{\partial^2 \Pi}{\partial x^2} .$$

This equation governs the fluctuations around the
macroscopic value $\Omega\varphi(t)$ found in the previous
step. It is similar to the equation for the
Ornstein-Uhlenbeck process, but its coefficients
depend on time. Yet it can be solved explicitly[6].

For $t = \infty$ the coefficients become constant
and the equation reduces to

$$\frac{\partial \Pi}{\partial t} = \frac{\partial}{\partial x} x\Pi + \frac{\partial^2 \Pi}{\partial x^2} .$$

This is the familiar (linear) Fokker-Planck equation for equilibrium fluctuations, as used in the
standard theory of Rayleigh, Einstein, Smoluchowski,
and many others.

The macroscopic equation also has the particular solution $\varphi(t) = 0$. The fluctuations
around it obey

$$\frac{\partial \Pi}{\partial t} = -\frac{\partial}{\partial x} x\Pi .$$

This leads to $\langle x \rangle_t = \langle x \rangle_0 e^t$, which demon-
strates that the solution $\varphi = 0$ is unstable.
Thus the non-existence of unicorns is an unstable
situation.

8.   Some critical remarks

We have found that the fluctuations are
governed in lowest order by a second order
equation, whose first coefficient is linear in
x and whose second coefficient is a constant.
It turns out that the next order adds a quadratic
term in the first coefficient, a linear term in
the second coefficient, and also a third order
derivative with a constant coefficient; and so
on[7].   It follows that it is inconsistent to use
the nonlinear Fokker-Planck equation without including at the same time higher derivatives. To
put it differently, the single parameter Ω
governs both the validity of the Fokker-Planck
approximation and the influence of the nonlinearity on the fluctuations; it is inconsistent to
add nonlinear terms without improving on the
Fokker-Planck approximation as well. In particular, it is incorrect to conclude from the first
two terms of the Kramers-Moyal series that the
equilibrium distribution is

$$P^{eq}(y) = \frac{const.}{\alpha_2(y)} \exp\left\{\int_0^y \frac{2\alpha_1(y')}{\alpha_2(y')}\,dy'\right\}.$$

Another way of describing fluctuations was origina-
ted by Langevin, who wrote for the velocity v of
a Brownian particle

$$\dot{v} = -\Gamma v + \eta(t) \ .$$

Γ is the friction coefficient and $\eta(t)$ is assumed to be a rapidly fluctuating force, due to the

random collisions of the gas molecules. The
rapidly fluctuating and irregular character of
$\eta(t)$ is taken into account by treating $\eta(t)$
as a random process having the properties

$$\langle \eta(t) \rangle = 0, \qquad \langle \eta(t_1)\, \eta(t_2) \rangle = c\, \delta(t_1 - t_2).$$

Of course this does not yet fully specify the
stochastic process; for this reason one sometimes completes the definition by stipulating
that $\eta(t)$ is a Gaussian process, but that is
an <u>ad hoc</u> assumption, and fortunately superfluous
for deriving the properties of Brownian motion.

The success of Langevin's treatment of
Brownian motion has given rise to the popular
idea that, for every macroscopically known phenomenon in nature, fluctuations can be taken into
account simply by supplementing the macroscopic
rate equation by a Langevin term, even though its
physical meaning may be obscure. Inasmuch as the
present Ω-expansion provides a systematic treatment of fluctuations, it is now possible to verify
this idea <u>ex post facto</u>. Our paradigm will serve
us to show that the Langevin assumption cannot be
correct in the case that the macroscopic equation
is nonlinear.

Following the popular idea we write the
"Langevin equation"

$$\frac{dn}{dt} = n - \frac{n^2}{\Omega} + \eta(t) .$$

Thus $\langle \eta(t) \rangle$ can be found from

$$\Omega \varphi' + \Omega^{\frac{1}{2}}\frac{d}{dt}\langle x \rangle = \Omega \varphi + \Omega^{\frac{1}{2}}\langle x \rangle - \frac{\langle \Omega \varphi + \Omega^{\frac{1}{2}}x)^2 \rangle}{\Omega} +$$
$$+ \langle \varkappa(t) \rangle \; .$$

On substituting the results of section 7 this reduces to

$$\langle \varkappa(t) \rangle \quad = \quad \langle x^2 \rangle \; .$$

The fact that $\langle \varkappa(t) \rangle$ does not vanish may not yet be so bad, but the fact that it depends on the solution itself is fatal to the Langevin approach. Even if one tries to save the Langevin assumption by noting that $\langle \varkappa(t) \rangle$ has one factor Ω less than the other terms in the above "Langevin equation", the same evil reappears when computing C.

9. Stochastic differential equations

This name denotes differential equations in which one or more of the coefficients are stochastic variables. Thus a stochastic differential equation stands for an ensemble of differential equations, each with some values for the coefficients. The solution of a stochastic differential equation with given initial condition (which may be stochastic too) is a stochastic function. The problem is to derive its stochastic properties from those of the coefficients of the equation.

An example is Langevin's differential equation for the Brownian particle, in which the

coefficient $\varkappa(t)$ is stochastic. The solution
is well known, and can rigorously be proved to
be a Wiener process when $\varkappa(t)$ is assumed to be
Gaussian. Note that, whereas the values of $\varkappa$
at two different times are uncorrelated, this
is not true for the Wiener process; thus the
solution of the equation involves more memory
than its stochastic coefficient. This is a
general feature of stochastic differential equations, since the solutions are obtained by integrations.

Other examples are: the current in an elec-
tric network due to a noisy source[8,9], the response of a control system to random perturbations[10],
line broadening[11], and the effect on a photoconductor of correlations in the incoming photon
beam[12]. Examples of stochastic _partial_ differential equations are: the Brownian motion of a
string or a drumhead[13] and the propagation of
sound waves or radio waves through a turbulent
medium[14]. The twinkling of stars and the propagation of ultrasound in the ocean have been treated
in the approximation of geometrical optics and
belong therefore to stochastic ordinary differential equations[15].

A rigorous solution of a stochastic equation
consists of two steps: first solve the equation
for y, with each possible value of the coefficients, and subsequently average the resulting
y and products thereof. This can be carried out
explicitly only in rare cases; one often has to

be satisfied with an approximate solution. A much
bolder approach to problems that cannot be solved
rigorously is the following.

First take an average of the equation as it
stands. In the Langevin case this yields a
closed equation for $\langle y(t) \rangle$, but in general
averages of products of higher powers of y and
of products of y and the coefficients will
arise. Next, guided by physical insight and
mathematical needs, assume that some of these
averages of products may be replaced with the
products of the averages, for example $\langle y^2 \rangle \rightarrow \langle y \rangle^2$.
In this way one arrives at a closed equation for
$\langle y(t) \rangle$, whose solution hopefully approximates
the actual $\langle y(t) \rangle$. Boltzmann's famous "Stosszahlansatz" and later forms of "molecular chaos"
or "random phase approximations" are assumptions
of that ilk [23]. However, we shall not use such
methods. Nor are we concerned with stochastic
boundary value problems and the probability distribution of their eigenvalues[22].

---

* J.B. Keller[15], has called this the "dishonest"
  method, but that term seems too disparaging for
  a method which has led to the greatest achieve-
  ments of statistical mechanics. I propose to
  restrict this qualification to work in which
  such an assumption is made but concealed.
  According to this definition linear response
  theory should be called dishonest[21].

10. <u>First example</u>

Consider the first order linear differential equation

$$\dot{y} \;=\; -ivy, \qquad\qquad y(0) \;=\; 1.$$

Let v be a random real constant with probability density $\varphi(v)$. This example describes a set of harmonic oscillators with different frequencies. For each v the solution is e^{-ivt}, and hence

$$\langle y(t) \rangle \;=\; \langle e^{-ivt} \rangle = \int_{-\infty}^{\infty} \varphi(v)\, e^{-ivt}\, dv \;.$$

It is instructive to look at a few special cases; in the following list v_0 and Υ are fixed parameters.

$$\varphi = \frac{\Upsilon/\pi}{\Upsilon^2 + (v-v_0)^2} \qquad\qquad \langle y \rangle \;=\; e^{-iv_0 t - \Upsilon t}$$

$$\varphi = (2\pi\Upsilon)^{-\frac{1}{2}}\, e^{-(v-v_0)^2/2\Upsilon} \qquad \langle y \rangle \;=\; e^{-iv_0 t - \Upsilon t^2}$$

$$\varphi = \tfrac{1}{2}\Upsilon\, e^{-\Upsilon|v-v_0|} \qquad\qquad \langle y \rangle \;=\; e^{-iv_0 t}\, \frac{\Upsilon^2}{\Upsilon^2 + t^2} \;.$$

In each case the averaged solution tends to zero as $t \to \infty$. This damping is due to the fact that the harmonic oscillators of the ensemble gradually lose the phase coherence they had at t = 0. In

plasma physics this is called "phase mixing",
in mathematics "Riemann-Lebesgue theorem". The
modulus of y is not subject to phase mixing and
does not tend to zero; in fact,

$$|y(t)| = 1, \quad \text{hence} \quad \langle|y(t)|^2\rangle = 1.$$

The form of the damping factor is determined by
φ. . Only for one particular φ does it have
the form that corresponds to a complex frequency.
It is therefore incorrect to make the Ansatz
$\langle y(t)\rangle = e^{-i\omega t}$ and search for eigenfrequencies
ω [6]. The problem is _not_ invariant for time
translations owing to the fixed initial value.

Note that $\langle y(t)\rangle$ is identical with the
characteristic function $\chi(t)$ of the distribution
$\varphi(v)$. Hence under rather weak restrictions it
can have almost any reasonable dependence on
t [17]. The fact that $\langle y(t)\rangle = \chi(t)$ suggests
the use of the cumulant expansion

$$\langle y(t)\rangle = \langle e^{-ivt}\rangle = \exp\left\{\sum_{m=1}^{\infty} \frac{(-it)^m}{m!} \overline{v^m}\right\} \quad,$$

where $\overline{v^m}$ stands for the m-th cumulant. We shall
make ample use of cumulant expansions, but one has
to bear in mind that not all probability distributions have one. For instance, the above formula suggests that there exists _no_ φ for which
$\langle y(t)\rangle$ has the form $e^{-iv_0t-\Gamma t}$; but we have seen
that that conclusion is wrong.

To characterize the stochastic process y

more fully we compute the higher moments,

$$\langle y(t_1)...y(t_n)\ y^*(s_1)....y^*(s_m)\rangle\ =$$

$$=\left\langle e^{-iv(t_1+...\ t_n-s_1\ -\ ...\ s_m)}\right\rangle=$$

$$=\ \chi\ (t_1+...\ t_n-s_1-\ ...\ s_m)\ .$$

Thus we have succeeded in this simple instance
in providing a complete solution of the problem
of deriving the stochastic properties of $y(t)$
from those of v.   The solution may be written
in a condensed form by means of a characteristic
functional $X[\vartheta]$ involving an auxiliary dummy
function $\vartheta(t)$,

$$X[\vartheta]\ \equiv\ \left\langle \exp\ i\int\left\{\vartheta(t)y(t)+\vartheta^*(t)y^*(t)\right\}\ dt\right\rangle=$$

$$=\ \left\langle \exp\ i\int\left\{\vartheta(t)e^{-ivt}+\vartheta^*(t)e^{ivt}\right\}\ dt\right\rangle.$$

11. Second example

Take the same equation with the same initial
condition, but let v be a stochastic function
of time. Then

$$\langle y(t)\rangle\ =\left\langle \exp(-i\int_0^t v(t')dt')\right\rangle\ .$$

This is the characteristic function belonging to
the stochastic variable represented by the integral. It can be written in the form of a cumulant
expansion

$$\langle y(t) \rangle =$$
$$= \exp\left\{ \sum_{m=1}^{\infty} \frac{(-i)^m}{m!} \int_0^t \overline{v(t_1)v(t_2)\cdots v(t_m)} \; dt_1 dt_2 \cdots dt_m \right\},$$

where the cumulants of $v(t)$ are given by the usual cluster expansion[18)]

$$\langle v(t_1) \rangle = \overline{v(t_1)}$$
$$\langle v(t_1)v(t_2) \rangle = \overline{v(t_1)}\;\overline{v(t_2)} + \overline{v(t_1)v(t_2)}$$
$$\langle v(t_1)v(t_2)v(t_3) \rangle =$$
$$= \overline{v(t_1)}\;\overline{v(t_2)}\;\overline{v(t_3)} + \overline{v(t_1)}\;\overline{v(t_2)v(t_3)} + \cdots$$
$$+ \overline{v(t_1)v(t_2)v(t_3)} \quad \text{etc.}$$

The following specialization of this general formula is often used. Suppose $v(t) = v_0 + \varepsilon v_1(t)$ with fixed constant v_0 and random $v_1(t)$ having $\langle v_1(t) \rangle = 0$. Then the first cumulant is simply equal to v_0, and

$$\langle y(t) \rangle =$$
$$= e^{-iv_0 t} \exp\left\{ -\frac{\varepsilon^2}{2} \iint_0^t \langle v_1(t_1)v_1(t_2) \rangle \; dt_1 dt_2 \cdots \right\}.$$

Note that the expansion in ε occurs in the exponent, which greatly diminishes the danger of secular terms.

If the functions $v_1(t)$ in the ensemble are all, or almost all, so slowly varying that they are practically constant during the time t we are interested in, the present case reduces to that of the previous section. Let us consider the opposite case of rapidly varying $v_1(t)$. That describes an harmonic oscillator whose basic frequency v_0 is perturbed by a rapidly varying influence, for

instance an atom undergoing collisions while emit-
ting light.

Accordingly we suppose that the correlation
between $v_1(t_1)$ and $v_1(t_2)$ depends on t_1-t_2 alone
and vanishes for $|t_1-t_2| > \tau$. Then for $t \gg \tau$ the
double integral is approximately equal to Dt
with[19)]

$$D = \int_{-\infty}^{\infty} \langle v_1(0) \; v_1(t) \rangle \, dt \; .$$

Hence in this case $\langle y(t) \rangle$ does have the form
$e^{-iv_0 t - \Gamma t}$, corresponding to a complex frequency
$v_0 - i\Gamma = v_0 - \frac{1}{2}i\epsilon^2 D$. This shows that short last-
ing collisions give rise to line broadening with
Lorentz profile, as was originally shown by Lorentz.
The higher orders alternately add corrections to
v_0 and Γ, but that is only true up to some order
n with $n\tau \ll t$.

A simple trick enables us to find higher
moments of y. One has, for instance, taking
$0 \leqslant t \leqslant s$,

$$\langle y(t)y(s) \rangle = \left\langle \exp\left\{ -i \int_0^t v(t')dt' - i \int_0^s v(t')dt' \right\} \right\rangle$$

$$= \left\langle \exp\left\{ -i \int_0^{\infty} \vartheta(t')v(t')dt' \right\} \right\rangle .$$

Here we have defined $\vartheta(t')$ by setting

$$\vartheta(t') \quad = 2 \qquad 0 < t' < t \, ,$$
$$= 1 \qquad t < t' < s \, ,$$
$$= 0 \qquad s < t' \, .$$

It is again possible to use the cumulant expansion;
if we also use an expansion in ϵ for convenience,

we find

$$\langle y(t)y(s)\rangle = e^{-iv_0 t - iv_0 s} \times$$

$$\times \exp\left\{-\frac{\varepsilon^2}{2}\iint_0^\infty \vartheta(t')\,\vartheta(t'')\,\langle v_1(t')v_1(t'')\rangle \,dt'dt''\right\}.$$

Similar equations hold for the other moments; in particular

$$\langle y(t)y^*(s)\rangle = \left\langle \exp\left\{-i\int_s^t v(t')dt'\right\}\right\rangle$$

$$= \exp\left\{-\frac{\varepsilon^2}{2}\iint_s^t \langle v_1(t')v_2(t'')\rangle dt'dt''\right.$$

12. The general formula for $\langle y(t)\rangle$.

As the final complication we suppose that y is a vector and take instead of v a random matrix $V(t)$

$$y = -i\,V(t)\,y, \qquad y(0) = a.$$

The initial value is some fixed vector a. For given $V(t)$ the solution is

$$y(t) = \left\{1 - i\int_0^t V(t_1)dt_1 - \int_0^t v(t_1)dt_1 \int_0^t V(t_2)dt_2 + \dots \right\}a.$$

In each term the factors V appear in chronological order with the latest time on the left. No error is made by writing the factors in arbitrary order, provided one adds a reminder that they should actually be read chronologically:

$$y(t) = \left\{ 1 - i \int_0^t V(t_1)dt_1 - \right.$$
$$\left. - \frac{1}{2} \int_0^t dt_1 \int_0^t dt_2 \ulcorner V(t_1)V(t_2) \urcorner + \ldots \right\} a \ .$$

where $\ulcorner\ \urcorner$ denotes the time-ordered product. More generally

$$y(t) = \left\{ 1 + \sum_{n=1}^{\infty} \frac{(-i)^n}{n!} \int_0^t dt_1 \ldots dt_n \ulcorner V(t_1) \ldots V(t_n) \urcorner \right\} a =$$
$$= \ulcorner \exp \left\{ -i \int_0^t V(t')dt' \right\} \urcorner a.$$

The last expression is nothing but a symbolic way of writing the previous one: the time ordering has to be performed after expanding the exponential.

Having written the solution of the ordinary differential equation in a suitable way we take the average,

$$\langle y(t) \rangle = \ulcorner \left\langle \exp \left\{ -i \int_0^t V(t')dt' \right\} \right\rangle \urcorner a =$$
$$= \ulcorner \exp \left\{ \sum_{m=1}^{\infty} \frac{(-i)^m}{m!} \times \right.$$
$$\left. \times \int_0^t dt_1 \ldots dt_m \ \overline{V(t_1) \ldots V(t_m)} \right\} \urcorner a,$$

where the bar denotes the cumulant. This is our general formula for the averaged solution of a stochastic linear differential equation. It is a condensed way of writing the result, but the actual evaluation of the right-hand side is not

always easy**＊**. It reduces to the result of the
previous section when all factors V commute, not
merely at all times but also for all samples in
the ensemble.

Let $V(t) = V_0 + \epsilon V_1(t)$ with fixed V_0 and
$\langle V_1(t) \rangle = 0$. Then the first cumulant can again
be written separately,

$$\langle y(t) \rangle = \lceil \exp\left\{-i \int_0^t V_0(t_1)dt_1 + \right.$$

$$+ \left. \sum_{m=2}^{\infty} \frac{(-i\epsilon)^m}{m!} \int_0^t dt_1 \ldots dt_m \, \overline{V_1(t_1)\ldots V_1(t_m)}\right\} \rceil a.$$

Although V_0 does not depend on time it has been
provided with a label t_1, which is required to
assign to each factor V_0 its proper place in the
time-ordered product. An equivalent expression
is[20)]

$$\langle y(t) \rangle =$$

$$= e^{iV_0 t} \lceil \exp\left\{ \sum_{m=2}^{\infty} \frac{(-i\epsilon)^m}{m!} \times \right.$$

$$\int_0^t dt_1 \ldots dt_m \, \overline{W_1(t_1)\ldots W_\perp(\tau_m)}\left.\right\} \rceil a \quad,$$

wnere $W_1(t)$ stands for the operator $V_1(t)$ in

interaction representation,

$$W_1(t) = e^{iV_0 t} V_1(t) e^{-iV_0 t}$$

Neglecting third and higher powers of ϵ in the
exponent

＊ The equation is mentioned by R. Kubo[11)] and a
modified form of it was used by J.T. Ubbink[12)].

$$\langle y(t) \rangle = \exp\left\{ -iV_0 t - \varepsilon^2 \int_0^t e^{iV_0 t_1} dt_1 \right. \times$$

$$\left. \times \int_0^{t_1} \left\langle V_1(t_1) e^{-iV_0(t_1-t_2)} V_1(t_2) \right\rangle e^{-iV_0 t_2} dt_2 \right\} a.$$

Actually the exponential should still be time-ordered, but that only affects higher orders in ε.

Suppose the correlation depends on t_1-t_2 alone, so that

$$\left\langle V_1(t_1) \, e^{-iV_0(t_1-t_2)} \, V_1(t_2) \right\rangle = R(t_1-t_2).$$

Moreover suppose that the correlation in practically zero for $|t_1-t_2| > \tau$; so that for $t \gg \tau$

$$\int_0^{t_1} R(t_1-t_2) \, e^{-iV_0 t_2} \, dt_2 =$$

$$= \int_0^{t_1} R(t') \, e^{-iV_0 t'} \, dt' . e^{-iV_0 t_1} \approx$$

$$\approx \int_0^\infty R(t') \, e^{-iV_0 t'} \, dt' . e^{-iV_0 t_1} = D \, e^{-iV_0 t_1} .$$

And, if t is much larger than the periods contained in V_0,

$$\int_0^t e^{iV_0 t_1} \, dt_1 \, D \, e^{-iV_0 t_1} = t \, \tilde{D} ,$$

where $\tilde{D}$ is the diagonal part of D with respect to V_0. Then

$$\langle y(t) \rangle = e^{-iV_0 t - \varepsilon^2 \tilde{D} t} a,$$

which is a superposition of waves with complex frequencies.

13.   <u>Application to the transmission of waves.</u>

An electromagnetic wave propagating in the
direction x through the troposphere may roughly
be described by the Helmholtz equation

$$\frac{d^2u}{dx^2} \; k^2 \, n^2 \, u = 0 \; .$$

k_o is the wave number for the average density of
air, and $n^2(x) = 1+\epsilon\mu(x)$ is the variation in the
refractive index due to density fluctuations. In
practice $\epsilon\mu \sim 10^{-6}$; the fluctuations extend over
some 50 meters and they vary slowly compared to
the passage time of a radio signal. The wave
length is supposed to be small compared to them,
because derivatives of n have been neglected in
the equation. This approximation amounts to taking into account the local variations of wave
length and amplitude, but not reflection (WKB
approximation). Hence one may consider separately
the wave going to the right; putting
$du/dx + ik_o nu = y$ one has in the same approximation

$$dy/dx \; = \; ik_o n(x)y.$$

One is interested in the effect of many signals
transmitted over a fixed distance x much longer
than 50 m. Hence $\mu(x)$ is a stochastic function
and the problem is of the type treated in section
11, except that x is substituted for t. The mean
amplitude is attenuated by a factor $e^{-\Gamma x}$ with

$$\Gamma = \frac{1}{2} \, \epsilon^2 \, k_o^2 \int_{-\infty}^{\infty} \langle \mu(0) \, \mu(x') \rangle \; dx' \; .$$

The intensity of the signal decreases proportionally

to $e^{-2\mathbf{\Gamma}x}$, the increase in noise is $1-e^{-2\mathbf{\Gamma}x}$.

Another problem is the passage of a wave through a space with randomly located scatterers. We take one dimension and write y_1, y_2 for the amplitudes of the waves going to the right and to the left. The scatterers are located at random positions ξ_s and are otherwise identical. Each of them is represented by a two-by-two matrix A such that

$$y(\xi_s + 0) = A \, y(\xi_s - 0) .$$

When we set $A = 1 - i\varepsilon B$ the equation for y is

$$\frac{dy}{dx} = \begin{pmatrix} ik_o & 0 \\ 0 & -ik_o \end{pmatrix} y - i\varepsilon \sum_s \delta(x-\xi_s) \, B \, y.$$

It can be verified that B is real when the scatterer is elastic and symmetric. The operator on the right has the form $-iV = -iV_o - i\varepsilon V_1$, where V_o is constant and $V_1(x)$ is random and depends on x. Hence the general formula says

$$\langle y(x) \rangle = \left[\exp\left\{ -i \int_0^x \left[V_o(x') + \varepsilon V_1(x') \right] dx' + \right. \right.$$
$$\left. \left. + \sum_{m=2}^{\infty} \frac{(-i\varepsilon)^m}{m!} \int_0^x \overline{V_1(x_1) \ldots V_1(x_m)} \, dx_1 \ldots dx_m \right\} \right].$$

In order to compute the cumulants we have to employ a suitable way of describing the random location of the scatterers[9]. Let $f_n(x_1, \ldots, x_n)$ be the probability density for finding a scatterer at each of the locations $x_1, \ldots, x_n$, regardless of all other scatterers. Then, if all $x_1, \ldots, x_n$ are different,

$$f_n(x_1,\ldots,x_n) = \left\langle \sum_{s_1,\ldots,s_n} \delta(x_1-\xi_{s_1})\ldots\delta(x_n-\xi_{s_n}) \right\rangle.$$

Subsequently the correlation functions g_n are defined by the usual cluster expansion

$$f_1(x_1) \;=\; g_1(x_1)$$
$$f_2(x_1,x_2) \;=\; g_1(x_1)g_1(x_2) + g_2(x_1,x_2), \quad \text{etc.}$$

We suppose that the distribution is homogeneous in space, so that $f_1 = g_1 = $ const. and f_2 and g_2 only depend on $x_1 - x_2$.

Returning to the general formula we first note

$$V_0 + \overline{eV_1(x)} \;=\; V_0 + \varepsilon\langle V_1(x)\rangle \;=\; V_0 + \varepsilon f_1 B.$$

Call this constant matrix W_0 and define the interaction representation by

$$W_1(x) \;=\; e^{iW_0 x} V_1(x) e^{-iW_0 x}.$$

The general formula may then be written to second order in ε

$$\langle y(x) \rangle = e^{-i(V_0 + \varepsilon f_1 B)x}$$
$$\exp\left\{ -\varepsilon^2 \int_0^x dx_1 \int_0^{x_1} dx_2\, \overline{W_1(x_1)W_1(x_2)} \right\}a.$$

The term $\varepsilon f_1 B$ is the averaged effect of single scatterers, while the double integral describes the additional effect of pairs. With some algebra one finds for it

$$\int_0^x dx_1 \int_0^{x_1} dx_2 \; g_2(x_1,x_2) e^{iW_o x_1} B \; e^{-iW_o(x_1-x_2)} \times$$

$$\times B \; e^{-iW_o x_2} =$$

$$= \int_0^x e^{iW_o x_1} \left\{ \int_0^{x_1} g_2(x') B \; e^{-iW_o x'} B \; e^{iW_o x'} dx \right\} \times$$

$$\times e^{-iW_o x_1} dx_1 .$$

It is by a similar calculation that Ubbink[12] managed to solve the problem of a photoconductor subject to randomly incident photons.

Various limiting cases of this formula may be recognized. First suppose that the positions are uncorrelated; then $g_2 = 0$ and the total effect of the scatterers is the sum of their individual effects. Next suppose that the range of $g_2(x)$ is much less than the period of $e^{-iW_o x}$ that is, the correlation length is small compared to the wavelength. Then the integral reduces to

$$\int_0^x e^{iW_o x} B^2 e^{-iW_o x_1} dx_1 \; . \int_0^\infty g_2(x') \, dx' \; .$$

This represents the coherent scattering by neighbouring pairs of scatterers. For large x this is proportional to $x \; \widetilde{B^2}$ where the tilde denotes the diagonal part of B^2 with respect to W^o. Finally let g_2 have a long range and vary smoothly compared to the wavelength. Then the integral reduces to incoherent scattering by pairs,

$$x(\widetilde{B})^2 \int_0^\infty g_2(x') \, dx' \; .$$

In all these cases $\langle y(x) \rangle$ depends exponentially on x owing to the assumption that x is much

larger than any of the other lengths involved in
the problem.

REFERENCES

1) A. Kolmogoroff, Grundbegriffe der Wahrschein-
   lichkeitsrechnung. Ergebn. Mathem. Grenzgeb.
   2, No. 3 (Springer, Berlin 1933; Chelsea,
   New York 1948).

2) A. Khitchine, Asymptotische Gesetze der Wahr-
   scheinlichkeitsrechnung. Ergebn. Mathem.
   Grenzgeb. 2, No. 4 (Springer, Berlin 1933).

3) A.T. Barucha-Reid, Elements of the theory of
   Markov processes and their applications.
   McGraw-Hill, New York 1960; I.I. Gichman
   und A.S. Skorochod, Stochastische Differen-
   tialgleichungen. (Akademie-Verlag, Berlin
   1971).

4) Reviewed in N.G. van Kampen, Fluctuations in
   nonlinear systems in Fluctuation Phenomena
   in Solids. (R.E. Burgess ed., Acad. Press,
   New York 1965).

5) A.J. Lotka, Elements of mathematical biology.
   (Dover Publ., New York 1956).

6) S. Chandrasekhar, Rev. Mod. Phys. 15, 1 (1943);
   reprinted in: Selected papers on noise and
   stochastic processes. (N. Wax ed., Dover
   Publ., New York 1954).

7) N.G. van Kampen, Can. Journ. Phys. 39, 551
   (1961).

8) See e.g. W.B. Davenport Jr. and W.L. Root,
   An introduction to the theory of random
   signals and noise. (McGraw-Hill, New York
   1958).

9) R.L. Stratonovic, Topics in the theory of
   random noise, I, II (Gordon and Breach,
   New York 1963, 1967).

10) W.M. Wonham, Random differential equations in control theory in: Probabilistic methods in applied mathematics 2 (A.T. Barucha-Reid, ed.,(Acad. Press, New York, 1970).

11) R. Kubo, J. Math. Phys. 4, 174 (1963).

12) J.T. Ubbink, Physica 52, 253 (1971).

13) G.A. van Lear Jr. and G.E. Uhlenbeck, Phys. Rev. 38, 1583 (1931).

14) V.I. Tatarski, Wave propagation in a turbulent medium.(McGraw-Hill, New York 1961).

15) J.B. Keller, Wave propagation in random media in: Proc.Symposium Applied Mathem. 13 (Amer. Math.Soc., Providence, R.I. 1962).

16) See a forthcoming paper by T. Chow and J.B. Keller.

17) See any textbook on probability, or in particular E. Lukacs, Characteristic functions. (Griffin, London 1960).

18) This has been called the "First Mayer Theorem" in G.E. Uhlenbeck and G.W. Ford, Lectures in Statistical Mechanics (Amer.Mathem.Soc., Providence, R.I. 1963).

19) G.E. Ihlenbeck and L.S. Ornstein, Phys.Rev.36 823 (1930); reprinted in: Selected papers on noise and stochastic processes. (N.Wax ed. Dover Publ., New York 1954).

20) R.P. Feynman, Phys. Rev. 84, 108 (1951).

21) N.G. van Kampen, Physica Norvegica 5,279(1971).

22) W.E. Boyce, Random Eigenvalue Problems in: Probabilisitic methods in applied mathematics 1 (A.T. Barucha-Reid ed., Acad. Press, New York 1968); Statistical theory of spectra (C.E. Porter ed., Acad. Press, New York,1965).

23) Webster's Dictionary.

HYDRODYNAMICS OF MAGNETIC CRYSTALS

C.P. Enz

Institut de Physique Théorique
Ecole de Physique, CH-1211 GENEVE 4

1.   Introduction: Systems with two-fluid hydrodynamics

It has been realized in recent times that a
number of many-body systems can be described by
two-fluid hydrodynamic equations of astonishing
similarity [1]. The class of these systems is characterized by a phase transition into a state in
which a condensed phase can be defined. In analogy
to superfluid Helium this condensed phase then
gives rise to a two-fluid description with its
characteristic hydrodynamic modes of propagation:
first and second sound. The principal systems
having this property are neutral superfluids
(Helium II), charged superfluids (superconductors),
dielectric crystals and magnetic crystals (ferromagnets and antiferromagnets). To a certain extent
also nematic liquid crystals fall into this class.
However, no second sound-type mode exists in the
latter system [1].

The main feature of this class of systems is
the existence of a broken symmetry. The symmetry
group which is broken in neutral and charged superfluids is the gauge group; in dielectric crystals
it is the translation (and rotation) group; in
magnetic crystals and in nematic liquid crystals
it is the rotation group. The breaking of these
symmetry groups is related by Noether's theorem
to non-conservation of the physical quantity acting
as generator of this group which is particle number
in superfluids, momentum in dielectric crystals and
angular momentum in magnetic crystals and in nematic

liquid crystals. From the statistical ensemble point of view this means that the system is described by a density matrix ϱ which although commuting with the Hamiltonian $\mathcal{H}$ of the system does not commute with this group generator G which, however, commutes with $\mathcal{H}$,

$$[\varrho,G] \neq 0, \quad [\varrho,\mathcal{H}] = 0, \quad [G,\mathcal{H}] = 0.$$

There exists a physical density $\eta(\vec{r})$ playing the role of an <u>order parameter</u>, the time dependence of which is given by the Heisenberg representation

$$\eta(\vec{r},t) = e^{i(\mathcal{H} - \lambda G)t} \eta(\vec{r}) e^{-i(\mathcal{H} - \lambda G)t}$$

where the term $-\lambda G$ accounts for conservation of G. Because of $[\varrho,G] \neq 0$ the average of η still depends on time

$$\langle \eta(\vec{r},t) \rangle = \text{Tr}(\varrho\, \eta(\vec{r},t)) = \langle e^{-i\lambda Gt} \eta(\vec{r}) e^{i\lambda Gt} \rangle$$

Above the phase transition, $T > T_c$, $[\varrho,G] = 0$ and $\langle \eta \rangle = 0$; below it, $T < T_c$, we have a two-fluid situation where $\langle \eta \rangle$ determines the hydrodynamic motion of the "second fluid" or "superfluid".

A ϱ for which $[\varrho,G] \neq 0$ can be obtained from the equilibrium density matrix $\varrho_{eq} = \exp[\beta(F - \mathcal{H} + \lambda G)]$ by adding to the Hamiltonian $\mathcal{H}$ a coupling to an external source ϕ_{ext}

$$\mathcal{H}' = \int d^3r\, M(\vec{r})\, \phi_{ext}(\vec{r}\ t)\, \eta(\vec{r}) + \text{h.c.}$$

Then $\langle \eta \rangle$ is given, in linear response to $\mathcal{H}'$, just by a Kubo formula. If both M and ϕ_{ext} are independent of $\vec{r}$ the above formulas hold with an $\vec{r}$-independent $\langle \eta \rangle$. If, however, M or ϕ_{ext} depend on $\vec{r}$ then the system will be close to local equilibrium as described by the density matrix

$$\rho_{loc\ eq} = \exp\left[\int \beta(\vec{r})(f(\vec{r}) - h(\vec{r}) + \lambda(\vec{r})g(\vec{r}))d^3r\right]$$

where $f(\vec{r})$, $h(\vec{r})$, $g(\vec{r})$ are the densities associated with the free energy F, the Hamiltonian $\mathcal{H}$ and the group generator G, respectively. This local equilibrium is perturbed by time-dependent variations of the external source coupling $\mathcal{H}'$. Then $\langle\eta\rangle$ also becomes a function of $\vec{r}$.

In the neutral superfluid η is the creation operator ψ^+ for particles, and ϕ_{ext} is the external annihilation operator ψ_{ext}. Then $\mathcal{H}'$, with an appropriate matrix element M, is just a tunneling Hamiltonian coupling the external particle source to the system. We start with an infinitely long wave coupling $\mathcal{H}'$. Then we have

$$\langle\psi^+(t)\rangle = \sqrt{n_0}\,e^{-i\varphi(t)}$$

where n_0 is the condensate density. G is the number operator N and λ the chemical potential $m\mu$ where m is the atomic mass. Since ψ^+ raises the particle number by one we conclude that

$$e^{-im\mu Nt}\psi^+ e^{+im\mu Nt} = \psi^+\, e^{im\mu t}$$

so that

$$\dot{\varphi} = -m\mu$$

If the coupling $\mathcal{H}'$ varies with $\vec{r}$ then n_0, φ and μ depend on the position $\vec{r}$ and there is a non-vanishing superfluid velocity defined by

$$\vec{v}_s = \frac{\hbar}{m}\nabla\varphi$$

From this follows the hydrodynamic equation of motion for the "second fluid" or superfluid,

$$\dot{\vec{v}}_S = -\nabla \mu$$

The cases of superconductors and of magnetic crystals, which we treat in detail below, are completely analogous; for the superconductor ψ^+ is replaced by the creation operator of a Cooper pair.

In the dielectric crystal η is the displacement operator $\vec{d}(\vec{R})$ at lattice position $\vec{R}$, G is the momentum operator and $-\lambda$ a velocity parameter $\vec{v}$. Then

$$\langle \vec{d}(\vec{r} + \vec{v}t) \rangle = \vec{u}(\vec{r} + \vec{v}t)$$

defines the elastic displacement field which satisfies the equation of motion of elasticity theory

$$\ddot{\vec{u}} = \frac{1}{\rho} \nabla \sigma$$

where σ is the stress tensor and ρ the mass density of the crystal. Defining a lattice velocity

$$\vec{v}_L = \dot{\vec{u}}$$

the elastic equation of motion takes a form analogous to that of the superfluid,

$$\dot{\vec{v}}_L = \frac{1}{\rho} \nabla \sigma$$

so that $\vec{v}_L$ here plays the role of the superfluid velocity and the elastic displacement field represents the condensed phase or "second fluid".

A similar situation holds for the nematic liquid crystal [1].

The two-fluid description is completed by the hydrodynamic equations for the transport density ρ_T, the momentum density $\vec{j}$ and the energy density per unit mass e;

$$\dot{\varrho}_T + \nabla \cdot \vec{j}_T = 0$$

$$\dot{\vec{j}} + \nabla \pi = - \frac{1}{\tau_J} \vec{j}$$

$$(\varrho e)^{\cdot} + \nabla \cdot \vec{J}_e = 0$$

where $\vec{j}_T$ is the transport current density, π the momentum flux, $\vec{J}_e$ the energy flux and τ_J is a relaxation timewhich accounts for momentum dissipation. In the neutral and charged superfluid and in the nematic liquid crystal $\varrho_T = \varrho$ is the total mass density and $\vec{j}_T = \vec{j}$ the total momentum density. In the dielectric crystal $\varrho_T = \varrho$, $\vec{j}_T = \vec{j}_L = = \varrho\vec{v}_L$, $\vec{j} = \vec{j}_p = \varrho_p \vec{v}_p$ is the momentum density of the phonon fluid, $\vec{v}_p$ is the phonon drift velocity and ϱ_p the associated excitation mass density. In the planar ferromagnet $\varrho_T = mM_z$, $\vec{j}_T = \vec{j}_s = \varrho_s\vec{v}_s$ and $\vec{j} = \vec{j}_M = \varrho_M v_M$ where the quantities will be defined below.

As will be seen below with the example of the magnetic crystal it is useful to transform the energy conservation into an entropy balance

$$(\varrho s)^{\cdot} + \nabla \cdot \vec{J}_s = \sigma$$

where s is the entropy per unit mass, $\vec{J}_s$ the entropy flux and σ the entropy production density.

The subsequent sections provide an improved version of an earlier treatment of the hydrodynamics of magnetic crystals[2] which was based on the work of Halperin and Hohenberg [3]. These authors were the first to give a systematic hydrodynamic description of ferromagnets and

antiferromagnets using purely hydrodynamic arguments
to derive the well-known spin-wave and diffusion
modes of such systems. Their method has also lead
to some new predictions with respect to the damping of the spin-wave modes of the different types
of magnetic crystals. These authors did not, however, introduce a local magnon drift $\vec{v}_M$ and, as a
consequence, did not obtain the analogue of second
sound. This "second spin-wave" or "second magnon"
has been discussed by Michel and Schwabl [4] starting from a Boltzmann equation approach. In subsequent papers [5] these authors derived the hydrodynamics from microscopic equations of motion.
This microscopic approach, which is not our aim
here, has given rise to an extended literature
for which we refer to refs. 3), 4), 5). More
recently Michel and Schwabl [6] and Forney [7]
have discussed the conditions under which a
second magnon could be detected by neutron
scattering.

In ref. 2) the introduction of a local magnon
drift has also been omitted, as in ref. 3). In
addition, ref. 2) contains an unfortunate confusion between the magnon drift velocity and the
"superfluid" velocity introduced in ref. 3) which
gave rise to erroneous conclusions with respect
to the comparison of refs. 3) and 4). This confusion will be completely clarified here by introducing a local magnon drift in close analogy to
the case of dielectric crystals [8]. Thereby we
arrive again at a two-fluid description where the

first or normal fluid is that of the thermal magnons while the second or super-fluid is that of
the c-number magnetization field. As we shall see,
however, the magnon drift $\vec{v}_M$ is a valid concept
only for a magnon spectrum ω_q such that $\omega_q > v_M q$
and, therefore, is zero for purely isotropic
ferromagnets where $\omega_q \propto q^2$, as well as for the
paramagnetic state where the magnon velocity
vanishes. This latter fact means that the acoustic
magnon is the soft mode of the magnetic phase
transition.

Strictly speaking, lattice displacements introduce variations in the coupling constants of the
interaction between the local spins and consequently lead to a coupling between the two magnetic
"fluids" and the two non-magnetic "fluids" formed
by the thermal phonons and the c-number displacement field. Neglecting this coupling the hydrodynamics of the magnetic and non-magnetic two-fluid
systems are completely independent, which means in
particular that both have independent local temperatures. We will follow here ref. 2), 3) and 4)
in adopting this simplified point of view, i.e.
discuss only the uncoupled magnetic two-fluid
system.

2. Hydrodynamics of the Planar Ferromagnet

As in ref. 3) we start with the case of the
planar ferromagnet because it contains all the
essential features which can be taken over to the
other cases with only slight modifications. A
planar ferromagnet is described by a Hamiltonian

$$\mathcal{K} = \mathcal{K}_0 - H_z \sum_i S_i^z$$

$$\mathcal{K}_0 = - \sum_{ij} \left\{ J_{ij}^z S_i^z S_j^z + J_{ij}^\perp (S_i^x S_j^x + S_i^y S_j^y) \right\} \tag{1}$$

with coupling constants $J_{ij}^\perp > J_{ij}^z > 0$.
(The external magnetic field $\vec{H} = (o, o, H_z)$ is measured in energy units and $\hbar = 1$). The isotropic or Heisenberg ferromagnet which is the special case $J_{ij}^z = J_{ij}^\perp$ is not a favourable starting point for deriving hydrodynamic equations because, as we mentioned, the spin-waves are not sound-like. As we shall see, however, the results for this case easily follow from those of the planar case.

The crucial feature of a ferromagnetic state is the existence of a non-vanishing average of the spin raising operator $S_i^+ = S_i^x + iS_i^y$. The ferromagnetic average has the property that it is taken with a density matrix which commutes with $\mathcal{K}_0$ but not with $\sum_i S_i^z$. Since S_i^+ increases the value of S_i^z by one and since $[\sum_i S_i^z, \mathcal{K}_0] = o$ the Heisenberg representation

$$S_i^+(t) = \exp\left\{ i\left[\mathcal{K}_0 - (\bar{H}_z - H_z)(-\sum_i S_i^z)\right] t \right\} S_i^+ \times$$

$$\times \exp\left\{ -i\left[\mathcal{K}_0 - (\bar{H}_z - H_z)(-\sum_i S_i^z)\right] t \right\}$$

(the field $\bar{H}_z - H_z$ plays the role analogous to a chemical potential, see Eq. (10) below) together with (1) imply the existence of a non-vanishing order parameter

$$\frac{1}{v}\langle S_i^+(t) \rangle = \frac{1}{v} e^{-i(\bar{H}_z - H_z)t} \langle S_i^+ \rangle = M_\perp e^{i\varphi(t)} = M_x + iM_y \tag{2}$$

where v is the volume of the unit cell, $M_\perp$ the (time-independent) perpendicular magnetization and $\varphi(t)$ the precession angle. From (2), which is just the time-integrated Bloch equation, we have

$$\varphi(t) = \varphi_0 - (\bar{H}_z - H_z)t \qquad (3)$$

Since the energy

$$\langle \mathcal{H}_0 \rangle = V \rho \, e \qquad (4)$$

(ρ is the mass density of the crystal) and the parallel magnetization

$$\frac{1}{v} \langle S_i^z \rangle = M_z \qquad (5)$$

are conserved quantities $M_\perp$ must be a function of e and M_z. As pointed out in ref. 3) this supposes the existence of a relaxation mechanism which brings $M_\perp$ to its value $M_\perp$ (e, M_z) in a microscopic time.

In thermal equilibrium at a temperature T below the Curie point T_c, e= e(M_z, T) so that

$$M_\perp = M_\perp (M_z, T) \qquad (6)$$

and

$$H_z = \rho \left(\frac{\partial e}{\partial M_z} \right)_s \equiv h_z (M_z, T) \qquad (7)$$

where s = s (M_z, T) is the entropy per unit mass. Thermal equilibrium is defined by an average with a canonical density matrix which commutes with $\mathcal{H}_0$ and with $\sum_i S_i^z$. Hence it follows from (2) that

$$\dot{\varphi}_{eq} = 0 \qquad (8)$$

i.e. there is no precession of $\vec{M}$. From (3) and (8) we conclude that $\bar{H}_z = h_z$ and obtain

$$\dot{\varphi} = -m\, \mu \tag{9}$$

with

$$\mu = \frac{1}{m}\,(h_z - H_z) + \mu' \tag{10}$$

(m is the atomic mass) where we have allowed for
a dissipative part μ'. Going over to local variables
$e(\vec{r},t)$, etc. defined as averages of density operators (see ref. 3)) we may define a velocity field

$$\vec{v}_s = \frac{1}{m}\,\nabla\varphi \tag{11}$$

which is exactly analogous to the superfluid velocity in Helium. Indeed, according to (9) $\vec{v}_s$
satisfies local conservation of circulation in
analogy to the superfluid,

$$\dot{\vec{v}}_s = -\nabla\mu \tag{12}$$

Local conservation of energy and of parallel
magnetization is expressed by

$$\rho\dot{e} + \nabla\cdot\vec{J}_e = 0 \tag{13}$$

and

$$mM_z + \nabla\cdot\vec{j}_s = 0. \tag{14}$$

In analogy to the case of phonons in dielectric
crystals [8] the magnon momentum density $\vec{j}_M$ and
local drift $\vec{v}_M$ define the magnon mass density
ρ_M through

$$\vec{j}_M = \rho_M\,\vec{v}_M \tag{15}$$

and the local magnon momentum balance equation
is, in analogy to the phonon case, [8],[9]

$$\dot{\vec{j}}_M + \nabla\pi_M = -\frac{1}{\tau_J}\,\vec{j}_M \tag{16}$$

Here

$$\pi_M = - \varrho f_M \mathbf{1} + \varrho_M (\vec{v}_M \otimes \vec{v}_M) + \pi'_M \qquad (17)$$

is the momentum flux density and f_M the magnon
free energy per unit mass where

$$d(\varrho f_M) = - \varrho s_M dT + \vec{v}_M \cdot d\vec{j}_M \qquad (18)$$

and s_M is the magnon entropy per unit mass. The
dissipative part π'_M satisfies a linear relation
analogous to the phonon case in which, however,
the cross term $\sim \nabla_k v_{s\ell}$ is neglected,

$$\pi'_{Mij} = - \sum_{k\ell} \gamma_{ij,k\ell} \nabla_k v_{M\ell} \qquad (19)$$

We convert the energy conservation equation
into an entropy balance equation in order to ob-
tain expressions for the currents $\vec{J}_e$ and $\vec{j}_s$. The
thermodynamic relation underlying this conversion
is obtained from (18) and the analogous expression
for the "second fluid",

$$d(\varrho fs) = h_z dM_z - \varrho s_s dT + \varrho_s \vec{v}_s \cdot d\vec{v}_s, \qquad (20)$$

together with $f = f_M + f_s = e - Ts$ and $s = s_M + s_s$,
viz.

$$d(\varrho e) = h_z dM_z + Td(\varrho s) + \vec{v}_s \cdot (\varrho_s d\vec{v}_s) + \vec{v}_M \cdot d\vec{j}_M \quad (21)$$

In order to avoid unessential complications we
will neglect the tensor character of ϱ_s as well
as of ϱ_M. Then ϱ_s must be positive; it has the
meaning of a stiffness constant and the dimension
of a mass density and is the analogue of the superfluid mass density in Helium. Apart from the last
term Eq. (21) has the same content as Eq. (2.28)
of ref. 3) with the identification of $-\vec{v}/m$ with
our $\vec{v}_s$ and of $m^2 \varrho_s$ with our ϱ_s. Eq. (21) implies

$$T \varrho \dot{s} = \varrho \dot{e} - h_z \dot{M}_z - \vec{v}_M \cdot \dot{\vec{J}}_M - \varrho_s \vec{v}_s \cdot \dot{\vec{v}}_s$$

Inserting from (10, 12-14, 16-18) we obtain with H_z = const.

$$\varrho \dot{s} = - \vec{\nabla} \cdot \left\{ \frac{1}{T}(\vec{J}_e - \frac{h_z}{m} \vec{J}_s) \right\} + (\vec{J}_e - \frac{h_z}{m} \vec{J}_s - \varrho_M^s T \vec{v}_M) \cdot \vec{\nabla} \frac{1}{T} -$$

$$- \frac{1}{Tm}(\vec{J}_s - \varrho_s \vec{v}_s) \cdot \vec{\nabla} h_z + \frac{\varrho_s}{T} \vec{v}_s \cdot \vec{\nabla} \mu' + \frac{1}{T}\vec{v}_M \cdot (\vec{\nabla} \pi_M') + \frac{\varrho_M}{T\tau_J} \vec{v}_M^2$$

$$(22)$$

In the absence of dissipation, entropy must be locally conserved which means that the second and third terms to the right must then vanish. Hence

$$\vec{J}_s = \varrho_s \vec{v}_s + \vec{j}_s' \tag{23}$$

and

$$\vec{J}_e = \frac{\varrho_s h_z}{m} \vec{v}_s + \varrho_s^s T \vec{v}_M + \vec{J}_e' \tag{24}$$

where $\vec{j}_s'$ and $\vec{J}_e'$ are the corresponding dissipative parts. The entropy balance equation now takes the form

$$\varrho \dot{s} + \vec{\nabla} \cdot \vec{J}_s = \sigma \tag{25}$$

where

$$\vec{J}_s = \varrho \, s_M \vec{v}_M + \vec{J}_s' \tag{26}$$

is the entropy current density

$$\vec{J}_s' = \frac{1}{T} (\vec{J}_e' - \frac{h_z}{m} \vec{j}_s') \tag{27}$$

its dissipative part and

$$\sigma = T\vec{J}_s' \cdot \vec{\nabla}\frac{1}{T} - \frac{1}{Tm} \vec{j}_s' \cdot \vec{\nabla} h_z + \frac{\varrho_s}{T}\vec{v}_s \cdot \vec{\nabla}\mu' +$$

$$+ \frac{1}{T} \vec{v}_M \cdot (\vec{\nabla} \pi_M') + \frac{\varrho_M}{T\tau_J} \vec{v}_M^2 \tag{28}$$

the entropy production density. The dissipative part $\vec{j}_s'$ describes spin diffusion; the analogous diffusion terms in superfluid Helium and in dielectric crystals are negligible [8]. Eqs. (25 - 28) are a generalization of eqs. (2.43, 44) of ref. 3) to the case with magnon drift.

For small perturbations the dissipative parts μ', $\vec{J}_e'$ and $\vec{j}_s'$ are linear functions of the gradients of the coefficients in eq. (21), i.e. of ∇h_z, ∇T, $\varrho_s \nabla_i v_{sk}$ and $\nabla_i v_{Mk}$. Time reversal and point group invariance then reduce these linear functions to

$$\mu' = - \varrho_s \nabla \cdot (\zeta \vec{v}_s) \qquad (29)$$

$$\vec{J}_e' = - \kappa \nabla T \qquad (30)$$

and

$$\vec{j}_s' = - m \, \xi \nabla h_z \qquad (31)$$

The tensor ζ is the analogue of a second viscosity in Helium and κ is the heat conductivity tensor of the magnon gas. Note that in (29), as in (19), a cross term, $\sim \nabla \cdot \vec{v}_M$, is neglected. Such cross terms would give rise to dissipative coupling between magnetic and thermal variables.

3. Hydrodynamic Modes of the Ferromagnet

The hydrodynamic equations (12, 14, 16, 25) are, in the linear approximation, and inserting from (10, 15, 17-19, 23, 26-31)

$$\dot{\vec{v}}_s + \frac{1}{m} \nabla h_z - \varrho_s \nabla (\nabla \cdot \zeta \vec{v}_s) = 0 \qquad (32)$$

$$\dot{M}_z + \frac{\varrho_s}{m} \nabla \cdot \vec{v}_s - \nabla \cdot (\xi \nabla h_z) = 0 \qquad (33)$$

$$\dot{v}_{Mi} + \frac{\rho}{\rho_M} s_M \nabla_i T - \frac{1}{\rho_M} \sum_{jk\ell} \Gamma_{ij,k\ell} \nabla_j \hat{v}_k v_{M\ell} + \frac{1}{\tau_J} v_{Mi} = 0$$

(34)

$$\dot{s} + s_M \nabla \cdot \vec{v}_M - \frac{1}{\rho T} \nabla \cdot (\kappa \nabla T) + \frac{h_z}{\rho T} \nabla \cdot (\xi \nabla h_z) = 0 \quad (35)$$

In the last equation we have omitted the term $\frac{\rho_s}{T} \vec{v}_{eq} \cdot \nabla \mu'$ coming from σ, assuming that the equilibrium value of $\vec{v}_s$, $\vec{v}_{eq} = \frac{1}{m} \nabla \varphi_{eq}$, is small which is the case for small h_z and for large stiffness constant (see eq. (71) below).

Choosing as independent variables T, M_z, $\vec{v}_s$ and $\vec{v}_M$ we have to lowest order

$$ds = \frac{c_M}{T} dT - \frac{\beta}{\rho} d M_z$$

$$dh_z = \beta dT + \frac{1}{\chi} dM_z$$

(36)

where c_M is the specific heat at constant magnetization per unit mass, χ the isothermal longitudinal susceptibility and

$$\beta = (\frac{\partial h_z}{\partial T})_{M_z} = -\rho (\frac{\partial s}{\partial M_z})_T$$

(37)

Here the second equality follows from (21).

Going over to plane waves and introducing

$$\lambda_N^2(\vec{k}, \omega)_{ij} = \frac{\tau_J}{\rho_M} \sum_{mn} \Gamma_{im,nj}(\vec{k}, \omega) \hat{k}_m \hat{k}_n$$

(38)

where $\hat{k} = \vec{k}/|\vec{k}|$ eqs. (32-35) become, after insertion of (36)

$$\omega \delta \vec{v}_s + i\vec{k} \rho_s (\vec{k} \cdot \zeta \delta \vec{v}_s) - \vec{k} \frac{1}{m} \frac{1}{\chi} \delta M_z - \vec{k} \frac{\beta}{m} \delta T = 0 \quad (39)$$

$$(\omega + ik^2 \frac{\xi_\ell}{\chi}) \delta M_z - \frac{\rho_s}{m} (\vec{k} \cdot \delta \vec{v}_s) + ik^2 \beta \xi_\ell \delta T = 0 \quad (40)$$

$$(-i\omega + \frac{1}{\tau_J} + \frac{1}{\tau_J} \lambda_N^2 k^2) \vec{v}_M + i\vec{k} \frac{\rho}{\rho_M} s_M \delta T = 0 \quad (41)$$

and

$$\left\{ \omega +ik^2 \frac{1}{\rho c_M} \left(\kappa_\ell - h_z \beta \xi_\ell \right) \right\} \delta T - \frac{s_M T}{c_M} (\vec{k}\cdot\vec{v}_M) -$$

$$- \frac{1}{\rho c_M} \left(\omega T \beta + ik^2 \frac{h_z \xi_\ell}{\chi} \right) \delta M_z = 0 \qquad (42)$$

ξ_ℓ and κ_ℓ are the longitudinal projections $\xi_\ell \equiv (\hat{k}\cdot\vec{\xi}\hat{k})$. Eqs. (40, 41, 42) are the same as eqs. (4.14a, b, c) of ref. 4)if there the external temperature and field variations and the coupling coefficient of the magnon drift to the magnetization are put equal to zero (Θ=const, h=const and C_{ij}^1 = 0. The latter could have been included by adding a term - $\rho_M C^1 dM_z$ to eq. (18) which, however, would lead to an additional coupling between magnetic and thermal variables). To solve the system (39-42) we first express the velocities

$$\delta\vec{v}_s = \frac{\vec{k}}{\omega m} (\frac{1}{\chi} \delta M_z + \beta \delta T)(1+i\frac{k^2}{\omega} \rho_s \zeta_\ell)^{-1} \qquad (43)$$

and

$$\vec{v}_M = - \frac{\varphi}{T\rho s_M} i\vec{k} \delta T \qquad (44)$$

where

$$\varphi(\vec{k}, \omega)= \frac{T\rho^2 s_M^2}{\rho_M}(-i\omega +\frac{1}{\tau_J} + \frac{1}{\tau_J} \lambda_N^2 k^2)^{-1} \qquad (45)$$

is the magnon convectivity tensor and ζ_ℓ is the longitudinal projection of $\vec{\zeta}$.

We first examine the special case of complete decoupling of the magnetic variables $\delta\vec{v}_s, \delta M_z$ and the thermal variables $\vec{v}_M, \delta T$. This decoupling is obtained in the approximation

$$h_z \simeq 0, \quad \beta \simeq 0. \qquad (46)$$

There now exists a purely magnetic mode ($\vec{v}_M=0, \delta T=0$)

and a purely thermal mode ($\delta \vec{v}_s = 0$, $\delta M_z = 0$). Indeed, with (46) and $\vec{v}_M = 0$, $\delta T = 0$ eqs. (41,44) are identically satisfied and (40,43) yield damped isothermal spin-waves,

$$\frac{\omega^2}{k^2} = c_1^2 (1 + i \frac{k^2}{\omega} \varrho_s \zeta_\ell)^{-1} - i\omega \frac{\xi_\ell}{\chi} \qquad (47)$$

with velocity

$$c_1 = \sqrt{\frac{\varrho_s}{m^2 \chi}} . \qquad (48)$$

Assuming all the imaginary parts to be small so that they need be retained only linearly, eq. (47) simplifies to

$$\omega = c_1 k - i k^2 \frac{1}{2} (\varrho_s \zeta_\ell + \frac{\xi_\ell}{\chi}) \qquad (49)$$

which is the same as eqs. (2.48,50) of ref. 3).

The thermal mode is obtained with (46) and $\delta \vec{v}_s = 0, \delta M_z = 0$, in which case eqs. (40, 43) are identically satisfied. (42,44) yield in analogy to the case of phonons in dielectric crystals [8,9)]

$$-i\omega + \frac{1}{\varrho c_M} (\varphi_\ell(\vec{k}, \omega) + \kappa_\ell(\vec{k}, \omega)) k^2 = 0. \qquad (50)$$

There exists a domain of second spinwaves (the second magnon of refs. 4)to 6) given by the conditions [8)]

$$1/\tau_J \ll \omega \ll 1/\tau_{eq}, \quad \omega \tau_J \gg k^2 \lambda_{Ni}^2. \qquad (51)$$

The first is the "window condition" familiar from the case of phonons in dielectric crystals (see ref. 9) for a historical survey of this case), τ_{eq} being the relaxation time responsible for thermal equilibrium of the magnon fluid. In the domain (51) we have from (45)

$$\varphi_\ell(\vec{k}, \omega) = \frac{i}{\omega} \frac{T\varsigma^2 s_M^2}{\varsigma_M} \left(1 - i \frac{1+(\lambda_N^2)_\ell\, k^2}{\omega\, \tau_J} \right) \qquad (52)$$

and inserting into (50)

$$\frac{\omega^2}{k^2} = c_2^2 \left[1 - i \frac{1+(\lambda_N^2)_\ell\, k^2}{\omega\, \tau_J} - i\,\omega\,\tau_2 \right]. \qquad (53)$$

Here

$$c_2 = \sqrt{\frac{\varsigma\, T s_M^2}{\varsigma_M c_M}} \qquad (54)$$

is the second magnon velocity which has exactly
the same form as the second sound velocity in the
phonon case [8] and

$$\tau_2(\vec{k}, \omega) = \frac{1}{\varsigma c_M c_2^2} \; K_\ell(\vec{k}, \omega) \qquad (55)$$

is the thermal conduction relaxation time which,
for a propagating second magnon, must satisfy the
additional condition

$$\omega\,\tau_2 \ll 1. \qquad (56)$$

The window condition of eq.(51)has been analysed by Michel and Schwabl[6] for the cases of the
antiferromagnets MnF_2 and K_2NiF_4(which is a planar
case). Their conclusions are favorable to the
existence of a second magnon at temperatures of
the order of 30° K for Mn F_2 and of 90° K for
K_2NiF_4. Detailed calculations of the window condition for the insulating ferromagnet Eu S have
been carried out by Forney[7] who finds an open
window below approximately 2°K (Fig. 1) The curves
of Fig. 1 represent the reciprocal relaxation
times due to normal (N), Umklapp (U) and impurity
(I) scattering. The calculation is based on a Heisenberg plus a dipole-dipole coupling, the first
giving rise to 4-magnon processes, the second to

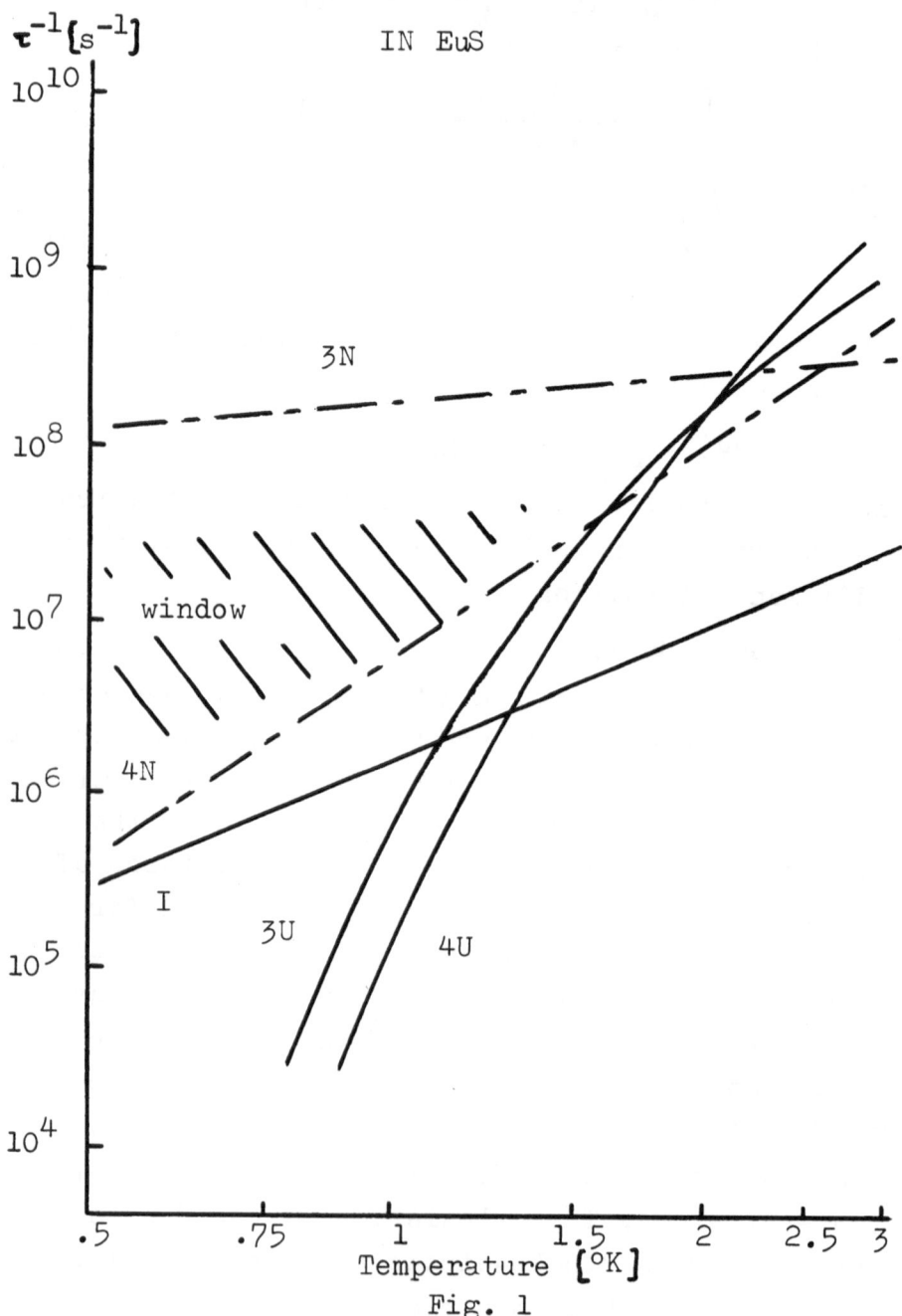

RATE OF RELAXATION OF MAGNONS
IN EuS

Fig. 1

3-magnon processes.

It is interesting to note that the conditions $\omega \tau_J \ll k^2 \lambda_{Ni}^2$, $k^2 \lambda_{Ni}^2 \gg 1$ which define the domain of Poiseuille flow in the phonon case[8] are also realizable for magnetic crystals of high magnon viscosity. To our knowledge this effect has never been looked for yet.

Heat diffusion by magnons is obtained in the domain

$$\omega \tau_J \ll 1 \ , \ k^2 \lambda_{Ni}^2 \ll 1 \tag{57}$$

where (50) reduces to

$$-i\omega + D_2 k^2 = 0 \tag{58}$$

with

$$D_2(\hat{k}) = \lim_{\omega \to o} \ \lim_{k \to o} \ \frac{1}{\varsigma c_v} (\varphi_\ell(\vec{k},\omega) + K_\ell(\vec{k},\omega)) \tag{59}$$

Neglecting here the convective contribution φ_ℓ we recover eq. (2.49) of ref. 3).

Taking now into account the coupling between the magnetic and the thermal fluids by relaxing the conditions (46) but making the simplifying a assumption that all dissipative terms be negligible insertion of (43) into (40) and of (44) into (42) yields with (48,54)

$$(\frac{\omega^2}{k^2} - c_1^2) \ \delta M_z - \frac{\varsigma_s \beta}{m^2} \ \delta T = 0 \tag{60}$$

and

$$(\frac{\omega^2}{k^2} - c_2^2) \ \delta T - \frac{\omega^2}{k^2} \ \frac{T\beta}{\varsigma c_M} \delta M_z = 0 \tag{61}$$

respectively. The solubility condition for this system of equations is completely analogous to that for superfluid Helium and for phonons in dielectric crystals [8)]

$$\left(\frac{\omega^2}{k^2} - (c_1^{ad})^2\right)\left(\frac{\omega^2}{k^2} - c_2^2\right) - \lambda = 0 \tag{62}$$

where c_1^{ad} is the adiabatic magnon velocity given by

$$(c_1^{ad})^2 = c_1^2 + \frac{\rho_s^T \beta^2}{m^2 \rho c_M} \tag{63}$$

and

$$\lambda = ((c_1^{ad})^2 - c_1^2) c_2^2 . \tag{64}$$

The paramagnetic state at $T > T_c$ is characterized by the fact that $\vec{v}_s$ is not a thermodynamic variable. Hence we see from (21,48,63) that in this case

$$\rho_s = 0 , \quad c_1 = c_1^{ad} = 0 \tag{65}$$

with c_1^{ad}/c_1 finite. Here the isothermal spin-wave mode (47) degenerates into the spin-diffusion mode

$$\omega = - ik^2 \frac{\xi_\ell}{\chi} \tag{66}$$

which is the same as eq. (2.51) of ref. 3). This implies of course that there is no thermal magnon fluid, $s_M = 0, \vec{v}_M = 0$.

In order to see more clearly the physical meaning of ρ_s and obtain an expression for $\vec{v}_{eq}$ we generalize (21) by allowing for a transverse component $H_\perp$ of the external magnetic field,

$$d(\rho f) = \vec{H} \cdot d\vec{M} - \rho s dT + \rho_s \vec{v}_s \cdot d\vec{v}_s + \vec{v}_M \cdot d\vec{j}_M \tag{67}$$

where $f = e - Ts$ is the free energy per unit mass and $\vec{H} = (0, H_\perp, H_z)$. From (2) we see that $\vec{M} = (M_\perp \cos\varphi, M_\perp \sin\varphi, M_z)$. Applying the equilibrium condition $df = 0, dT = 0, d\vec{j}_M = 0$ to (67) we obtain for the Fourier components, in view of (6,11),

$$(H_z + H_\perp \frac{\partial M_\perp}{\partial M_z} \sin \varphi) dM_z + (H_\perp M_\perp \cos \varphi - k^2 \frac{\varsigma_s}{m^2} \varphi) d\varphi = 0.$$

Since M_z and φ are independent variables this equation implies for the local equilibrium values $\varphi = \varphi_{eq}$, $\vec{H} = \vec{h}$,

$$\varphi_{eq} = \frac{m^2}{k^2 \varsigma_s} M_\perp h_\perp + O(h_\perp^3) \tag{68}$$

$$h_z = - \frac{m^2}{k^2 \varsigma_s} M_\perp \frac{\partial M_\perp}{\partial M_z} h_\perp^2 + O(h_\perp^4).$$

Inserting the expression for φ_{eq} we also find

$$M_x = M_\perp + O(h_\perp^2) \tag{69}$$

where 3) $\quad M_y = \chi_\perp h_\perp$

$$\chi_\perp = \frac{m^2}{k^2 \varsigma_s} M_\perp^2 + O(h_\perp^2) \tag{70}$$

is the isothermal transverse susceptibility. From (68,11) we further conclude that

$$\vec{v}_{eq} = \pm \hat{k} (h_z M_\perp / \varsigma_s \frac{\partial M_\perp}{\partial M_z})^{\frac{1}{2}}. \tag{71}$$

This shows that in equilibrium, magnetization and magnetic field in the (x, y)-plane tend to be orthogonal; this tendency increases with increasing stiffness constant ς_s. However, as shown explicitly in ref. 2) M_x and M_y are not conserved. For $H_z \neq h_z$ M_x and M_y essentially precess with the angle $\varphi(t)$ while for $H_z = h_z$ they essentially follow the motion of M_z.

It should be remarked that rigorously (2) holds only in a local ferromagnetic equilibrium state. When dissipation is taken into account the magnetization could have the more general form

$$\vec{M} = (M_\perp \cos \varphi + M_x', \; M_\perp \sin \varphi + M_y', \; M_z)$$

where M_x' and M_y' are small dissipative corrections. Since there is no equation of motion for M_x and M_y analogous to (14) M_x' and M_y' cannot be determined from hydrodynamics. Presumably therefore these dissipative corrections relax to zero in a microscopic time.

In the case of the isotropic ferromagnet without external magnetic field, rotational symmetry implies that

$$\chi = \chi_\perp = \frac{m^2 M_\perp^2}{k^2} \varrho_s$$

so that (48,72) give

$$c_1 = \frac{\varrho_s}{m^2 M_\perp} k \qquad\qquad (73)$$

This combined with (49) leads to the well-known quadratic dispersion $\omega_q = Dk^2$ of the Heisenberg ferromagnet [10]. Such a dispersion law does not allow for a magnon drift since the magnon occupation number

$$n_{\vec{q}} = (\exp[\beta(\omega_{\vec{q}} - \vec{v}_M \cdot \vec{q})] - 1)^{-1}$$

is negative for sufficiently small $\vec{q}$ with $\vec{v}_M \cdot \vec{q} > 0$. It can be shown, however (see, e.g., ref. 7)) that inclusion of a dipole-dipole interaction in the Hamiltonian of eq. (1) gives rise to an energy gap in the magnon spectrum, $\omega_q = \omega_0 + Dq^2$, and that in this case the magnon drift exists for $|\vec{v}_M| < 2\sqrt{\omega_0 D}$. Similar considerations hold in the case of an axial ferromagnet for which $J_{ij}^z > J_{ij}^\perp > 0$, where the anisotropy energy also gives rise to a gap [10].

The effect of isotropy on the linear relations (29,31) is more delicate. Omitting again a cross term $\sim \nabla \cdot \vec{v}_M$ μ' can have terms proportional to ∇h_z and $\nabla \cdot \vec{v}_s$ with coefficients transforming like $h_z \vec{j}_s$ and h_z^2, respectively. Since these relations must be linear in h_z and $\vec{v}_s$ the only vector which in the coefficients can be associated to $\hat{h}$ and $\vec{j}_s$ in order to have the correct transformation properties is ∇. Hence

$$\mu' = - \frac{\eta}{m} \nabla_z \nabla^2 h_z - \varsigma_s \varsigma' \nabla_z^2 (\nabla \cdot \vec{v}_s) \qquad (74)$$

Similarly $\vec{j}_s'$ can have such terms with coefficients transforming like $\vec{j}_s^2$ and $\vec{j}_s h_z$. Since $\vec{j}_s^2$ is a scalar, this implies the form

$$\vec{j}_s' = - m \, \xi \, \nabla h_z - \varsigma_s \, \eta \, \nabla_z \nabla (\nabla \cdot \vec{v}_s) \qquad (75)$$

where the appearance of the same coefficient η in (74) and (75) expresses Onsager's reciprocity principle.

If we put $\eta = 0$ we recover all the former equations with the only modification in (29) that according to (74) now

$$\xi = \xi' \nabla_z^2 \qquad (76)$$

With (72,73,76) the decoupled isothermal spin-wave mode (49) becomes

$$\omega = \frac{\varsigma_s}{m^2 M_\perp} k^2 - ik^4 \frac{\varsigma_s}{2} \left(\frac{\xi}{m^2 M_\perp^2} - \hat{k}_z^2 \xi' \right) \qquad (77)$$

Here the real part is in agreement with eq. (7.9) of ref. 3) while the imaginary part proportional to to k^4 corresponds to the expectation expressed in this reference.

The plane wave representation (40-43) for $\eta \neq 0$ is

modified as follows:

$$\delta\vec{v}_s = \frac{\vec{k}}{\omega m}\left(\frac{1}{\chi}\,\delta M_z + \beta\;\delta T\right)(1+ik_z k^2 \eta)\left(1-i\frac{k_z^2 k^2}{\omega}\,\varsigma_s \zeta'\right)^{-1} \tag{78}$$

$$(\omega +ik^2\frac{\xi}{\chi})\delta M_z - \frac{\varsigma_s}{m}(1+ik_z k^2 \eta)(\vec{k}\cdot\delta\vec{v}_s)+ik^2\beta\xi\,\delta T = 0 \tag{79}$$

$$(-\,i\omega +\frac{1}{\tau_J} + \frac{1}{\tau_J}\lambda_N^2 k^2)\vec{v}_M + i\vec{k}\,\frac{\varsigma}{\varsigma_M}\,s_M\,\delta T = 0 \tag{80}$$

$$\left[\omega + ik^2\,\frac{1}{\varsigma\,c_M}\,(\kappa - h_z\beta\,\xi)\right]\delta T - \frac{s_M T}{c_M}(\vec{k}\cdot\vec{v}_M) -$$

$$-\frac{1}{\varsigma\,c_M}(\omega\,T\,\beta + ik^2\,\frac{h_z\,\xi}{\chi})\,\delta M_z + ik_z k^2\,\frac{h_z\,\varsigma_s\,\eta}{m\varsigma\,c_M}(\vec{k}\cdot\delta\vec{v}_s)=0 \tag{81}$$

In the decoupling approximation (46) we recover the two modes (47) for δM_z and (50) for δT, the former being modified as follows:

$$\frac{\omega^2}{k^2}=c_1^2(1+ik_z k^2 \eta)^2(1-i\frac{k_z^2 k^2}{\omega}\,\varsigma_s \zeta')^{-1}-i\omega\frac{\xi}{\chi} \tag{82}$$

Linearizing the imaginary parts and taking into account (72,73) leads again to eq. (77), the contribution of the term proportional to η being of fifth order in k.

4. Hydrodynamics of the Antiferromagnet

The antiferromagnet is characterized by the existence of two sublattices A and B such that the coupling constants J_{ij}^z and $J_{ij}^{\perp}$ are positive or negative if the positions i and j are on the same or on different sublattices, respectively. Defining numbers [3)]

$$\eta_i = \begin{cases} +\,1;\ \text{if i on A} \\ -1\ ;\ \text{if i on B} \end{cases} \tag{83}$$

the operators $\eta_i S_i^x$, $\eta_i S_i^y$, S_i^z satisfy the same
commutation relations as S_i^x, S_i^y, S_i^z and $\eta_i \eta_j J_{ij}^{\perp}$
is ferromagnetic. Hence the Hamiltonian (1)
written in these variables describes ferromagnetism
in the (x,y)-plane. Defining the sublattice magne-
tizations by

$$\vec{M}_A = \frac{1}{v} \langle \vec{S}_i \rangle \quad ; \text{ i on A}$$
$$\vec{M}_B = \frac{1}{v} \langle \vec{S}_i \rangle \quad ; \text{ i on B} \tag{84}$$

the total magnetization is

$$\vec{M} = \frac{1}{v} \langle \sum_i \vec{S}_i \rangle = \vec{M}_A + \vec{M}_B \tag{85}$$

and

$$\vec{N} = \frac{1}{v} \langle \sum_i \eta_i \vec{S}_i \rangle = \vec{M}_A - \vec{M}_B \tag{86}$$

is the staggered magnetization. Hence for the planar antiferromagnet

$$N_x + iN_y = N_o e^{i\varphi} \tag{87}$$

is equivalent to $M_x + iM_y$, eq. (2), and in an external magnetic field $\vec{H} = (0, 0, H_z)$ M_z is conserved
together with the energy e, eq. (4).

On the other hand, N_z is not conserved but is
supposed to relax to zero in a microscopic time.
Since (85, 86) imply

$$\vec{M} \cdot \vec{N} = 0 \tag{88}$$

we then can combine (87, 88) in writing

$$\vec{M} = (- M_o \sin \varphi, \; M_o \cos \varphi, \; M_z)$$
$$\vec{N} = (N_o \cos \varphi, \; N_o \sin \varphi, \; 0) \tag{89}$$

where M_o and N_o are supposed to relax to values
M_o (e, M_z) and N_o (e, M_z), respectively, in
microscopic times.

In thermal equilibrium at temperature T below
the Néel temperature T_N, $e = e(M_z, T)$ so that

$$M_o = M_o (M_z, T)$$
$$N_o = N_o (M_z, T)$$

(90)

As can be seen from the generalization of
(84-86) space-time dependent quantities, (88) and
hence (89) are still valid in local antiferromagne-
tic equilibrium.

As is shown in ref. 2), in the case of the
antiferromagnet the condition $h_z(k=0) = 0$ is necessary for M_z to be conserved in thermal equilibrium.
For $h_z \stackrel{\sim}{=} 0$ the dynamics of the planar antiferromagnet is exactly the same as that of the planar
ferromagnet described in Section 3, if there $M_\perp$
is replaced by N_o. In particular, in the decoupling approximation (46) the spin-wave mode (49)
with velocity (48) and the thermal mode (50) are
recovered.

It should be remarked that rigorously the
orthogonality (88) only holds in a local antiferro-
magnetic equilibrium state while with dissipation,
deviations from (88) are possible. Hence (89)
should be generalized to the form

$$\vec{M} = (-M_o \sin \varphi + M_x', \; M_o \cos \varphi + M_y', \; M_z)$$
$$\vec{N} = (N_o \cos \varphi + N_x', \; N_o \sin \varphi + N_y', \; N_z')$$

where the primed quantities are small dissipative
corrections. This does not, however, affect the
above conslusions since in a planar antiferromagnet, even without external magnetic field, M_x and
M_y are not conserved. Hence there is no equation

of motion for these quantities analogous to (14)
which would allow to determine hydrodynamically
the motion of M_x' and M_y'. Presumably therefore the
above primed quantities relax to zero in microscopic times.

In the isotropic antiferromagnet without external magnetic field all three components of $\vec{M}$
are conserved. However, the main direction of the
staggered magnetization introduces an asymmetry
into the problem. Assuming this direction to be
again the x- direction there is complete equivalence between the y- and z- directions. Hence the
dynamics requires two precession angles φ and ψ describing the motion of N_y and N_z relative to N_x,
respectively. The orthogonality (88) then implies
the form, for small φ and ψ ,

$$\vec{M} = (- M_y \varphi - M_z \psi, \; M_y, \; M_z)$$
$$\vec{N} = (N_x, \; N_x \varphi, \; N_x \psi) \tag{91}$$

where we have assumed that dissipative corrections
analogous to those above are again negligible.

The equations of motion for φ and ψ are,
for $\vec{H} = 0$,

$$\dot{\varphi} = - m \mu = - h_z - m \mu'$$
$$\dot{\psi} = - m \nu = - h_y - m \nu' \tag{92}$$

Defining two velocity fields by

$$\vec{v}_s = \frac{1}{m} \nabla \varphi \tag{93}$$

$$\vec{w}_s = \frac{1}{m} \nabla \psi$$

we have local conservation of circulation

$$\dot{\vec{v}}_s = - \nabla \mu \tag{94}$$
$$\dot{\vec{w}}_s = - \nabla \nu$$

Local conservation of M_y and M_z is expressed by

$$m \dot{M}_z + \vec{\nabla} \cdot \vec{j}_s = 0$$
$$m \dot{M}_y + \vec{\nabla} \cdot \vec{\ell}_s = 0 \tag{95}$$

In ref. 3) an analogous equation for M_x is added to eqs. (95) while the orthogonality (88) is not used. Here the latter implies the form (91) so that M_x is not an independent variable, and such an equation is redundant.

The thermodynamic relation (21) is now generalized to

$$d(\varrho\, e) = h_z dM_z + h_y dM_y + Td(\varrho\, s) +$$

$$+ \varrho_s(\vec{v}_s \cdot d\vec{v}_s + \vec{w}_s \cdot d\vec{w}_s) + \vec{v}_M \cdot d\vec{j}_M \tag{96}$$

where $\vec{j}_M$ is determined by eqs. (15-17, 19) and we neglect the fact that there is actually a distinct magnon fluid for each polarization. Inserting (16-18, 92-95) and the local energy conservation (13) into

$$T\varrho\, \dot{s} = \varrho\dot{e} - h_z \dot{M}_z - h_y \dot{M}_y - \varrho_s(\vec{v}_s \cdot \dot{\vec{v}}_s + \vec{w}_s \cdot \dot{\vec{w}}_s) - \vec{v}_M \cdot \dot{\vec{j}}_M \tag{97}$$

we obtain

$$\varrho\dot{s} = - \vec{\nabla} \cdot \left\{ \frac{1}{T}\left(\vec{J}_e - \frac{h_z}{m}\vec{j}_s - \frac{h_y}{m}\vec{\ell}_s\right)\right\} +$$

$$+ \left(\vec{J}_e - \frac{h_z}{m}\vec{j}_s - \frac{h_y}{m}\vec{\ell}_s - \varrho\, s_M T\, \vec{v}_M\right) \cdot \vec{\nabla}\frac{1}{T} -$$

$$- \frac{1}{Tm}\left\{(\vec{j}_s - \varrho_s\vec{v}_s)\cdot\vec{\nabla} h_z + (\vec{\ell}_s - \varrho_s\vec{w}_s)\cdot\vec{\nabla} h_y\right\} +$$

$$+ \frac{\varrho_s}{T}(\vec{v}_s\cdot\vec{\nabla}\mu' + \vec{w}_s\cdot\vec{\nabla}\nu') + \frac{1}{T}\vec{v}_M\cdot(\vec{\nabla}\pi_M') + \frac{\varrho_M}{T\tau_J}\vec{v}_M^{\,2}.$$

$$\tag{98}$$

As before, absence of dissipation implies local
entropy conservation so that

$$\vec{j}_s = \rho_s \vec{v}_s + \vec{j}_s{}'$$
$$\vec{l}_s = \rho_s \vec{w}_s + \vec{l}_s{}' \tag{99}$$

and

$$\vec{J}_e = \frac{\rho_s}{m}(h_z \vec{v}_s + h_y \vec{w}_s) + \rho s_M T \vec{v}_M + \vec{J}_e{}' \tag{100}$$

The entropy balance (25) then holds with

$$\vec{J}_s - \rho s_M \vec{v}_M = \vec{J}_s{}' = \frac{1}{T}\left(\vec{J}_e{}' - \frac{h_z}{m}\vec{j}_s{}' - \frac{h_y}{m}\vec{l}_s{}'\right) \tag{101}$$

and

$$\sigma = T\vec{J}_s{}' \cdot \nabla \frac{1}{T} - \frac{1}{Tm}(\vec{j}_s{}' \cdot \nabla h_z + \vec{l}_s{}' \cdot \nabla h_y) +$$

$$+ \frac{\rho_s}{T}\vec{v}_s \cdot \nabla \mu' + \frac{\rho_s}{T}\vec{w}_s \cdot \nabla \nu' + \frac{1}{T}\vec{v}_M \cdot (\nabla \pi_M{}') + \frac{\rho_M}{T \tau_J}\vec{v}_M{}^2 \tag{102}$$

Rotational invariance in the (y, z)- plane implies
for the dissipative parts, to lowest order in the
derivatives,

$$\mu' = - \rho_s(\nabla \cdot \zeta \vec{v}_s + \nabla \cdot \eta \vec{w}_s)$$
$$\nu' = - \rho_s(\nabla \cdot \eta \vec{v}_s + \nabla \cdot \zeta \vec{w}_s) \tag{103}$$

and

$$\vec{J}_e{}' = - \kappa \nabla T$$
$$\vec{j}_s{}' = - m \xi \nabla h_z \tag{104}$$
$$\vec{l}_s{}' = - m \xi \nabla h_y$$

The hydrodynamic equations (94,95,25) now are, in
linear approximation, inserting from (92,99,101-104),

$$\dot{\vec{v}}_s + \frac{1}{m}\nabla h_z - \rho_s \nabla \left[\nabla \cdot (\zeta \vec{v}_s) + \nabla \cdot (\eta \vec{w}_s)\right] = 0 \tag{105}$$
$$\dot{\vec{w}}_s + \frac{1}{m}\nabla h_y - \rho_s \nabla \left[\nabla \cdot (\eta \vec{v}_s) + \nabla \cdot (\zeta \vec{w}_s)\right] = 0$$

$$\dot{M}_z + \frac{\varrho_s}{m} \vec{\nabla} \cdot \vec{v}_s - \vec{\nabla} \cdot (\xi \vec{\nabla} h_z) = 0$$

$$\dot{M}_y + \frac{\varrho_s}{m} \vec{\nabla} \cdot \vec{w}_s - \vec{\nabla} \cdot (\xi \vec{\nabla} h_y) = 0 \tag{106}$$

$$\dot{s} + s_M \vec{\nabla} \cdot \vec{v}_M - \frac{1}{\varrho T} \vec{\nabla} \cdot (\kappa \vec{\nabla} T) + \frac{1}{\varrho T} \Big[h_z \vec{\nabla} \cdot (\xi \vec{\nabla} h_z) +$$

$$+ h_y \vec{\nabla} \cdot (\xi \vec{\nabla} h_y) \Big] = 0 \tag{107}$$

On generalizing eqs. (36) to

$$ds = \frac{c_M}{T} dT - \frac{\beta}{\varrho} (dM_z + dM_y)$$

$$dh_z = \beta \, dT + \frac{1}{\chi} dM_z \tag{108}$$

$$dh_y = \beta \, dT + \frac{1}{\chi} dM_y$$

and inserting plane waves, eqs. (105-107) become

$$\omega \delta \vec{v}_s + i\vec{k} \, \varrho_s \Big[\vec{k} \cdot (\xi \, \delta \vec{v}_s) + \vec{k} \cdot (\eta \, \delta \vec{w}_s) \Big] -$$

$$- \vec{k} \frac{1}{m} \frac{1}{\chi} \delta M_z - \vec{k} \frac{\beta}{m} \delta T = 0$$

$$\omega \delta \vec{w}_s + i\vec{k} \, \varrho_s \Big[\vec{k} \cdot (\eta \, \delta \vec{v}_s) + \vec{k} \cdot (\xi \, \delta \vec{w}_s) \Big] -$$

$$- \vec{k} \frac{1}{m} \frac{1}{\chi} \delta M_y - \vec{k} \frac{\beta}{m} \delta T = 0 \tag{109}$$

$$(\omega + ik^2 \frac{\xi_\ell}{\chi}) \, \delta M_z - \frac{\varrho_s}{m} (\vec{k} \cdot \delta \vec{v}_s) + ik^2 \beta \xi_\ell \, \delta T = 0$$

$$(\omega + ik^2 \frac{\xi_\ell}{\chi}) \, \delta M_y - \frac{\varrho_s}{m} (\vec{k} \cdot \delta \vec{w}_s) + ik^2 \beta \xi_\ell \, \delta T = 0 \tag{110}$$

$$\Big[\omega + \frac{ik^2}{\varrho c_M} (\kappa_\ell - (h_z + h_y) \beta \xi_\ell) \Big] \delta T - \frac{s_M T}{c_M} (\vec{k} \cdot \vec{v}_M) -$$

$$- \frac{1}{\varrho c_M} (\omega T \beta + ik^2 \frac{h_z \xi_\ell}{\chi}) \, \delta M_z - \frac{1}{\varrho c_M} (\omega T \beta + ik^2 \frac{h_y \xi_\ell}{\chi}) \, \delta M_y = 0$$

$$\tag{111}$$

These equations together with (41) are the

generalization of eqs. (39-42) to the case of the isotropic antiferromagnet.

In eqs. (109) $\delta\vec{v}_s$ and $\delta\vec{w}_s$ are coupled through the term proportional to η. Although these equations are readily solved exactly we wish to simplify the problem by making the decoupling assumption

$$\eta \cong 0 \qquad (112)$$

Then (43) holds together with

$$\delta\vec{w}_s = \frac{\vec{k}}{\omega m} \left(\frac{1}{\chi} \delta M_y + \beta\, \delta T\right)\left(1 + i\, \frac{k^2}{\omega}\, \varsigma_s \zeta_\ell\right)^{-1} \qquad (113)$$

Furthermore we consider only the approximation analogous to (46)

$$h_z \cong 0,\ h_y \cong 0,\ \beta \cong 0 \qquad (114)$$

which decouples the motion of δM_z, δM_y and δT. In this case we recover exactly the spin-wave mode (47) for δM_z and for δM_y and the thermal mode (50) for δT. Apart from the convection term φ_ℓ in (50) this result is in agreement with that of ref. 3).

The motion of $M_x = - M_y \varphi - M_z \psi$ may be obtained with the help of (92, 106). In plane wave representation with the decoupling approximation (112, 114) which also implies $M_z \cong 0$, $M_y \cong 0$, and making use of (43,108,48), we find it to be

$$\frac{\omega^2}{k^2}\, \delta M_x = -\left\{c_1^2\left(1 + i\, \frac{k^2}{\omega}\, \varsigma_s \zeta_\ell\right)^{-1} - i\omega \frac{\xi_\ell}{\chi}\right\} \cdot$$

$$\cdot\left(\varphi_{eq}\, \delta M_y + \psi_{eq}\, \delta M_z\right) \qquad (115)$$

The equilibrium values of φ and ψ are found in the same way as in Section 3 from the condition df=0, dT=0, d$\vec{j}_M$=0 . With an external magnetic field $\vec{H}$ = (H$_x$, H$_y$, H$_z$) and with (91,93) we find from (96) for the Fourier components

$$(-H_x \varphi + H_y) \, dM_y + (- H_x \psi + H_z) dM_z -$$

$$- (h_x M_y + \frac{k^2 \rho_s}{m^2} \varphi) d\varphi - (h_x M_z + \frac{k^2 \rho_s}{m^2} \psi) d\psi = 0$$

This determines the equilibrium values $\vec{H}=\vec{h}$, $\varphi = \varphi_{eq}$, $\psi = \psi_{eq}$ as follows:

$$\varphi_{eq} = \frac{h_y}{h_x} \; ; \; \psi_{eq} = \frac{h_z}{h_x} \tag{116}$$

and

$$\pm i \frac{kc}{h_x} = \sqrt{\frac{M_y}{\chi^{h_y}}} = \sqrt{\frac{M_z}{\chi^{h_z}}} \tag{117}$$

With (116,108) we obtain

$$\varphi_{eq} \, \delta M_y + \psi_{eq} \, \delta M_z = \frac{\chi}{2h_x} \, \delta(h_y^2 + h_z^2)$$

and assuming that $\vec{h}^2$ relaxes to a value $\vec{h}^2(e, \vec{M}^2)$ in a microscopic time so that it is a constant of the motion, we find with $\delta M_x = \chi_{\|} \delta h_x$

$$\varphi_{eq} \, \delta M_y + \psi_{eq} \, \delta M_z = -\frac{\chi}{\chi_{\|}} \, \delta M_x \tag{118}$$

Defining the isothermal spin-wave velocity parellel to the main direction of the staggered magnetization by

$$c_{\|} = \sqrt{\frac{\rho_s}{m^2 \chi_{\|}}} \tag{119}$$

eq. (115) becomes

$$\frac{\omega^2}{k^2} \, \delta M_x = \left\{ c_{\|}^2 \, (1 + i \frac{k^2}{\omega} \rho_s \zeta_\ell)^{-1} - i\omega \frac{\xi_\ell}{\chi_{\|}} \right\} \delta M_x \, . \tag{120}$$

Thus the motion of M_x is that of a damped spin-wave of the form (47 but with c_1 replaced by $c_\parallel$ and χ by $\chi_\parallel$. This result is in disagreement with eq. (6.9b) of ref. 3) which leads to a diffusive mode.

REFERENCES

1) C.P. Enz, "Two-Fluid Aspects of Condensed Matter", International Seminar on Statistical Mechanics and Field Theory, Haifa, 1971, to be published.

2) C.P. Enz, Phys. Kondens. Materie 12, 262(1971).

3) B.I. Halperin and P.C. Hohenberg, Phys. Rev. 188, 898 (1969). P.C. Hohenberg, "Hydrodynamics of Systems with Broken Symmetry". Lectures for the Tokyo Summer Institute on Quantum Fluids, 1970.

4) K.H. Michel and F. Schwabl, Solid State Comm. 7, 1781 (1969), Phys. Kondens. Materie 11, 144 (1970).

5) F. Schwabl and K.H. Michel, Phys. Rev. B2, 189 (1970); K.H. Michel and F. Schwabl, Z. Phys. 238, 264 (1970); 240, 354 (1970) and Proc. of the 1970 International Conference on Magnetism, Grenoble (to be published), MeG 11, 12.

6) K.H. Michel and F. Schwabl, Phys. Rev. Letters 26, 1568 (1971).

7) J.J. Forney, Doctoral Thesis (Geneva University, 1972) and to be published.

8)   C.P. Enz, "Unified hydrodynamic description of ordered systems", to be published.

9)   C.P. Enz, Ann. Phys.(N.Y.) <u>46</u>, 114 (1968)

10)   See e.g., C. Kittel, <u>Quantum Theory of Solids</u> (Wiley, New York, 1963).

SYMMETRY BREAKING INSTABILITIES

M.G. Velarde

Departamento De Fisica
Universidad Autonoma
Canto Blanco (Madrid-34) Spain

Introduction

I.   The Bénard-Rayleigh Problem.

II. Soret-driven Instability.

III. Final Comments.

INTRODUCTION

We shall discuss in these lectures a few
macroscopic aspects of irreversibility and non-
linear eigenvalue problems. We shall remain on
a rather phenomenological level but hopefully
these problems will be taken up in the near
future from the more sophisticated many body or
kinetic theory point of view.

The general features of these problems are
as follows. Suppose that we have, with suitable
boundary conditions (b.c.) and initial conditions
(i.c.), a non-linear equation of the form

$$\mathcal{N}(\vec{v}, R) \quad = \quad 0 \qquad\qquad (1)$$

where $\mathcal{N}$ is some non-linear operator, depending
on a parameter R, which operates on the unknown
function $\vec{v}$. The parameter could be a Rayleigh
number (to be defined below), a Reynolds number,
etc. in thermodynamic problems. The first question is whether or not equation (1) has any
solution $\vec{v}$ for a given value of R, say R_o.
If a solution exists, the next question is how
many solutions it has branched at R_o. Among all
the solutions, one next asks, which one is the
one to be realizable in a particular experiment.
From the mathematical point of view the realizable one should be the stable one. Another
relevant question is how the solutions depend
on $[R - R_o]$ in a right- or left-hand neighbourhood of R_o. One might consider analytic
solutions or perhaps non-analytic solutions,

as R_0 could be called a "critical" point. We shall denote the first branching or bifurcation point, or critical point, of our problem by R_c.

Let us illustrate a little bit more the qualitative features of the non-linear eigenvalue problem. Let $\mathcal{L}$ be the linearized operator associated with $\mathcal{N}$ and such that for the chosen b.c. and i.c. the corresponding linearized eigenvalue problem is

$$\mathcal{L}\vec{v} = R\vec{v} \tag{2}$$

In equation (2) $\mathcal{L}$ is a linear operator acting on functions $\vec{v}$ belonging to some normed vector space. R is always in our cases a real number. For every value of R the trivial solution is always a solution to problem (2)

$$\vec{v} = \vec{0} \tag{3}$$

Usually one has a sequence of eigenvalues $R_1 < R_2 < \ldots$ with corresponding eigenvectors $\vec{v}_1 < \vec{v}_2 < \ldots$ such that

$$\mathcal{L}\vec{v}_i = R_i\vec{v}_i \quad \text{and} \quad \|\vec{v}_i\| = 1, \ i = 1,2,\ldots \tag{4}$$

If λ is a real number, other solutions of (2) are indeed

$$\vec{v} = \lambda\vec{v}_i \quad i = 1,2,\ldots \tag{5}$$

One notices that the solution (3) is normalized to zero whereas the norm of solutions (5) is λ. At the points R_i the solution $\vec{v} = \vec{0}$

splits into two branches.

Now from the linearized problem one goes
over to the non-linear one. A few of the relevant
features of the non-linear branching are the
following:
(i)   the non-linear branch emanating or bifurcating from an eigenvalue of the linearized problem may be quite complicated, indeed much more
so than the simple branching [5].
(ii) there may be no non-linear bifurcation from
an eigenvalue of the linearized problem.
(iii) there may be several solutions bifurcating
from an eigenvalue of the linearized operator.

Other possibilities do exist but we shall
not enter into any more details here as we shall
concentrate on type (i) problems.

Often the solution of a non-linear eigenvalue problem is intractable in practice and in
general is only an intermediate step in obtaining
physically relevant information. Because of the
intractability of the problem one then turns to
some perturbative analysis of the solution, thus
providing some approximate but relevant physical
predictions whose validity is to be tested in an
experiment. For instance, let $\vec{v}_j$ be a convective solution in a thermohydrodynamical problem. In general, convective problems are intractable in practice. However, a perturbative
analysis might lead in the first few orders to
some relevant conclusion regarding, for instance,
convective heat flux. The heat flux is something

that can be observed and one hopes that the
mathematical error is indeed covered by the experimental one. One may thus get some good
agreement between theory and experiment, but one
should be careful, as perturbation theory only
provides suggestions but not proofs of anything.

Our purpose is to illustrate all these
questions by looking at two different physical
problems. Because of the surprisingly good
agreement between theory and experiment I hope
someone of you will become interested enough to
go into a deeper analysis.

In Section I we shall discuss the so-called
Bénard-Rayleigh problem, a rather old problem.
Section II is devoted to a Soret-driven convective instability quite recently discovered.
In the last section we shall give a few general
comments. In fact at the very end of the last
section one might start again with a new series
of lectures, but that is not my purpose here.

I.   The Bénard-Rayleigh Problem

Let us consider a shallow horizontal layer
of a single component fluid, subject to an adverse
vertical temperature gradient, and suppose the
Gibbs local equilibrium description to be
valid [1,2,3].

Among the local fields are the mass density
ρ , the momentum density $\rho\vec{v}$, and the temperature
T (or else the internal energy density ρu) of a

macroscopic element of fluid. The postulated
differential equations for the time evolution of
these fields are a consequence of the basic
principles of thermodynamics and are well known.
First is the continuity equation

$$\partial_t \varrho \quad = \quad - \, \nabla \cdot \, \varrho \vec{v} \qquad\qquad (I.1)$$

where ∂_t stands for the partial derivative with
respect to time. The second equation is the
Navier-Stokes equation,

$$\partial_t \varrho \vec{v} = \, - \, \nabla \cdot (\varrho \vec{v} \, \vec{v} + \vec{P}) + \, \varrho \vec{F} \qquad\qquad (I.2)$$

where $(\varrho \vec{v} \, \vec{v} + \vec{P})$ is the momentum flow tensor.
$\vec{P}$ is the total pressure $P_{ij} = p \, \delta_{ij} + \pi_{ij}$
with p the equilibrium pressure and π the
viscous stress tensor, here assumed to be symmetric. $\varrho \vec{F}$ is the external force field, say
gravitation ($\vec{F} = 0,0,-g$). The third equation
is Fourier's equation:

$$\varrho c_v \, \partial_t \, T + \varrho c_v \, \vec{v} \cdot \nabla T \quad = \quad \lambda \nabla^2 \, T \qquad\qquad (I.3)$$

In equation (I.3) we have eliminated, for simplicity, viscous dissipation and any internal
heat sources, and use has been made of an
equation of state $\varrho = \varrho(u, T)$. In fact we
shall take

$$\varrho = \quad \varrho^*(1 - \alpha [T - T^*]) \qquad\qquad (I.4)$$

where ϱ^* is the density corresponding to some
reference temperature T^*. α is the volumetric
expansion coefficient of the fluid, c_v the

specific heat at constant volume, and λ is the
heat conductivity of the fluid.

In order to simplify our mathematical description we shall take the Oberbeck-Boussinesq
model and reduce equations (I.1) and (I.2) to
the following ones,

$$\nabla \cdot \vec{v} = 0 \qquad\qquad\qquad\qquad (\text{I.1 bis})$$

$$\varrho^* \partial_t \vec{v} + \varrho^* \vec{v} \cdot \nabla \vec{v} = -\nabla p + \mu \nabla^2 \vec{v} - \varrho(T) \vec{F}$$
$$(\text{I.2 bis})$$

where μ is the shear viscosity of the fluid;
this and the other material parameters of the
fluid will be considered as constants. The density will be considered as constant everywhere
except in the buoyancy term $\varrho\vec{F}$ where only its
temperature dependence will be retained, according to the equation of state (I.4).

The Oberbeck-Boussinesq model has been de-
fended as the relevant ad hoc approximation for
describing a shallow layer of fluid with small
density differences, vanishingly small compressibility and thermohydrodynamic velocities small
enough compared to the speed of sound in the
fluid. For more details concerning the validity
of this model the reader is referred to Mihaljan's
paper[4] or to Chandrasekhar's book[5] or else to
a forthcoming 1972 review paper[6].

The macroscopic local fields are "averages"
of some underlying microscopic quantities and
hence are subject to fluctuations. To be realizable in an experiment any solution say $\vec{v} = \vec{0}$,

bifurcating or not, of the system (I.1) to (I.3)
must be stable in the presence of these flucta-
tions.

Experimentally a shallow horizontal fluid-
layer heated from below will remain at rest provided the imposed adverse temperature gradient
is less than a certain critical value. The
temperature difference is usually expressed in
terms of a Rayleigh number defined as

$$R \quad = \quad \frac{g \; \alpha \; d^3 \; \Delta T}{\kappa \; \nu} \tag{I.5}$$

Here   d   is the depth of the fluid layer, κ the
thermal diffusivity ($\kappa = \lambda / \varrho \, c_v$)   and ν the
kinematical viscosity   ($\nu = \mu / \varrho$).

Equations (I.1 bis), (I.2 bis), (I.3) and
(I.5) with suitable b.c. constitute the non-linear
eigenvalue problem with parameter   R.

For some critical difference of temperature,
say R_c,   a spontaneous motion of the fluid is
observed in which distinct and regular steady
cell patterns can be discerned (see, for instance,
the forthcoming 1972 review paper by Koschmieder [7]).
For higher differences of temperature, $R > R_c$,
other time-dependent bifurcating solutions arise,
finally giving turbulent behaviour. Here we shall
concentrate on the first steady bifurcating solution.

The first experimental work seems to be due
to Bénard at the beginning of this century and the
first theoretical explanation to Rayleigh in 1916,
although later it appeared that Rayleigh's explanation was not in fact directly related to Benard's

results, but provided the prediction of a new
kind of instability. We shall not enter into
details concerning this point and the reader is
again referred to ref. 6.

The intuitive rationale of the problem that
we shall call the Bénard-Rayleigh problem is as
follows. The fluid adjacent to the lower heated
plate becomes warmer and so less dense. Then
the equal pressure, equal temperature, equal
density horizontal lines are distorted, thus
provoking horizontal motion. Lastly, because of
the buoyancy effect $g\vec{F}$, the fluid at the bottom
starts rising. Viscous forces tend to resist
this buoyancy but, at the critical point, the
buoyancy force overcomes the stabilizing viscosity dissipation, and convection begins. Notice
the striking similarity of this phenomenon with
the gas-liquid equilibrium phase transition where
surface tension is the stabilizing mechanism.
The first liquid droplet disappears unless the
fluctuation, i.e. the droplet, is big enough to
overcome the gas-liquid surface tension.

The problem can in fact be recast into an
energy variational or entropy variational problem for the fluctuations. In terms of energy,
see [5] for more details, one could say that
"Instability, or first bifurcation, occurs at
the minimum temperature gradient for which a
balance can be steadily maintained between the
kinetic energy dissipated by viscosity and the
internal energy released by the buoyancy force."

In more sophisticated terms it has been
shown by Glansdorff and Prigogine[2] that
"Instability, or first bifurcation occurs at
the minimum temperature gradient at which a
balance can be steadily maintained between the
entropy generated through heat conduction by the
temperature fluctuations and the corresponding
entropy carried away by the velocity fluctuations."

   We shall not have space here to discuss
further the variational approaches to the problem.
Should the fluid layer be unstable for infinitesimal fluctuations it will be unstable for any
fluctuation of finite amplitude. If all fluctuations are assumed to be Fourier analysable then
a sufficient condition for instability is that
any one of the normal modes be allowed to start
growing at the critical point. From another
point of view, some correlation length must diverge as the system approaches the bifurcation
point.

   We must remark that, unfortunately, sufficient conditions for instability sometimes
turn out to be of no use as happens in some
cases of turbulence in pipes. Obviously a system might start being unstable to finite amplitude fluctuations much earlier than to infinitesimal ones. This clearly shows the need for
some complementary analysis. For instance we
might need the use of a variational technique
that could provide an upper bound for stability
in the parameter R. Even though such variational

methods cannot with certainty conclude instability, the comparison of the upper bound for
stability with the sufficient condition for instability yields the range of values of the
relevant stability parameter, in which subcritical instabilities (i.e. instability to
finite amplitude fluctuations) are possible.

An energy method has been developed for the
Bénard-Rayleigh problem. It has been shown[8]
that no subcritical instabilities are possible
for the single component, one adverse temperature gradient problem described above.

Let us consider the shallow fluid layer
confined between two rigid horizontal boundaries
and let us assume that temperature is homogeneous-
ly fixed at the top (T_2) and bottom $(T_1, T_1 > T_2)$.
For small enough values of R the following equations describe a steady solution of the thermohydrodynamic balances (Oberbeck-Boussinesq model)

$$\vec{v}^{\,s} = \vec{0} \qquad\qquad (1.6a)$$

$$T^s = T_1 - \Delta T \frac{z}{d} \qquad\qquad (1.6b)$$

$$\rho^s = \rho_1[1 - \alpha(T-T_1)] \qquad\qquad (1.6c)$$

This is indeed the first solution of the linearized eigenvalue problem for small enough values of
R.

Let then δT, $\delta\vec{v}$, $\delta\rho$, be the macroscopic
fluctuations of the local steady fields. Should
the solution (I.6) be unstable for some $R_c > 0$,

then we will have a bifurcation at R_c.

Let these fluctuations be Fourier-Laplace analysable. One then has for a generic normal mode

$$\begin{pmatrix} \delta V_3 \\ \delta T \end{pmatrix} \sim \begin{pmatrix} A_k & W(z) \\ B_k & \Theta(z) \end{pmatrix} e^{i\vec{k}\cdot\vec{r}} e^{\sigma t} \quad . \quad (I.7)$$

Here we have kept only the vertical velocity component and the temperature fluctuations, as they are the only quantities of interest to us. $\vec{r}$ is a horizontal two-component vector in configuration space and $\vec{k}$ is the horizontal two-component wave-number of the fluctuations. At this point we must remark that we have taken a shallow layer of fluid in the limit $d \ll L$ where L is some horizontal dimension. For $L \rightarrow \infty$ the Fourier spectrum is indeed continuous, but in experimental situations $L < \infty$ and the spectrum will rather be discrete. The quantity σ is the time constant, which, in general, may be complex. The sign of its real part determines stability. Should $Re[\sigma]$ be positive (negative) the fluctuations start growing (decaying). According to linear theory these fluctuations will go to infinity (zero) for $t \rightarrow \infty$. In fact, the fluctuations never go to infinity for $Re[\sigma] > 0$ since, once the bifurcate state is formed, the non-linear coupling in the equations is a stabilizing mechanism in some neighbourhood of the critical point. At $Re[\sigma] = 0$ (if at the same time $Im[\sigma] = 0$)

we have neutral stability. A fluctuation then
cannot start growing or decaying.

One can show that, for R > 0, the problem
(I.1 bis), (I.2 bis), (I.3) for the fluctuations
(I.7) around the steady state (I.6), with the
appropriate boundary conditions on rigid boun-
daries(for the dimensionless fields)

$$\delta V_3 = \frac{\partial}{\partial z} \delta V_3 = \delta T = 0 \quad \text{at} \quad z = \pm \frac{1}{2}$$

$$(\text{I}.8)$$

is real; one has, for R > 0, Im $[\sigma]$ = 0 if
Re $[\sigma]$ = 0. For R < 0 the fluid layer is
always stable and bifurcation is not possible.

In more precise terms, and for infinitesimal
fluctuations, one is able to reduce the original
problem to the simpler one

$$(D^2 - k^2)(D^2 - k^2 - \sigma)(D^2 - k^2 - \frac{\kappa}{\nu}\sigma)W =$$
$$= - Rk^2 W \qquad\qquad (\text{I}.9)$$

with b.c. on rigid plates

$$W = DW = (D^2 - k^2)W = 0 \quad \text{at} \quad z = \pm \frac{1}{2} . \quad (\text{I}.10)$$

Here   D $\equiv \frac{d}{dz}$.   We also have dropped the symbol
δ as we are now dealing only with fluctuations,
and there is no possibility of confusion. The
problem (I.9) and (I.10) is not only real but
selfadjoint. The scalar product will be taken
with integration of the two appropriate functions
over the entire layer. Care has been taken to
make quantities dimensionless; in doing so, the

vertical gap is measured in units of d and the
origin on the vertical axis is taken at the
middle of the layer. More technical details can
be found in Chandrasekhar's book[5)].

From equation (I.9), and setting $\sigma = 0$,
one gets solutions of the type

$$W = A_o \begin{Bmatrix} \cos \\ \sin \end{Bmatrix} (q_o\, z) + A \begin{Bmatrix} \cosh \\ \sinh \end{Bmatrix} (qz) + c.c.$$

(I.11)

The parameters $\pm iq_o$, $\pm q$ and $\pm q^*$ are the three
pairs of solutions of $(q^2 - k^2)^3 = - Rk^2$. We
shall take $\mathrm{Re}[q] = q_1$, $\mathrm{Im}[q] = q_2$.

The operator (I.9) (with $\sigma = 0$) indeed
admits separately even and odd solutions and so
do the b.c. (I.10). As solutions (I.11) must
satisfy the b.c. and the problem is homogeneous
we end up with a 6×6 homogeneous linear set
of algebraic equations. This 6×6 system breaks
into two 3×3 parts. The necessary and sufficient condition for a non-trivial solution is
given by Rouché-Froebenius theorem. This condition is, for the even modes, a 3×3 vanishing determinant. Explicitly, it leads to the
following transcendental equation

$$-q_o \tan \left(\tfrac{1}{2}q_o\right) = \frac{(q_1 + q_2\sqrt{3})\sinh q_1 + (q_1\sqrt{3} - q_2)\sin q_2}{\cosh q_1 + \cos q_2}$$

(I.12)

A similar condition holds for odd modes but we
shall not consider it here.

Equation (I.12) is then solved by trial and
error. All points satisfying relation (I.12)

provide a non-trivial solution to the linear
perturbation problem and so lead to bifurcation.
The minimum temperature difference that provides
a non-trivial solution corresponds to the first
even mode at k = 3.117 and R $\simeq$ 1,708. We shall
call this the actual critical point. According
to Joseph[8], and some other people, this is also
the value given by a finite amplitude analysis,
and so there exists one and only one solution
for $0 \leqq R < R_c \simeq 1,708$.

Let us now see the relevance of this bifurcating solution to the physical problem. To
do so we shall concentrate on some measurable
quantity, e.g. the convected heat flux. To
evaluate how much heat flux is convected by the
bifurcating solution, one must know something of
the non-linear eigenvalue problem. At this stage
we shall make a few comments on this.

Firstly, the linear problem does not say
anything about the bifurcating solution; it
just tells us that there will exist at least one
at the critical point R_c = 1,708 with wavelength
λ_c = 2π/3.117. Both numbers, R_c = 1,708 and
λ_c = 2π/3.117, are in quite good agreement
with experimental results for shallow layers.
But $\vec{k}_c$ (or $\vec{\lambda}$) has infinitely many ways of
being decomposed into two components, so we
know nothing about the actual structure that is
formed at the critical point.

The proper approach to the non-linear eigenvalue problem would be to decide first if any

realizable set of i.c. and b.c. will lead to a
steady convective state. This presumably would
permit us to ascertain the variety of possible
states which can be found and possibly one of
these would be the only stable one. Unfortunately this problem is still open, though discussed
in a remarkable paper by Newell, Lange and
Aucoin[9].

As a matter of fact, it has thus far been
proved[10] (i) that the cell pattern of solutions
of the boundary value problem discussed above is
ensured by invariance under rotation by an angle
$\omega = \pi/N$,   N = 1,2,3   of the horizontal projection
of the velocity field. Rolls, squares and horizontal hexagons are obtained for N = 1,2,3 respectively. (ii) Furthermore, by requiring that
the normal component of the horizontal velocity
field vanishes on the boundaries of each cell,
it has been shown:

For $(k_x = k_y\sqrt{3},\ k_y = \tfrac{1}{2}k_c)$ $\left[k_x = k_y = k_c/\sqrt{2}\right]$

$$\left\{k_x = k_c,\ k_y = 0\right\}$$

there exist (two) [a unique] {a unique} real
branching solution(s) of the eigenvalue problem
with (hexagonal) [square] {roll} patterns for
values of R belonging to a sufficiently small
right-hand neighbourhood of R_c = 1,708. These
solutions can be developed in a power series of
$\left[R - R_c\right]^{1/2}$. There are no real branching
solutions with (hexagonal) [square] {roll}

pattern for $0 \leqslant R < R_c$.

These are rigorous results. As yet, unfortunately, nothing is known about relative stability of the different solutions, and so these theorems are of very little physical use. Quite a number of papers have indeed appeared discussing stability by a perturbative analysis of the non-linear problem, usually up to the first few orders. Our modest opinion on this point is that a perturbative stability analysis does not provide any real advance. The non-linear stability problem is thus entirely open to more sophisticated, and possibly new, mathematical techniques.

Now, to avoid the mathematical difficulty, we go back to experimental results. The solution of rolls is the one found by Koschmieder for rigid-rigid plates and normal fluids (for instance, silicone oil)[*]. All we can possibly do with perturbation theory is just obtain some qualitative insight into the non-linear problem by computing a physical quantity assuming that rolls are the realizable solution.

One defines a series expansion, assumed to be convergent in a small righthand neighbourhood of R_c,

$$R = R_c + \sum_{i=1}^{\infty} \epsilon^{(i)} R^{(i)} \qquad (I.13)$$

for ϵ small enough. We also define a power series

---

[*] One must emphasize how much Koschmieder's work is a definitive piece of experimental evidence.

expansion of the local fields W and T, assumed
to be steady solutions of the branching problem.
By keeping just second order terms one then com-
putes $\langle \epsilon^2 \ WT \rangle$, where the bracket here means
a normalized integration over a wavelength. This
provides us with some estimate of the convected
heat flux. The details of the calculations can
be found in a paper by Segel[11]. One obtains a
convected heat flux in quite reasonable agreement
with experimental results[7]. A much more complete approximation scheme has been developed by
different authors (see for instance ref. 12) but
here we shall leave this question open (see also
ref. 6).

II. Soret-driven Instability

We shall turn now to our second problem.
Let us take a shallow horizontal two-component
fluid layer subjected to a vertical temperature
gradient, adverse or otherwise, and again assume
to be applicable the Oberbeck-Boussinesq model
described in the previous section. We must sup-
plement the continuity equation with

$$\partial_t \, N_i + \vec{v} \cdot \nabla N_i \quad = - \frac{1}{\rho} \nabla \cdot \vec{J}_i \quad . \qquad (II.1)$$

Here $N_i = \rho_i / \rho$; $\rho_1 + \rho_2 = \rho$. Subscripts
refer to component one and two; we shall consider
as one the denser component. $\vec{J}_1 = \rho \, (\vec{v}_1 - \vec{v})$

and $\vec{v}$ is the local barycentric velocity of a
fluid element. For small enough inhomogeneities
one knows [1,2])

$$\vec{J}_1 = N_1 N_2 \, D' \, \nabla T + D \, \nabla N_1 \qquad (II.2)$$

where D is the normal diffusion coefficient and
D' is the thermal diffusion coefficient, defined
here positive if the more dense component migrates
towards the cold plate.

In equation (II.2) we have incorporated the
so-called Soret effect. There is mass flow, not
only due to an actual concentration gradient but
also to an Onsager cross-effect because of the
thermal gradient. In the heat equation (I.3)
we shall disregard the complementary effect, i.e.
the heat flux due to a concentration gradient
(Dufour effect). For more simplicity we shall
take the product $N_1 N_2$ as a constant equal to
its average value $N_1^* N_2^*$. For details concerning the relevance of these approximations see
refs. 1, 6, 13.

We shall discuss now the non-linear eigenvalue problem for both $R > 0$ and $R < 0$. We
shall see that in both cases bifurcation is
indeed predicted, depending upon the sign of D'.

We shall take as b.c. for our problem,
fixed temperatures on the two rigid plates.
These plates will also be taken as impervious
to mass flux, i. e. $\vec{J}_1 = 0$ on the top and bottom.
Because of this latter condition the original
temperature gradient builds up an actual concentration gradient.

Let us guess, as in the previous section, the intuitive rationale of the instability to be expected. We assume that the heating is from above $(R < 0)$ and the more dense component migrates towards the top plate $(D' < 0)$. A fluid particle taken from the bottom (top) to the top (bottom) plate finds itself in a warmer (colder) but lighter (denser) neighbourhood. As in a liquid one has $\varkappa/D \gtrsim 100$, then the temperature fluctuation relaxes one hundred times faster than the concentration fluctuation, thus providing a mechanism for vertical convective instability. In the case of heating from below (and, say, $D' > 0$) in order to rationalize possible convective instability we just take over the argument developed for the single Bénard-Rayleigh problem.

For small enough values of R we have a steady solution of the thermohydrodynamic equations through supplementing equations (I.6) with

$$\frac{\partial}{\partial z} N_1^S = \frac{D' N_1^* N_2^* \Delta T}{D} = \text{constant} \qquad (II.3a)$$

$$\varrho^S = \varrho_1 \left[1 - \alpha (T^S - T_1) + \Gamma (N_1^S - N_1^*) \right] \qquad (II.3b)$$

By defining the following separation coefficient,

$$S = \Gamma \frac{D' N_1^* N_2^*}{\alpha D} \qquad (II.4)$$

where Γ is the corresponding partial derivative of ϱ with respect to N_1 at constant temperature; one can thus write

$$\frac{\partial}{\partial t} \rho^{s} = \alpha \rho_{1} \Delta T [1+S] \quad . \qquad (II.5)$$

Notice that the overall density gradient might
be made negative by having $-1 < S < 0$. The interest of the stability problem to be discussed
lies in the fact that we shall find bifurcation and so convective instability even in the
region $-1 < S < 0$ where the overall density
favours stability.   This is a most remarkable
and intuitively unexpected situation, showing
indeed that a simple mechanical interpretation
of the stability problem is no longer possible.
One needs a deep analysis of the non-equilibrium
fluctuations problem.   We shall come back to this
point in the next section.

Now equations (I.1 bis), (I.2 bis), (I.3)
and (II.1) with the corresponding b.c. and i.c.
lead to a complex eigenvalue problem.   That is
very bad as, in general, it is quite difficult
to handle complex eigenvalue problems (indeed
non-selfadjoint problems).   We may turn around
this mathematical difficulty by taking advantage
of the experimental fact $\varkappa \gg D$ and go to the
limit $D/\varkappa \rightarrow 0$.   To do so we shall take a suitable transformation to non-dimensional quantities
in the following way

$$\tau = \frac{tD}{d^{2}} \qquad \vec{V} = \frac{\vec{v}d}{D} \quad . \qquad (II.6)$$

Here instead of t a new time scale τ is chosen

with the time unit being the relaxation time of
a concentration fluctuation. We expand all
macroscopic fields and take the limit $D/\varkappa \longrightarrow 0$
in a formal way. This corresponds to taking
first order perturbation theory in the small
parameter $D/\varkappa$. The reader can find complete
details concerning this subtle point in ref. 14.

If now Θ , W, Γ denote respectively the
temperature, the z-component of velocity and the
concentration fluctuations, one has $\Theta = 0$ for
the chosen b.c. Indeed one is not able to detect
a temperature fluctuation with time units one
hundred bigger than its relaxation time.

A similar analysis to the one already given
for the simple Bénard-Rayleigh problem leads us
to the following real, but not self-adjoint,
problem

$$\left[(D^2 - k^2) - \sigma \right] \Gamma(z) - \frac{P \varkappa \nu R}{\alpha g d^3 D} W(z) = 0 \qquad (\text{II.7a})$$

$$\frac{g \varkappa d^3}{\nu D} k^2 \Gamma(z) + \left[D^2 - k^2 \right]^2 W(z) = 0 \qquad (\text{II.7b})$$

subject to the b.c.

$$W = \frac{dW}{dz} = \frac{d\Gamma}{dz} = 0 \quad \text{at} \quad z = \pm \frac{1}{2} \quad . \qquad (\text{II.8})$$

The neutral stability is defined by setting
$\sigma = 0$ in equations (II.7). One then gets the
simpler problem

$$\left[D^2 - k^2 \right]^3 W = - \tilde{R} k^2 W \qquad (\text{II.9})$$

where $\tilde{R} = RS \frac{\varkappa}{D}$. Surprisingly enough equation

(II.9) is just (I.9). Even more, the two problems are mathematically one and the same if, for the simple Bénard-Rayleigh problem, one fixes on the boundaries not the temperature but the heat flux.

We shall not give here any details of the linearized eigenvalue problem but just give the figures corresponding to the critical point. Notice at this stage that problem (II.9) is symmetric in ΔT and so both the heating from below and the heating from above problems are symmetrically described.

The minimum, $\tilde{R}$, again corresponds to the first even mode. We get $\tilde{R}_c = 720 = RS \frac{\varkappa}{D}$ for $k_c = 0$. For fluids with $\frac{\varkappa}{D} \approx 100$ one gets the hyperbolic relationship $RS = 7.20$.

A striking result is that $k_c = 0$. We must remember that this mathematical result belongs to the case $(^d/L) \rightarrow 0$ i.e. to a horizontally unbounded shallow layer of fluid. A few remarks are needed to explain the model. Firstly, in an experimental set-up the container, or the "boat" as it is called by workers in the field, is always horizontally bounded. For this very reason the Fourier spectrum of the fluctuations will be discrete rather than continuous. Secondly, as $k_c = 0$ means $\lambda_c = \infty$, in a realistic case λ_c will be very large but not infinite. One would expect to have something like $\lambda_c \sim L/d$. Beyond an intuitive argument, the correctness of this suggestion remains open.

Obviously instability is predicted for the region $R < 0$ and $-1 < S < 0$ contrary, as pointed out earlier, to intuitive arguments.

Now experimentalists, studying the Soret separation using the so-called parallel plate method, have traditionally assumed that convective instability cannot occur if the fluid layer is heated from above[16]. Consequently I am not aware of any work reporting either critical point parameters (say R_c or k_c) or pattern development for a Soret-driven convective instability. The problem is totally open for experimental research. A comprehensive list of experiments to be performed, guided by this new predicted instability, is given in ref. 13.

From the mathematical point of view we are in a much poorer state than for the simple Bénard-Rayleigh problem. We do not have at the present time any existence (and non-uniqueness) theorem describing bifurcation. Yet we would like to explore some qualitative and quantitative aspects of the non-linear eigenvalue problem and get some prediction of physical relevance. We shall take a natural extrapolation of Koschmieder's experimental results for the single component problem. Let us assume that rolls are a realizable bifurcating structure at the critical point. As in the case discussed in the previous section, the first relevant physical quantity should be the convected heat flux. However it has been shown that in the limit $D \ll \varkappa$ there is a vanishing

convected heat flux[14]. So we turn to another
relevant physical quantity. Let us consider the
Soret separation between the two-components.

As was indicated earlier in these lectures,
a power series expansion, assumed to be convergent in some small right-hand neighbourhood
of $\tilde{R}_c$ = 720, is taken for all local steady
fields. As k_c = 0 in the exact case, we shall
consider k_c small but not vanishing. Up to
second order perturbations around $\tilde{R}_c$ one gets

$$\langle \epsilon^2 W\, N_1 \rangle \approx - 0.21\ k^{2/3}\ [\tilde{R} - \tilde{R}_c] \frac{D\,\nu}{gd^3} \quad . \qquad (II.10)$$

This is the predicted convected mass flow
that reduces the Soret separation. Needless to
say the predicted value is not a rigorous result.
Yet surprisingly good agreement is found on com-
paring with some available experimental data.

We select a very interesting and unique
experimental work on LiI - H$_2$O. From data reported by Agar and Turner[17] one gets

$$\langle \epsilon^2 W\, N_1 \rangle_{exp} \approx -1.25 \times 10^{-3} \qquad\qquad (II.11)$$

whereas a numerical estimate for their set-up is
(unfortunately for k_c = 2π)

$$\langle \epsilon^2 W\, N_1 \rangle_{theory} \approx -1.48 \times 10^{-3} \qquad\qquad (II.12)$$

Also, the predicted convective velocities
for their experiment are of the order of $W \sim 10^{-3}$
to 10^{-2} cm/sec. This is a very small velocity

and great care will be required for its observation. Indeed, this datum is not reported by
Agar and Turner. Again we refer to ref. 13
for a more detailed comparison between theory
and experiments.   Much work, both theoretical
and experimental, remains to be done. It should,
for instance, be interesting to ascertain if the
influence of the b.c. is so strong as to lead to
a pattern of its own, possibly by forcing rolls
every time that rigid-rigid boundaries are considered. We have in mind only Newtonian normal
fluids and the Oberbeck-Boussinesq model.

III. Final comments

We have shown with two rather simple hydrodynamic examples how one gets disorder-order non-
equilibrium phase transitions. Though some intuitive rationale has been given for these instabilities, we have seen that in certain cases
the problem is well beyond a naive and simple
explanation.

Among others, Prigogine and co-workers[2,18]
[and references quoted therein; see also ref. 19]
have shown that very similar transitions arise
in purely non-convective, dissipative systems,
such as chains of autocatalytic, non-linear
chemical reactions. At bifurcation points the
system may evolve to states showing spatial,
temporal or functional ordering, which give rise
to dissipative structures[2,18].

In every case the local equilibrium assumption[3] turns out to be a relevant starting point.
This can be expressed by assuming a Maxwell-
Boltzmann momentum probability density. However,
this does not prevent the system from reaching
an overall state which is very far from equilibrium, owing to external macroscopic constraints
such as thermal or concentration gradients as
discussed earlier in these lectures. It seems
fairly well established[2] that the local equilibrium assumption is useful if the system is dense
enough, when _elastic_ collisions continuously restore the local Maxwell-Boltzmann distribution.
At the same time this state is perturbed by macroscopic transport processes or _reactive_ collisions.
Thus systems which in the small scale are purely
dissipative, and so tend to damp out all thermal
fluctuations, undergo large scale instabilities
which drastically change their macroscopic state.
One might say that when a finite amplitude emerges,
a local inhomogeneity, a "droplet" say, would at
first sight tend to be damped by dissipative
mechanisms. However, in a close neighbourhood of
a critical point, diffusion or heat flow should
rather provide a mechanism for propagating such
fluctuations over macroscopic distances.

The response of a system to a fluctuation
bringing the system back to its initial state is
indeed deterministic. But fluctuations constitute
a purely stochastic process. It is at the moment

that a fluctuation reaches a macroscopic level that
coupling appears between the deterministic and the
stochastic aspects of the problem. This is given
in terms of non-linear equations.

The approaches to the fluctuation problem can
be summarized as follows. (i) There is the Langevin-
Landau approach which amounts to introducing some
a priori stochastic terms in the evolution equations. But this procedure compromises a priori the
very understanding of the emergence of fluctuations
(see for more detailed comments refs. 2), 18)).
(ii) Another way of looking at the problem consists in deriving stochastic master equations which
underlie, for instance, certain markovian reaction
equations in chemical problems. (iii) Still another
possibility is a kinetic theory approach. One looks
for a (generalized) Boltzmann equation whose moments will contain the usual macroscopic equations
and incorporate,presumably,some macroscopic correlation of fluctuations[20]. We shall not enter here
into any details concerning these approaches.

I would like to end my lectures by calling
your attention to the related, and rather ambitious,
programme started many years ago (1945) by
Prigogine. This is the question of generalizing
Hamilton-Lagrange classical dynamics to dissipative
systems. One would like to have some "Lagrangian"
and thus some "action principle" for dissipative
systems. Unfortunately, evidence exists that such
a programme is far from being "universally" valid.

One can say something about steady states by using Prigogine's minimum entropy production theorem (over a volume). Unfortunately, if the phenomeno-logical coefficients[1,2] are not strictly constant and convective terms are not negligibly small, the entropy production does not suit the purpose. This is the case for many natural irreversible processes and, indeed, for non-linear phenomena.

Prigogine and coworkers[2] have recently over-come this limitation as follows. Let P be the entropy production per unit time in an element of volume. One has from linear theory

$$P = \sum_i J_i X_i \; . \tag{III.1}$$

Here J_i denotes some flux and X_i generalized forces[1,2]. From (III.1) one gets

$$P = \sum_i (\delta J_i) X_i + \sum_i J_i (\delta X_i) \; . \tag{III.2}$$

Define

$$\delta_x P = \sum_i J_i (\delta X_i) \; . \tag{III.3}$$

They show that for at least time-independent b.c. and when no convective terms are present, a univer-sal evolution criterion should be

$$\delta P \leq 0 \tag{III.4}$$

This inequality has been generalized to include convective terms but the thing becomes too heavy to have time to discuss it again here[2].

One notices that $\delta_x P$ is not a total

differential and this very fact justifies indeed
the difficulty in finding a true "action principle" for far from equilibrium problems. Yet
some <u>local potential</u> was still available[2].
Obviously in linear problems $\delta_x P$ covers as a
particular case the minimum entropy production
theorem.

At the same time Prigogine and coworkers
turned to Einstein-Landau fluctuation theory.
By putting together fluctuation theory and their
"extended" non-equilibrium thermodynamics[2] they
came up with a general stability criterion for
non-equilibrium states.

Let S_0 be, for instance, the steady state
entropy and let S be the non-equilibrium one
due to fluctuations. Let us again forget about
convective problems. For an equilibrium state
one knows that a fluctuation regresses according
to the Einstein law

$$P \sim e^{(S-S_0)/k} \qquad\qquad (III.5)$$

Here, up to second order terms,
$\Delta S = S-S_0 \approx \frac{\delta^2 S}{2} < 0.$ k is the Boltzmann's
constant. Some evidence exists that equation
(III.5) with $(\delta^2 S)$ should be valid for non-
equilibrium states[2]. Now starting with $\delta^2 S < 0$
around the reference state, the system will
fluctuate to finally decay to this state, which
will correspond to the most probable configuration,
as long as $\partial_t(\delta^2 S) > 0$ $(\geqslant 0)$. If $\partial_t(\delta^2 S) < 0$,
then the system will evolve to a new regime and

a dissipative structure could be formed. The
central point is thus the balance equation for
the quantity $(\delta^2 S)$ which measures the fluctuations.

This criterion associates then $\delta^2 S$ with
a Liapounov's function[21] (up to the sign). One
can thus take over the powerful direct method of
Liapounov. This fascinating approach could indeed be the starting point for another series of
lectures. We should start from this very last
section finally to end up with the hydrodynamic
examples, discussed earlier in these lectures,
as particular applications of general criteria.

REFERENCES

1) S.R. de Groot and P. Mazur, Non-equilibrium Thermodynamics, North Holland, Amsterdam, 1962.

2) P. Glansdorff and I. Prigogine, Thermodynamic Theory of Structure, Stability and Fluctuations, Wiley, Interscience, N.Y. 1971. There also exists a French edition Structure, Stabilité et Fluctuations, Masson, Paris, 1971.

3) G. Nicolis, J. Wallenborn and M.G. Velarde, Physica 43 (1969), 263.

4) J.M. Mihaljan, Astrophys. J. 136 (1962), 1126.

5) S. Chandrasekhar, Hydrodynamic and Hydromagnetic Stability, Clarendon Press, Oxford, 1961.

6) M.G. Velarde and R.S. Schechter – a review paper in preparation for Advances in Chemical Physics (1972-73).

7) E.L. Koschmieder, Advances in Chemical Physics, a review paper to appear.

8) D.D. Joseph and C.C. Shir, J. Fluid Mech. 26 (1966), 753 and references quoted therein.

9) A.C. Newell, G.G. Lange and P.J. Aucoin, J. Fluid Mech. 40 (1970), 513.

10) H. Görtler, K. Kirchgässner and P. Sorger, Deutsche Versuchsanstalt für Luft- und Raumfahrt E.V. (Sondersruck aus Problems of Hydrodynamics and Continuum Mechanics) 901, 353.

11) L.A. Segel in Non-equilibrium Thermodynamics, Variational Techniques and Stability, R.J. Donnelly, R. Herman and I. Prigogine, editors, University of Chicago Press, Chicago 1966. On general grounds this is a very interesting book.

12) A. Schluter, D. Lortz and F. Busse, J. Fluid Mech. 23 (1965), 129.

13) M.G. Velarde and R.S. Schechter, Chem. Phys. Letters, 12 (1971), 312.

14) M.G. Velarde and R.S. Schechter, Phys.
     Fluids, in press.

15) M.G. Velarde and R.S. Schechter, in Modern
     Developments in Thermodynamics, B. Gal-
     Or, editor, Haifa, 1972, to appear.

16) A well documented and excellent account of
     experimental techniques on thermal dif-
     fusion can be found in H.J.V. Tyrrell,
     Diffusion and Heat Flow in Liquids,
     Butterworths, London, 1961.

17) J.N. Agar and J.C.R. Turner, Proc. Roy. Soc.
     A255 (1960), 307. See also P.N. Snowdon
     and J.C.R. Turner, Trans. Faraday Soc.
     59 (1960), 1409, 1812.

18) (a) I. Prigogine and G. Nicolis, in 3rd
     Int. Conf. "Theoretical Physics and
     Biology", Versailles, France (1971), to
     appear.
     (b) G. Nicolis, Fluctuations around non-
     equilibrium states in open, non-linear
     systems, preprint available (1971).

19) Bifurcation theory and non-linear eigen-
     value problems, Keller and Antman,
     editors, Benjamin, N.Y. 1969.

20) Though the turbulence problem is far more
     complicated and may or may not. depend-
     ing on your taste, bear similarity
     with the simple instability problems,
     we mention an approach from first prin-
     ciples: R. Balescu and A. Senatorsky,
     Ann. Phys. (N.Y.) 58 (1970), 587.

21) See e.g. J. La Salle and S. Lefshetz,
     Stability by Liapounov's Direct Method,
     Academic Press, N.Y., 1961.

BROWNIAN MOTION AND THE
STOCHASTIC THEORY OF QUANTUM MECHANICS

E. Santos

Facultad de Ciencias
Universidad de Valladolid
Valladolid, Spain

Introduction

Introduction

The purpose of this lecture is to give the principles of the simplest stochastic mechanics in configuration space. The work simplest will be explained later; configuration space is the space of the position coordinates of the particles and stochastic mechanics is defined as follows: We call stochastic mechanics to any dynamical theory in which the equations of motion are not fully known. The partial ignorance about the equations of motion is taken into account by introducing stochastic parameters in it. The typical example of stochastic mechanics is the theory of Brownian motion. We distinguish stochastic from statistical mechanics by defining statistical mechanics as a theory in which the differential equations of motion are in principle known even though the initial conditions are not completely specified.

The theory of Brownian motion started in 1905 with the theory of Einstein, who studied the motion of a Brownian particle in ordinary space in absence of external forces. The forces were included afterwards by Smoluchowski. This theory is imperfect in the sense that it introduces infinite velocities. The difficulty was eliminated by Uhlenbeck and Ornstein in 1930 by developing the theory in phase space. The theory of Einstein - Smoluchowski became

a limiting case of that by Uhlenbeck and Ornstein
for large enough time intervals.

The analogy between the theory of Brownian
motion and quantum mechanics was studied in 1933
by Fürth, who derived for the Brownian particles
some uncertainty relations similar to those of
Heisenberg. The analogy is also evident in the
path-integral formulation of quantum mechanics due
to Feynmann. Nevertheless, the analogy was considered purely formal until 1952, when Fenyes showed
that Schrödinger equation might be derived from a
classical stochastic theory. This idea has been
developed from that time by a number of people,
mainly L. de la Peña-Auerbach, whose formulation
will be followed here.

1.   KINEMATICS OF STOCHASTIC (MARKOVIAN) MOTION

For simplicity we will consider a single particle in one dimension, though the generalization
to N particles is easy. The basic idea of the
theory is that the path of the particle, given by
$x=x(t)$, is a stochastic process. This function
$x(t)$ will not be necessarily differentiable, although this might be considered a purely mathematical idealization. So, we need some substitute
for the ordinary time derivatives in order to be
able to define velocities and accelerations. If
the particle is in the point x at time t, a stochastic theory cannot tell us in which point will
be the particle at time $t+\Delta t$. We can only ask for
the probability distribution of the position at

that time. Now, a probability distribution is fully
characterized by the moments, which will be denoted
by $E\{[x(t+\Delta t)-x(t)]^n\}$ (E means expectation value
of what follows). In principle the moments for all
Δt would be necessary to characterize fully the
probability distribution, but we start simplifying
by assuming that only the values for small Δt are
needed. So we define the function

$$S_n^+(x,t) = \lim_{\Delta t \to 0} \frac{E\{[x(t+\Delta t) - x(t)]^n\}}{\Delta t},$$

assuming that they exist. In the same way we could
ask for the probability distribution of the position at time $t-\Delta t$ if we know the position at time
t. This leads us to define the functions

$$S_n^-(x,t) = \lim_{\Delta t \to 0} \frac{E\{[x(t) - x(t-\Delta t)]^n\}}{\Delta t}$$

We make a further simplification of the theory
by assuming that as many functions $S_n(x,t)$ vanish
as it is possible. It is easy to see that the functions S_2 cannot vanish because then there is no
stochasticity at all. In fact, $S_2 \Delta t$ is a measure
of the variance of the probability distribution
for small enough Δt. If S_2 were zero, the motion
would be deterministic. Then, the simplest assumption is that all S_n vanish for $n > 2$. Also, let us
assume that S_2^+ and S_2^- are constant, and indeed
the same constant 2D. This is a rather strong
assumption which is by no means evident. Its intuitive meaning is that the stochasticity of the motion is the same for all places and all times.

Besides S_2, only the functions $S_1^{\pm}$ remain. We shall
denote it by $v_{\pm}(x,t)$ in the following. It is easy
to see that the functions represent mean forward
and backward velocities.

It can be shown that the hypotheses made up
to now are sufficient to characterize the stochastic process $x(t)$ as Markovian. Roughly speaking,
a stochastic process is a Markov process if there
is not memory in it. A more precise definition
can be found in the references. It is interesting
to note that the hypotheses which we have made
(i.e. the Markovian character) imply the existence
of infinite actual velocities for the particle. In
fact, for S_2 to be finite, $x(t+\Delta t)-x(t)$ must be
of the order of $\sqrt{\Delta t}$ for small enough Δt. This
shows that the average velocity in the time inter-
val Δt is of order $1/\sqrt{\Delta t}$, which goes to infinity
as Δt goes to zero. Nevertheless, this fact may be
considered a consequence of the idealizations involved rather than a physical fact.

Now, we are able to define the time derivatives of any function $f(x,t)$. All which we need
is to assume that f is an analytic function of x
and t. Then, we define a forward time derivative
by

$$\mathcal{D}_+\, f(x,t) \equiv \lim_{\Delta t \to 0} \frac{E\{f(x(t+\Delta t),t+\Delta t)\} - f(x(t),t)}{\Delta t}$$

A series expansion of f in the neighborhood of
$(x(t),t)$ leads to

$$\mathcal{D}_+ f(x,t) = \frac{\partial f}{\partial t} + v_+(x,t)\frac{\partial f}{\partial x} + D\frac{\partial^2 f}{\partial x^2}\ .$$

In a similar way, the backward time derivative
is

$$\mathcal{D}_- f(x,t) \equiv \lim_{\Delta t \to 0} \frac{f(x(t),t)-E\{f(x(t-\Delta t),t-\Delta t)\}}{\Delta t} =$$

$$= \frac{\partial f}{\partial t} + v_-(x,t)\frac{\partial f}{\partial x} - D\frac{\partial^2 f}{\partial x^2}$$

From these definitions trivially follows that

$$v_\pm(x,t) = \mathcal{D}_\pm x .$$

After that, we can define the accelerations which
will be the time derivatives of the velocities.
As there are two velocities, there are four
accelerations:

$$\mathcal{D}_+ v_+, \mathcal{D}_+ v_-, \mathcal{D}_- v_+, \mathcal{D}_- v_-.$$

It is convenient to define two new velocities
as linear combinations of the earlier ones. These
are

$$v \equiv (v_+ + v_-)/2 , \quad u \equiv (v_+ - v_-)/2.$$

In the same way, we define two new operators of
time derivation:

$$\mathcal{D}_c = (\mathcal{D}_+ + \mathcal{D}_-)/2 = \frac{\partial}{\partial t} + v\cdot\nabla \quad \mathcal{D}_s \equiv (\mathcal{D}_+ - \mathcal{D}_-)/2 \equiv u\cdot\nabla + D\nabla^2 .$$

The operator $\mathcal{D}_c$ (c for current or systematic) is
seen to coincide with the total time derivative
of hydrodynamics. The second operator $\mathcal{D}_s$ is a
purely stochastic derivative. It is interesting
to note that time reversal changes v_+ into v_- and
$\mathcal{D}_+$ into $\mathcal{D}_-$. On the other hand v and $\mathcal{D}_c$ are
time reversal invariant while u and $\mathcal{D}_s$ change sign
under time reversal.

2. EQUATION OF CONTINUITY

Let us consider any conserved quantity $f(x,t)$, that is, any function whose value does not change with time during the motion of the particle. Then it follows that

$$\mathcal{D}_{\pm} \, f = 0.$$

At this moment, we introduce the probability density for the position of the particle, $\varrho(x,t)$. From the previous equation it follows

$$\int \varrho(x,t) \, \mathcal{D}_{\pm} \, f(x,t) \, dx = 0.$$

Now, performing integrations by parts and assuming that ϱ vanishes rapidly enough at infinity it follows that

$$\frac{d}{dt} \int \varrho f \, dx = \int f \left[\frac{\partial \varrho}{\partial t} + \nabla (\varrho v_{\pm}) \mp D \, \nabla^2 \varrho \right] dx.$$

The left hand side of this equation vanishes because f was assumed to be a conserved quantity. The vanishing of the right hand side for any conserved quantity f leads to the vanishing of the term in brackets, which leads to

$$\frac{\partial \varrho}{\partial t} + \nabla (\varrho v) = 0 \, , \quad \nabla (\varrho u) = D \nabla^2 \varrho \, , \left(\nabla \equiv \frac{\partial}{\partial x} \right).$$

The first equation is the continuity equation. The second one can be integrated and gives

$$u = D \, \text{grad} \, \log \varrho \, .$$

This is straightforward in one dimension. In three dimensions a term of the form $\varrho^{-1} \text{rot} \, \vec{w}(x,t)$ might

be added, which we neglect. This last relation
allows us to write

$$v_{\pm} = v \pm D \text{ grad log } \varphi ,$$

a relation which we shall use later. A relation
between the accelerations can be found if it is
assumed that $v(x,t)$ is the gradient of some function $S(x,t)$. (This condition is trivially always
true in one dimension but not in three). The relation, which follows straightforwardly from the
continuity equation, can be written in either one
of the two forms:

$$\mathcal{D}_c u + \mathcal{D}_s v = 0 \quad , \quad \mathcal{D}_+ v_+ = \mathcal{D}_- v_- .$$

3. DYNAMICS OF BROWNIAN MOTION

We have developed a quite general theory of
stochastic Markovian motion in configuration
space. Now, we can particularize by introducing
additional postulates. For example, let us consider a Brownian particle in absence of external
fields of force. Then, there are no privileged
directions in space. This means that, if the particle is in point x at time t, the center of the
probability distribution will be also x for any
future time. This is equivalent to saying that v_+
vanishes, which leads to

$$v = - D \text{ grad log } \varphi .$$

Taking into account the continuity equation,
this leads to the diffusion equation

$$\frac{\partial \varphi}{\partial t} = D \nabla^2 \varphi .$$

This was just the relation derived by Einstein in
his theory of Brownian motion. What we have shown
is that it is the simplest equation for the evolution of the probability density of a Brownian
particle. If the Brownian particle actually obeys
it or not is a question to be answered by experiments or by a detailed, microscopic theory, which
we do not intend to develop here.

Let us derive the Brownian motion equation in
the presence of forces. In this case it is obvious
that v_+ must be different from zero. If we assume
that Brownian motion is a dissipative phenomenon
we must have a proportionality between the force
and the velocity, that is

$$f = m \beta v_+ ,$$

where β is some constant and m the mass of the
particle introduced here to agree with standard
notation. Hence, the Smoluchowski equation is easily obtained

$$\frac{\partial \varphi}{\partial t} = - \frac{1}{m \beta} \nabla (f \varphi) + D \nabla^2 \varphi .$$

This is well known to be the equation of Brownian
motion in the presence of forces. Clearly, this
and the diffucion equation can be used only to
follow the evolution to the future. If we know the
position of the particle at a time and we ask for
the previous positions, we must use a backward
equation, derived from the condition

$$f = m \beta v_- .$$

4. STOCHASTIC THEORY OF QUANTUM MECHANICS

Let us try to derive now a non-dissipative theory based upon the previous formalism. The non-dissipative character is necessary if we wish to arrive at the Schrödinger equation because, as it is well known, Schrödinger equation conserves the mean energy. At this point we already depart from the theory of Brownian motion. For non-dissipative forces it seems that the forces must be proportional to the accelerations. As we have four of these, it seems that there are four independent constants in the theory. Nevertheless, we must take into account the time reversal properties for a non-dissipative theory. As only two of the accelerations are time reversal invariant and the external force is assumed to be also invariant, the following relation is the most general which we can write

$$f = m\left(\mathcal{D}_c v + \lambda \mathcal{D}_s u\right).$$

The parameter m is the mass of the particle, because in absence of stochasticity $\mathcal{D}_s u$ vanishes and $\mathcal{D}_c v$ equals the total time derivative. The parameter λ characterizes the type of stochastic motion. It is an open problem that of studying the compatibility of this equation with the previous hypotheses. It can be proved that the value $\lambda = 0$ is not compatible with $D \neq 0$ and probably only the values ± 1 are possible.

Now, it is straightforward to write the dynamical equation in terms of time and spatial derivatives of v and φ. If this is done, a rather

cumbersome equation appears which we shall not
write here. This and the continuity equation are
the basic equations of our theory. They are a
pair of coupled, non-linear partial differential
equations which seems very difficult to handle.
However, this is not so. It seems almost magic
that, if we assume that v is a gradient and the
force field f derives from a potential, the equations can be both, decoupled and linearized. In
fact, introducing the new functions

$$\psi_{\pm} = \sqrt{\rho}\, \exp\left[\mp S/2\sqrt{\lambda}\, mD\right] \ , \quad mv \equiv \mathrm{grad}\, S,$$

the equations became

$$\pm\, 2\sqrt{\lambda}\, mD\frac{\partial \psi_{\pm}}{\partial t} = 2\lambda mD^2\, \nabla^2 \psi_{\pm} + V\psi_{\pm} \ ; \quad f \equiv -\,\mathrm{grad}\, V .$$

These equations are quite different whether
λ is positive or negative. If λ is negative and we
introduce the new parameter $\hbar$ by

$$4\lambda m^2\, D^2 = -\hbar^2,$$

the equations became

$$i\hbar\, \frac{\partial \psi_{+}}{\partial t} = -\frac{\hbar^2}{2m}\, \Delta \psi_{+} + V\psi_{+}$$

and its complex conjugate. This is just Schrödinger
equation. Indeed, from the definition of ψ_{+} it
follows the usual interpretation of $|\psi_{+}|^2$ as the
probability density for the particle having the
position x at time t.

These facts are the basis for the stochastic
interpretation of quantum mechanics. According
to this theory the electrons, for example, would

be point-like, classical, particles subject to
some random motion. This random motion would give
to the electrons a wave-like character in some
conditions.

At this moment, a number of questions remain
open. For example, as it was pointed out, it is
not obvious that negative values for λ are compatible with all the previous assumptions of the
theory and general physical principles. It is to
be noted that the choice $\lambda > 0$ gives rise to equations of diffusion type, which are very different
from Schrödinger equation. Also, a theory in phase
space would be more satisfactory and the physics
behind it would be more transparent. Some attempts
have been made to generalize the theory to the
relativistic domain and to introduce the spin.
Also, the Lamb shift has been calculated with
rather straigthforward generalizations of this
theory, with partial success. It is quite obvious
that the theory is far from compete with usual
quantum mechanics at this moment.

A problem which remains is to find the cause
of the random motion if it is not to be interpreted
as a basic fact of nature. The more attractive theory along this line is the so-called stochastic
electrodynamics. The basic idea of this theory is
to assume a random background radiation in all
space, which would be the cause of the random motion of material particles. A number of interesting
results has been derived from this theory which
can be seen in the references.

References

Brownian Motion and Stochastic Processes

N. WAX, ed.: Selected papers on noise and stochastic processes, Dover, 1954.

Stochastic Theory of Quantum Mechanics

E. NELSON: Phys. Rev. 150, 1079 (1966) and references herein.

E. SANTOS: Nuovo Cimento, 59B, 65 (1969).

L. de la PEÑA-AUERBACH: J. Math. Phys. 10, 1620
(1969), 12, 453 (1971).

L. de la PEÑA-AUERBACH and A.M. CETTO: Phys. Rev.
D3, 795 (1971).

Stochastic Electrodynamics

M. SURDIN: Ann. Inst. H. Poincaré 15, 203 (1971),
a review article.

E. SANTOS: Lett. Nuovo Cimento, to appear about
June 1972.